GEOSYNTHETIC CLAY LINERS FOR WASTE CONTAINMENT FACILITIES

# Geosynthetic Clay Liners for Waste Containment Facilities


*Editors*

Abdelmalek Bouazza
*Department of Civil Engineering, Monash University, Melbourne, Victoria, Australia*

John J. Bowders
*Department of Civil and Environmental Engineering, University of Missouri, Columbia, Missouri, USA*


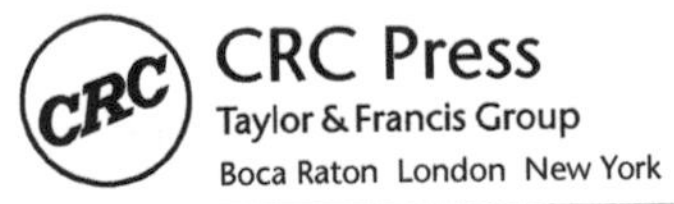

CRC Press
Taylor & Francis Group
Boca Raton London New York

CRC Press is an imprint of the
Taylor & Francis Group, an **informa** business
A BALKEMA BOOK

Published by: CRC Press/Balkema
P.O. Box 447, 2300 AK Leiden, The Netherlands
e-mail: Pub.NL@taylorandfrancis.com
www.crcpress.com – www.taylorandfrancis.co.uk – www.balkema.nl

First issued in paperback 2020



ISBN-13: 978-0-367-57722-3 (pbk)
ISBN-13: 978-0-415-46733-9 (hbk)



**Visit the Taylor & Francis Web site at**
**http://www.taylorandfrancis.com**

**and the CRC Press Web site at**
**http://www.crcpress.com**

Cover photo courtesy of Geofabrics Australasia Pty. Ltd.

*Library of Congress Cataloging-in-Publication Data*

Geosynthetic clay liners for waste containment facilities / edited by
Abdelmalek Bouazza & John J. Bowders.
p. cm.
Includes bibliographical references.
ISBN 978-0-415-46733-9 (hardback : alk. paper) – ISBN 978-0-203-85523-2 (e-book) 1. Sanitary landfills–Linings. 2. Clay soils–Permeability. 3. Geosynthetics. 4. Engineered barrier systems (Waste disposal) I. Bouazza, Abdelmalek. II. Bowders, John J. (John Joseph), 1957–
TD795.7.G4475 2010
628.4′4564—dc22

2009037858

Typeset by Macmillan Publishing Solutions, Chennai, India

# Table of Contents

# Preface

Geosynthetic clay liners have taken a prominent role in civil engineering infrastructure, in particular, in the waste containment industry. Increasingly stringent regulation of pollution and waste production worldwide has driven the need to isolate contaminants posing a threat to human and environmental health by using engineered barrier systems. Over the past two decades, geosynthetic clay liners have gained widespread acceptance for use in such barrier systems. Nowadays, they are often used as a component of primary and secondary base liners or final cover systems in municipal solid-waste landfills as well as in regulated industrial storage and mining waste-disposal facilities.

This book gives a comprehensive and authoritative review of the current state of practice on geosynthetic clay liners in waste containments. It provides an insight into individual materials (bentonite and the associated geosynthetics) and the manufacturing process. This is followed by the coverage of important topics such as hydraulic conductivity, chemical compatibility, contaminant transport, gas migration, shear strength and slope stability, and field performance.

The idea of this book gelled when the first editor, Bouazza, spent a sabbatical with Bowders at the University of Missouri in 1999. We originally thought our profession lacked a comprehensive document on the background, behaviour and applications of these relatively new low-permeability barriers. This book has been a long time in coming and many developments in our understanding of geosynthetic clay liners and in the products themselves have occurred so we have asked some of the leading people in geosynthetic clay liners research and applications to contribute their expertise for this treatise, which we hope will constitute a valuable reference in the years to come for practitioners and manufacturers as well as researchers. We are extremely indebted and much appreciative of the time and effort of each of the contributing authors. Only through their willingness to share their knowledge is this book made possible. Finally, we wish to express our appreciation and thanks to the reviewers for their highly competent efforts. Last but not least, we would like to gratefully acknowledge the long-suffering, patience and encouragement of our editor, Mr. Janjaap Blom of Taylor and Francis publishers.

Abdelmalek Bouazza, Monash University, Melbourne, Victoria, Australia
John J. Bowders, University of Missouri, Columbia, Missouri, USA

# Contributors


**CRAIG. H. BENSON**, Professor, Department of Civil and Environmental Engineering, University of Washington, Seattle, USA

**ABDELMALEK BOUAZZA**, Associate Professor, Department of Civil Engineering, Monash University, Melbourne, Victoria, Australia

**JOHN J. BOWDERS Jr**, Professor, Department of Civil and Environmental Engineering, University of Missouri, Columbia, USA

**ANDREA DOMINIJANNI,** Research Fellow, DITAG, Politecnico di Torino, Torino, Italy

**WILL P. GATES,** Research Fellow, Department of Civil Engineering, Monash University, Melbourne, Australia

**ROBERT B. GILBERT,** Professor, Department of Civil, Architectural and Environmental Engineering, University of Texas, Austin, USA

**Y. GRACE HSUAN,** Professor, Department of Civil and Environmental Engineering, Drexel University, Philadelphia, USA

**TAKESHI KATSUMI,** Associate Professor, GSGES, Kyoto University, Kyoto, Japan

**EDWARD KAVAZANJIAN Jr,** Associate Professor, Department of Civil, Environmental and Sustainable Engineering, Arizona State University, Tempe, USA

**GEORGE R. KOERNER**, Director Designate, Geosynthetic Institute, Folsom, USA

**ROBERT M. KOERNER,** Professor, Director, Geosynthetic Institute, Folsom, USA

**CRAIG B. LAKE,** Associate Professor, Department of Civil and Resource Engineering, Dalhousie University, Halifax, Canada

**WILLIAM J. LIKOS,** Associate Professor, Department of Civil and Environmental Engineering, University of Missouri, Columbia, USA

**MARIO MANASSERO,** Professor, DITAG, Politecnico di Torino, Torino, Italy

**JOHN S. McCARTNEY**, Assistant Professor, Department of Civil, Environmental and Architectural Engineering, University of Colorado, Boulder, USA

**R. KERRY ROWE,** Professor, Department of Civil Engineering, Queen's University, Kingston, Canada

**J. SCALIA,** Department of Civil and Environmental Engineering, University of Wisconsin, Madison, USA

**STEPHEN G. WRIGHT,** Professor, Department of Civil, Architectural and Environmental Engineering, University of Texas, Austin, USA

**JORGE G. ZORNBERG,** Associate Professor, Department of Civil, Architectural and Environmental Engineering, University of Texas, Austin, USA

# CHAPTER 1

# Background and overview of geosynthetic clay liners

R.M. Koerner & G.R. Koerner
*Geosynthetic Institute, Folsom, USA*

## 1 INTRODUCTION

A geosynthetic clay liner (GCL) is defined as follows: A factory-manufactured hydraulic or gas barrier consisting of a layer of bentonite or other very low permeability material supported by geotextiles and/or geomembranes, mechanically held together by needling, stitching, or chemical adhesives. GCLs are also known as Geosynthetic Barriers-Clay (GBR-C) by the International Standards Organization.

Since their introduction in the mid-1980's, GCLs have been utilized greatly as the lower portion of geomembrane/GCL composites in both landfill liners and final covers. They have also been used in other containment applications and by themselves as single barrier systems when modified with a geofilm or polymer coating within the cover geotextile. In most applications they have served as replacement materials for the more traditional compacted clay liners (CCLs). As such, GCLs must be comprehensively identified and characterized so as to make a technical equivalency of one material versus another. This opening chapter is meant to provide the necessary information. It is an abbreviated form of Chapter 6 in Koerner (2005).

The background section describes the individual materials (bentonite and the associated geosynthetics) and some insight into the manufacturing processes. This is followed by a description of the current GCL products, both nonreinforced and reinforced types. A section on test methods and properties is subdivided into physical, hydraulic, mechanical, and endurance properties. Using this information, a generic manufacturing quality control specification is presented. Finally, equivalency issues of GCLs versus CCLs are compared and contrasted for both landfill liners and final covers. A brief summary concludes the chapter which includes a vista into the subsequent chapters in the book.

## 2 BACKGROUND

Since Geosynthetic Clay Liners, or GCLs, are composite materials consisting of bentonite clay contained within, or on, geotextiles or geomembranes, it is necessary to describe each of the component parts before considering the manufactured product. For perspective, however, it should be mentioned that GCLs have had different names in the past and even presently, e.g., the International Standards Organization (ISO) refers to them as Geosynthetic Barriers-Clay (GBR-C), Zanzinger et al. (2002).

### 2.1 *Bentonite*

Bentonite is a well-known member of the clay family of geologically cohesive materials having extremely low hydraulic conductivity, or permeability, to liquids and gases. Of the various types, sodium bentonite is known to have the lowest permeability of any naturally occurring geologic material. It is only available in large quantities in Wyoming and North Dakota in the U.S. The transportation costs of moving bentonite from these locations to worldwide factories for GCL manufacturing are relatively high. An alternative can be found in the large natural deposits of higher permeability calcium bentonite, which is much more available on a worldwide basis. By using sodium hydroxide to treat the calcium bentonite, a replacement of the calcium ions occurs, decreasing the permeability to that of the naturally occurring sodium bentonite. The process is sometimes called *peptizing* and is worldwide in its usage. Although not fully established, this trend will probably increase in the future.

Regarding identification of the bentonite, X-ray diffraction is a precise method of determining the composition of clays. However, the test is costly, and relatively few geosynthetics testing laboratories are capable of performing it. Although not as accurate, the American Petroleum Institute's methylene blue analysis is easy to perform and is thought to give conservative results. When using such a test, a montmorillonite content of at least 70% is felt to be required to yield adequate swell and permeability values for use in GCLs. This value is approximately equivalent to an X-ray diffraction value of 90% (Heerten et al. 1993). A more complete treatment of bentonite properties is treated in this book by Likos et al.

### 2.2 *Geotextiles*

Geotextiles are generally used as the carrier material beneath the bentonite and the cap (or cover) fabric above it. The geotextiles can be nonwoven, woven, or composite nonwovens with a woven scrim. Presently, most of the geotextiles are made from polypropylene resins with minor additions of additives such as high temperature processing aids, ultraviolet light stabilizers, and long-term durability additives. The paper by Hsuan and Koerner in this book will focus on the importance of the geotextile polymer formulation. When considering needlepunched reinforced GCLs, at least one of the geotextiles must be of the needlepunched variety. Woven geotextiles are either of the slit-film or spunlaced variety which have excellent strength and stiffness characteristics, but must be of a sufficiently tight weave so as not to allow bentonite to squeeze through the openings.

The properties of the geotextiles are an important consideration. In addition to opening size, the mass per unit area, tensile strength, tensile elongation, and installation survivability properties are all important.

### 2.3 *Geomembranes*

As will be seen in the manufacturing section to follow, one product uses a geomemrane as a substrate and adheres bentonite to it. Of course, an adhesive must be blended with the bentonite for this reason. The geomembrane can be of any type, thickness, or surface feature. Textured high density polyethylene (HDPE) geomembranes are quite common in the 1.0 to 1.5 mm thickness range. The product can be installed in three ways;

- geomembrane up; bentonite down
- geomembrane down; bentonite up
- geomembrane down; bentonite up with an additional covering geomembrane above the bentonite

### 2.4 *Other associated materials*

In addition to the possible adhesives added to the bentonite as just mentioned, very thin polymer films ($\simeq$0.10 mm thick), i.e., geofilms, have been added to the composite material for further reduction of permeability. Such films are placed either above or below the cap geotextile to lower its permeability from the as-manufactured condition.

A different approach in achieving very low permeability than using a geofilm is to polymer treat the cap geotextile to lower its permeability from the as-manufactured condition. Both of these approaches are often recommended when using GCLs as a sole barrier material for noncritical and/or temporary applications. (For landfill and hydraulic applications, GCLs are invariably covered by a geomembrane).

Lastly, it is also possible to polymer modify the bentonite particles themselves within their molecular structure or adhere polymers to the outside of the platlets. This subset of nano-technology gives rise to the terms of internal or external bentonite modification, respectively Jeon (2009). Such processes, however, are not widespread and presently the low permeability component of GCLs is entirely bentonite. Perhaps the future will even see an all-polymer product, but that remains for extended research and development.

### 2.5 *GCL manufacturing*

There are currently two structurally different GCL types distinguished by the method of manufacturing the composite material: nonreinforced and (internally) reinforced. The two forms are based

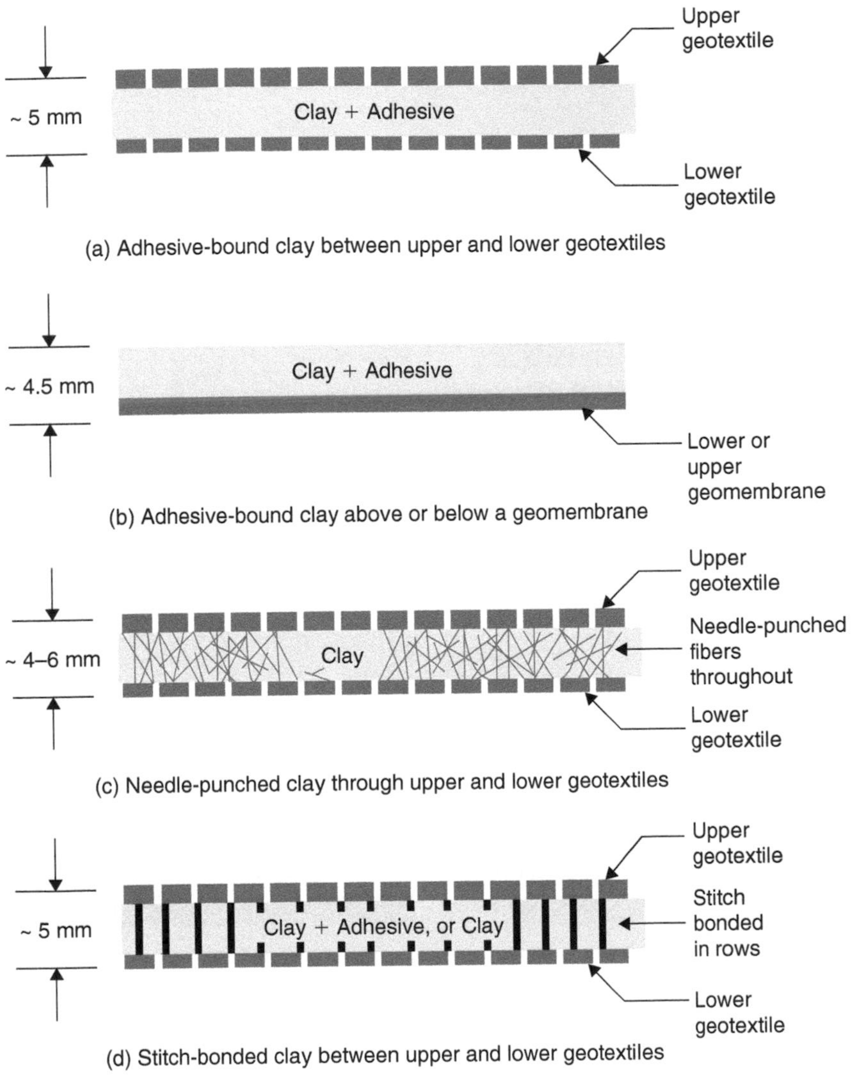

(a) Adhesive-bound clay between upper and lower geotextiles



(b) Adhesive-bound clay above or below a geomembrane



(c) Needle-punched clay through upper and lower geotextiles



(d) Stitch-bonded clay between upper and lower geotextiles

Figure 1. Cross sections of currently available GCLs, after Koerner (2005).

on the realization that hydrated bentonite is extremely low in its shear strength. Thus, nonreinforced GCLs are used on very flat surfaces and reinforced GCLs (by needle punching or stitch bonding) are used on relatively steep slopes. Each type has a number of subvariations.

Nonreinforced GCLs are either geotextile-related, geotextile/polymer-related, or geomembrane-related. The geotextile-related types have geotextiles on both surfaces and an adhesive mixed with the bentonite for bonding purposes, as shown in Figure 1a. The geotextile/polymer-related types are similar but have a polymer impregnated in the upper geotextile to decrease the permeability lower than the bentonite itself. The geomembrane-related types have the bentonite adhesively bonded to a geomembrane (see Figure 1b). The geomembrane can be of any type, thickness, or surface texture.

More common than the above are reinforced GCLs. The usual method of reinforcement is by needle punching from a nonwoven cap geotextile through the bentonite and opposing geotextile, which creates a labyrinth of fibers throughout (see Figure 1c). Alternatively, we can stitch bond between two woven geotextiles through the sandwiched bentonite layer (see Figure 1d). Further variations of reinforced GCLs are geotextile-related, geotextile/polymer-related, and geotextile/geofilm-related. The geotextile-related types have various geotextiles on both surfaces. The geotextile/polymer-related types have a polymer impregnated in the upper geotextile to lower the permeability beyond

(a) Claymax from CETCO

(b) Bentofix from BTI, Naue GmbH, and Terrafix

(c) Bentomat from CETCO

(d) Gundseal from GSE Lining Technology

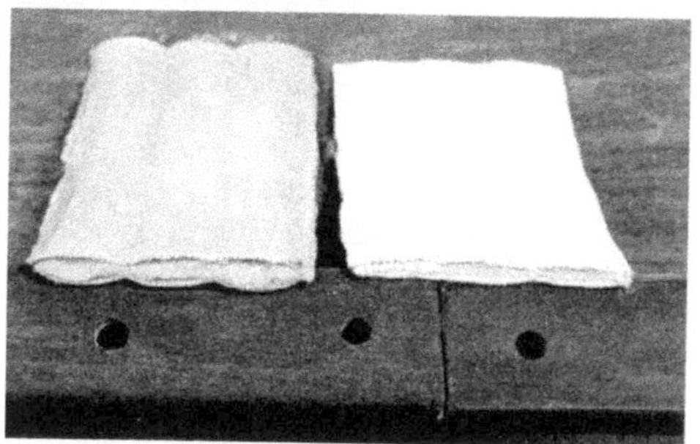

(e) NaBento from Huesker

Figure 2. A selection of commercially available GCLs with as-received product shown on right side and hydrated product on left side, after Koerner (2005).

the bentonite itself. The geotextile/film-related types have a thin plastic film either above or beneath the cap geotextile. After needle punching, this film becomes part of the composite material and decreases the permeability much lower than the bentonite itself.

## 3 CURRENT GCL PRODUCTS

GCLs being among the newest of the geosynthetic family of materials are still seeing new products and variations of existing products. Figure 2 shows the most widely used GCLs in their as-received and hydrated states. Figures 2a and e are stitch-bonded reinforced GCLs; Figures 2b and c are needlepunched reinforced GCLs; and Figure 2d is a nonreinforced geomembrane-related GCL. Without the stitch bonding, Figure 2a is a nonreinforced geotextile-related GCL.

Table 1 presents company names and trademarks of the various GCLs but for the most current information, additional details are available from the respective manufacturers either by brochure or their internet websites.

### 3.1 *Nonreinforced GCLs*

As mentioned previously, bentonite has the lowest permeability of any naturally occurring geologic material which makes it ideal as a barrier, or containment, material. Unfortunately, when hydrated,

Table 1. Currently available GCLs.

| Manufacturer | Trademark | Substrate | Infill | Superstrate | Comments |
|---|---|---|---|---|---|
| CETCO | Claymax | geotextile | bentonite | geotextile | can include thin plastic film |
| CETCO | Bentomat | geotextile | bentonite | geotextile | can include thin plastic film |
| Naue/BTI | Bentofix | geotextile | bentonite | geotextile | can include polymer treatment |
| GSE | Gundseal | geomembrane | bentonite | none | can include a covering geomembrane |
| Huesker | NaBento | geotextile | bentonite | geotextile | stitch bonded |
| Rawell | unknown | geotextile | bentonite (polymer modified) | none | considerable moisture |
| Geosynthetics | Equiva-Seal | geotextile | bentonite within a geonet | geotextile | thermal bonding |
| Laviosa | Modulo-Geobent | geotextile | bentonite | geotextile | |
| Aashi | unknown | geotextile | bentonite | geotextile | |
| GID | Trisoplast | soil subgrade | bentonite (polymer modified and sand filler) | none | in-situ fabrication (not a GCL, per se) |

it also has the lowest shear strength of any naturally occurring geologic material. Furthermore, the suction capacity of bentonite is enormous and it will readily hydrate when a GCL is placed on, or covered with, a moist soil, Daniel et al. (1993). Thus, unreinforced GCLs must only be used on very flat subgrades. With the free swell friction angle of bentonite being as low as 6 deg., this becomes a most important consideration.

The possible exception to this is the geomembrane-related GCL with a covering geomembrane placed above thus sandwiching the bentonite between two geomembranes. The purpose of this being that the relatively dry bentonite has an significantly higher shear strength, than when it is hydrated, Thiel et al. (2004).

### 3.2 *Reinforced GCLs*

By needle punching or stitch-bonding from cap (cover)-to-carrier geotextiles one has created a reinforced GCL. This is clearly within the expertise of textile engineering and relies on the connecting fibers or yarns for internal strength of the composite material. In so doing, reinforced GCLs can be used on slopes in direct relationship to the adequacy of this complicated fiberous network joining the composite system together. The mechanism is called internal strength when the potential shear plane passes within the GCL structure. It is called external, or interfacial, strength when either the cap or carrier GCL surfaces are involved. In this book, Zornberg and McCartney describe the requisite shear strength testing and Gilbert and Wright discusses slope stability designs incorporating the subsequent test properties.

## 4 TEST METHODS AND PROPERTIES

There is considerable ongoing activity in characterizing and evaluating GCLs. Many test methods have been approved by ASTM and ISO, and still others are in various stages of development. The most important ones are described in this section which is an abbreviated form of that given in Koerner (2005).

### 4.1 *Physical properties*

*Thickness* of GCLs in the as-manufactured state (or dry, as an index value) is a difficult property to measure and can be considered at best to be a quality-control property for manufacturing. It is usually not included in a product specification. Where thickness is relevant in a performance role is in permeability testing to convert a flow rate (or flux) value to a hydraulic conductivity or permeability value. Here the thickness of the hydrated test specimen is required and the issue is quite controversial. What pressure to apply and how to deduct for the geosynthetics are unresolved issues.

The measurement of *mass per unit area of a composite* GCL follows ASTM D5993. It is somewhat subjective for reasons of thickness measurement difficulties as just mentioned. In addition, cutting out a GCL test specimen, containing powdered or granular bentonite without losing material is very difficult. Moisture has been added around the edges before cutting with little success, and the mass of the adsorbed water must be deducted, which is difficult if a water spray is used.

Concerning the *mass per unit area of the bentonite* (without the associated geotextiles or geomembranes) difficulties arise with respect to sampling, the removal of the geotextiles or geomembrane, and the deduction for the amount of adhesive (if present). Unfortunately, the GCLs that are easiest to sample are those that contain adhesives, which is a difficult situation if we desire a separate mass per unit area for the bentonite component by itself. From a manufacturing perspective, most GCLs are targeted to have 3.7 $kg/m^2$ of bentonite.

Bentonite is a very hydrophilic mineral. As such it will generally have a measurable *moisture content* at all times. The value can be as high as 10%, approximately the shrinkage limit, yet this is still considered to be the as-received, or dry, condition. The situation is farther complicated by those GCLs that contain adhesives in the bentonite. Generally, some adhesive in liquid form remains after oven heating. This, along with the humidity adsorption of the bentonite, can lead to a total moisture content of up to 15% as the product leaves the manufacturing facility. For GCLs which have water purposely added to the bentonite during manufacturing, the moisture content will be even higher, e.g., 30 to 40%. The measurement of moisture content is straightforward via ASTM D5993 and is defined as the moisture content divided by the oven-dry weight of the specimen, expressed as a percentage. Some manufacturers base the moisture content on the wet weight of the test specimen, which results in lower values.

### 4.2 *Hydraulic properties*

Since GCLs are used in their primary function as liquid barriers, hydraulic properties are critically important. This section only treats the topic superficially since other chapters in this book are focused entirely on specific hydraulic issues, e.g., Katsumi, Dominijanni and Manassero, and Benson and Scalia.

Bentonite, the essential low permeability component of GCLs, is known to hydrate differently depending upon the nature of the *hydrating liquid.* It is also known to hydrate differently as a function of the applied normal stress. Koerner (2005) illustrates that hydration occurs in decreasing magnitude with respect to the following;

- distilled water
- tap water
- mild leachate
- harsh leachate
- diesel fuel

The test is formalized as ASTM D5887 using distilled water and a suggested testing protocol for index permeability or flux values. The test can be modified for site specific conditions as desired.

The amount of *swelling* of bentonite under zero normal stress has been formalized in a test known as the swell index, and designated as ASTM D5890. In this test, a graduated cylinder is filled with 100 ml of water, to which 2.0 g of bentonite is added. The bentonite is milled to a powder and added to the water slowly so as to allow the clay to flocculate and settle to the bottom of the cylinder. After leaving the cylinder undisturbed for 24 hours, the volume occupied by the clay is measured and a recommendation is given. Heerten et al. (1993) recommend a minimum swell index value of 24 ml per 2.0 g of bentonite.

The fact that the bentonite in GCLs can readily *absorb* water from the adjacent soil has been shown by Daniel et al. (1993). They placed samples of GCLs on sand soils of varying water contents from 1 to 17% and measured the uptake of water in the GCL. Figure 3 shows the resulting curves. Two important messages stem from this data: soils as dry as 1% can result in GCL hydration to 50% and the time for hydration is quite rapid (e.g., within 5 to 15 days). For a laboratory determination of absorption, some GCL manufacturers report a plate water absorption test, performed in accordance with ASTM E946, to determine the volumetric increase of a clay sample as it draws water from an underlying saturated porous stone.

Another index text now focused on the *fluid loss* of the bentonite tested under pressure is ASTM D5891. It is an indirect measure of the adhesive characteristics of the pore liquid to the clay particles. In this test a carefully prepared bentonite slurry is poured into an assembled cell which has a filter paper suspended on a support screen in its base. The cell is then placed in a filter press that is pressurized to 700 kPa. The clear water collected from the base outlet of the cell between 7.5 and 30 min. is the fluid loss. A maximum value of 18 ml is often specified for this test.

Although the proper term is hydraulic conductivity, we will continue to use the word *permeability* since it is embedded in the GCL literature at this time. As with CCLs, the permeability of a GCL should be evaluated under field-simulated pressure conditions in a flexible wall permeameter. Rigid wall permeameters cannot be used, see Koerner (2004). The general performance test for GCLs is ASTM D5887, which results in a "flux" value in units of $m^3$/sec-$m^2$.

A 100 mm diameter GCL test specimen is placed in a rubber membrane that is then placed in a triaxial permeameter. It is subjected to a total stress of 550 kPa and then back pressure saturated at 515 kPa with deionized water for 48 hr. Permeation through the specimen is initiated by raising the pressure on the influent side of the test specimen to 530 kPa. Permeation is continued until inflow and outflow are equal to ±25%, or until the flow rate is sufficiently low to ensure conformance with a required value. The final value of flux or permeability is then obtained and reported.

The test can also be conducted using site specific conditions set forth by the parties involved. Permeants which are potentially incompatible with the bentonite can also be evaluated, e.g., permeants which cause a cation exchange resulting in an increase in permeability have been reported

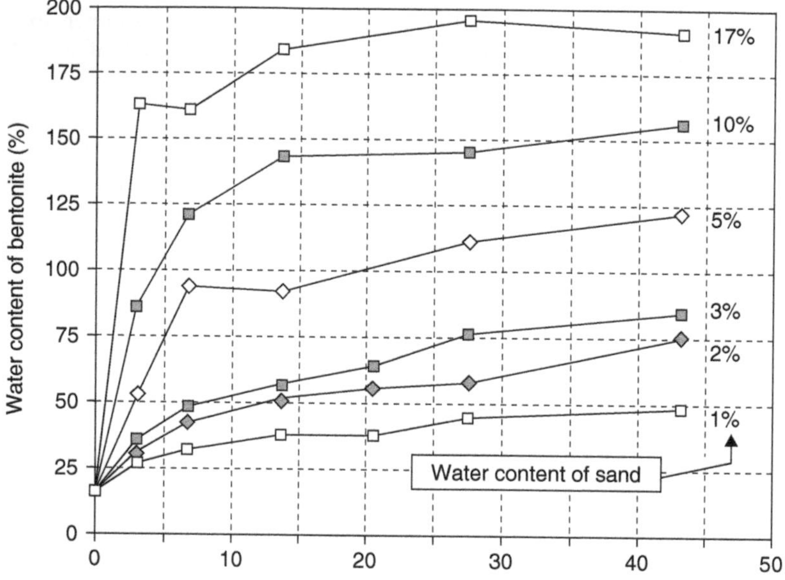


Figure 3. Water content versus time for GCL samples placed in contact with sand at various water content. (After Daniel et al., 1993)

by Kolstad, et al. (2004), and others. ASTM D6766 provides such a test protocol in which two variations are mentioned. One is with initial GCL saturation using water, the second saturates the GCL with a specific test liquid. Both are followed by permeation with the specific test liquid. The second option is the most aggressive.

Because GCL roll edges and ends are placed in the field using an overlap configuration, it is important that flow does not occur between the upper and lower GCL panels. Using a large laboratory test tank measuring 2.4 m long × 1.2 m wide × 0.9 m high with an overlap seal along the long direction of the tank, Estornell and Daniel (1992) measured the permeability in the overlap region. The amount of overlap was 150 mm. Above the GCLs 300 mm of gravel supplied the applied normal stress. Within experimental error, the overlap permeability was as low as with the control sample having no overlap seam. This study, in general, confirms the general recommendation of a minimum overlap requirement of 150 mm. A smaller version of this test is also available.

In addition to liquid permeation through GCLs, the issue of contaminant transport has arisen. Lake and Rowe address this topic in the book. Lastly, gas permeability and gas diffusion are topics of interest and they are addressed in this book by two separate chapters by Bouazza.

### 4.3 *Mechanical properties*

GCLs placed on side slopes, under high shear stresses, adjacent to rough or yielding subgrades, under thermal stresses, and so on can readily challenge the individual product's mechanical properties thereby affecting its functionality as a hydraulic or gas barrier. Invariably, some aspect of tensile stress will be involved. In addition, the critically important aspect of shear strength will be addressed in this section.

Using ASTM D6768, a GCL can be evaluated for its *wide width tensile behavior*. Since the clay component has little tensile strength, either dry or saturated (in comparison to the geosynthetics), the recommended manner of testing is the dry state. The resulting strength will be essentially that of the geotextiles or geomembrane involved. See Koerner (2005) for data of this type. The GCL should be tested as a composite, however, and not as individual geosynthetics or according to the published values of the individual as-manufactured geosynthetic materials. Thus, fugative bentonite all over the testing machine should be anticipated. It is possible to seal the test specimen edges with hot glue, but if this is done carelessly it could affect the tensile strength response.

The question frequently arises as to the *axi-symmetric tensile behavior* of GCLs, particularly when used in landfill closure applications. Koerner et al. (1996) have used an axi-symmetric tension test setup for geomembranes and modified it for GCLs. The test setup used had a very flexible geomembrane over the GCL test specimen. Hydrostatic pressure is applied until failure occurs, which is always in the GCL before the geomembrane. The load taken by the geomembrane is deducted (via a separate test on the geomembrane by itself), and the stress-versus-strain behavior of the GCL is obtained. In general, the failure strain for the geotextile-related GCLs is from 10 to 19%, and for HDPE geomembrane-related GCLs from 15 to 22%. Such values are orders of magnitudes higher than CCL's, and in keeping with stiffer geomembranes like HDPE and scrim reinforced geomembranes (e.g., fPP-R, CSPE-R and EPDM-R). Geomembranes with the highest flexibility, like LLDPE, fPP, and PVC, strain considerably farther (50 to 100%) in this type of test.

Other than the hydraulic conductivity tests mentioned earlier, a major focus of GCL testing has been on the *internal shear strength* of both reinforced and nonreinforced GCLs. The test method used currently is ASTM D6243. The suggested shear box is 300 mm-by-300 mm with sufficient displacement to obtain large deformation (i.e. residual) strength as well as peak strength. The general behavior of GCLs is as follows:

- GCLs are strongest in the dry condition and weakest in the free swell condition. The results from constrained swell conditions are intermediate between the two extremes.
- The type of hydrating liquid affects shear strength, but to a lesser extent than other factors. Hydration with distilled water is the worst-case shear strength condition in this regard.
- GCLs fabricated by needle punching between two geotextiles required much larger displacements than unreinforced GCLs to reach their limiting shear strength condition.
- Needle punching significantly increases the shear strength under all conditions. Stitch bonded GCLs result in similar and even greater improvements.

Depending upon the nature of the upper and lower surfaces of the GCL on the adjacent soil or other geosynthetic materials, separate *interface shear tests* are needed. These interfaces must be evaluated with respect to the adjacent site-specific materials: typically they are either geotextiles, geogrids, geonets, geomembranes or any type of soil. Also note that the interface surface may be considerably changed from the as-received or as-constructed materials, due to hydrated bentonite intruding into nonwoven needle punched geotextiles or extruding out of the woven geotextiles into the interface of concern. Woven slit film, spun laced, and monofilament geotextiles with even the slightest open area between fibers or yarns are all of concern in this regard. Product-specific simulated testing is called for in most circumstances. The open literature is particularly strong with respect to GCL shear strength testing.

ASTM D6496 addresses the *peel strength* between upper and lower geotextiles of reinforced GCLs. In this test, a 100 mm wide specimen has its cap and carrier geotextiles gripped individually in opposing tensile grips and pulled at a constant rate of extension by a tensile testing machine until the layers of the specimen separate. The reinforcing fibers or yarns are successfully placed in tension and have the effect of bundling against one another. The average peel strength is calculated and reported in units of kN/m. The test is an index test used to evaluate the quality of the reinforcement process. Numerous attempts have been made at relating peel strength with internal direct shear strength but without complete success, von Maubeuge & Lucas (2002) and Zornberg et al. (2005).

Due to the relative thinness of GCLs compared to CCLs, *puncture* concerns are understandably often voiced. There are a number of puncture tests that can be used with GCLs, including: ASTM D4883, which uses a 8.0 mm puncturing probe; ASTM D6241, which uses a CBR probe of 50 mm diameter; and ISO 12236, which also uses a 50 mm diameter probe. Although all of these tests are straightforward to perform, it is important to recognize the self-healing puncture characteristics of GCLs which contain bentonite. No puncture test by itself can reproduce this self-sealing mechanism, since the GCL is being used as a hydraulic barrier and puncture per se may not be a defeating, or even limiting, phenomenon.

*Lateral squeezing*, however, can occur if a uniform and constant load is stationed on a GCL with insufficient cover soil. The degree of squeezing is dependent on the bentonite's initial moisture content, the type of GCL, and the cover soil thickness, Koerner & Narejo (1995).

### 4.4 *Endurance properties*

Since the soil component of the barrier material in a GCL is clay, its long-term integrity is generally assured. However, the *liquid that activates and permeates the bentonite*, resulting in its low permeability, is certainly an issue insofar as moisture barrier endurance is concerned. It should be noted that a permeant other than distilled water can be used per ASTM D6766. The permeant suggested is a 0.1 M calcium chloride solution so as to assess the possible ion exchange that might occur during service conditions. Site-specific liquids can also be used as decided upon by the parties involved.

The central property of a hydrated GCL insofar as *freeze-thaw behavior* is concerned is its permeability. Daniel et al. (1997) used a rectangular laboratory flow box and subjected the entire assembly to 10 freeze-thaw cycles. The permeability showed a slight increase from $1.5 \times 10^{-9}$ to $5.5 \times 10^{-9}$ cm/sec. Kraus et al. (1997) report no change in flexible wall permeability tests of the specimens evaluated after 20 freeze-thaw cycles. While the moisture in the bentonite of the GCL can freeze, causing disruption of the soil structure, upon thawing the bentonite is very self-healing and apparently returns to its original state. In this regard, it is fortunate that most GCLs have geotextile or geomembrane coverings so that adjacent soil particles cannot invade the bentonite structure during the expansion cycle.

The behavior of *alternate wet and dry cycles* insofar as a GCL's permeability is important in many circumstances, particularly so when the duration and intensity of the dry cycle is sufficient to cause desiccation of the clay component of the GCL. Boardman & Daniel (1996) evaluated a single, albeit severe, wet-dry cycle on a number of GCLs and found essentially no change in the permeability. The results are encouraging and mimic the freeze-thaw results, but the results of numerous wet-versus-dry cycles awaits further investigation. Perhaps more significant than change in permeability is that shrinkage can cause loss of overlap and (combined with other mechanisms) can possibly cause separation at the roll edges or ends. If this occurs in the field, friction with the underlying surface will prevent expansion back to the original overlapped condition. Thus cover soil, placed in a timely manner and sufficiently thick to resist shrinkage, is necessary. Alternatively, one can compensate by increasing the usual overlap distance or physically bond the opposing geotextiles together, Rowe, et al. (2009).

The *adsorptive capacity* of GCLs is important when they are used for landfill liners and interface with the various leachates that they are meant to contain. Both organic and inorganic solutes are of concern. The situation is described in Koerner & Daniel (1995), particularly in comparison to CCLs and addressing the issue of making an equivalency assessment. The cation exchange capacity of the bentonite clay must be determined and, along with its thickness, such a comparison can be made. It is in this particular instance that GCLs usually are not considered to be equivalent to the much thicker CCLs. For this reason there is a tendency to use a three-component composite liner (i.e. GM/GCL/CCL). The CCL component, however, can be significantly higher in its permeability than the usual regulated value. The hydraulic conductivity of the GCL/CCL system has been analytically investigated by Giroud et al. (1997).

*Water breakout* time is of particular interest for GCLs used in landfill closures. It is this point that steady-state seepage will occur through the GCL and into the underlying solid waste. The data can be obtained from a permeability test, as described previously, but now starting with the as-received dry GCL instead of starting with a fully saturated test specimen.

For a GCL placed beneath a landfill or surface impoundment, it is the *solute breakout time* (rather than water) that is of concern. The test method is again the permeability test, but now with the liquid of concern (e.g. with the leachate), as the permanent. This is an area where research seems to be warranted, particularly in light of showing the equivalency of GCLs to CCLs.

The *durability* of geotextile coverings of GCLs, as well as needle-punched fibers or sewing yarns providing internal reinforcement, is critically important for the long-term performance of a liner system. This applies to liner systems as well as cover systems. Thus, recent efforts have focused on a geotextile fiber and fabric lifetime (Hsuan & Koerner 2002). The approach is similar to that of geomembrane lifetime prediction with the exception that the simulated laboratory incubation setup is very challenging. The topic is treated in this book by Hsuan and Koerner.

### 4.5 *Generic specification for GCLs*

There has been a considerable effort focused on the development of a generic specification for GCLs. Table 2 presents such a document. It is crafted according to the GCL types described in the introduction to this chapter and includes many of the test methods and limiting properties just

Table 2. GRI-GCL3 generic specification for geosynthetic clay liners (GCLs).

| Property | ASTM test method | Reinforced GCL | | | Non-reinforced GCL | | | Testing frequency |
|---|---|---|---|---|---|---|---|---|
| | | GT-related | GT polymer-related | GM-GF-related | GT-related | GT Polymer-related | GM-GF-related | |
| **Clay (as received)** | | | | | | | | |
| Swell index (ml/2g) | D5890 | 24 | 24 | 24 | 24 | 24 | 24 | 50 tonnes |
| Fluid loss (ml)(1) | D5891 | 18 | 18 | 18 | 18 | 18 | 18 | 50 tonnes |
| Geotextiles (as received) | | | | | | | | |
| Cap fabric (nonwoven) – mass/unit area ($g/m^2$)(2) | D5261 | 200 | 200 | 200 | 70 | 100 | n/a/70 | 20 000 $m^2$ |
| Cap fabric – (woven) – mass/unit area ($g/m^2$) | D5261 | 100 | 100 | 100 | – | – | – | 20 000 $m^2$ |
| Carrier fabric (nonwoven composite) – mass/($g/m^2$)(2) | D5261 | 240 | 240 | 240 | 90 | 100 | n/a/90 | 20 000 $m^2$ |
| Carrier fabric (woven) – mass/unit area ($g/m^2$) | D5261 | 100 | 100 | 100 | – | – | – | 20 000 $m^2$ |
| Coating – mass/unit area ($g/m^2$)(3) | D5261 | n/a | 100 | n/a | n/a | 100 | n/a | 4 000 $m^2$ |
| **Geomembrane/Geofilm (as received)** | | | | | | | | |
| Thickness(5) (mm) | D5199/D5994 | n/a | n/a | 0.40/0.50/0.1 | n/a | n/a | 0.40/0.75/0.1 | 20 000 $m^2$ |
| Density (g/cc) | D1505/D792 | n/a | n/a | 0.92 | n/a | n/a | 0.92 | 20 000 $m^2$ |
| Break tensile strength, MD&XMD (kN/m) | D6693 | n/a | n/a | n/a | n/a | n/a | 6.0 | 20 000 $m^2$ |
| Break tensile strength, MD (kN/m) | D882 | n/a | n/a | 2.5 | n/a | n/a | 2.5 | 20 000 $m^2$ |
| **GCL (as manufactured)** | | | | | | | | |
| Mass of GCL ($g/m^2$)(6) | D5993 | 4000 | 4050 | 4100 | 4000 | 4050 | 4100 | 4 000 $m^2$ |
| Mass of bentonite ($g/m^2$)(6) | D5993 | 3700 | 3700 | 3700 | 3700 | 3700 | 3700 | 4 000 $m^2$ |
| Moisture content(1) (%) | D5993 | (4) | (4) | (4) | (4) | (4) | (4) | 4 000 $m^2$ |
| Tensile str., MD (kN/m) | D6768 | 4.0 | 4.0 | 4.0 | 4.0 | 4.0 | 4.0 | 20 000 $m^2$ |
| Peel strength (N/m) | D6496 | 360 | 360 | 360 | n/a | n/a | n/a | 4 000 $m^2$ |
| Permeability(1) (m/sec), "or" | D5887 | $5 \times 10^{-11}$ | n/a | n/a | $5 \times 10^{-11}$ | n/a | n/a | 25 000 $m^2$ |
| Flux(1) ($m^3/sec\text{-}m^2$), | D5887 | $1 \times 10^{-8}$ | n/a | n/a | $1 \times 10^{-8}$ | n/a | n/a | 25 000 $m^2$ |
| GCL permeability(1),(7) (m/sec) (max. at 35 kPa) | D6766 | $1 \times 10^{-8}$ | n/a | n/a | $1 \times 10^{-8}$ | n/a | n/a | yearly |
| GCL permeability(1),(7) (m/sec) (max. at 500 kPa) | D6766 mod. | $5 \times 10^{-10}$ | n/a | n/a | $5 \times 10^{-10}$ | n/a | n/a | yearly |
| **Component Durability** | | | | | | | | |
| Geotextile and reinforcing yarns (8) (% strength retained) | See § 5.6.2 | 65 | 65 | n/a | 65 | 65 | n/a | yearly |
| geomembrane | See § 5.6.3 | n/a | n/a | GM Spec(9) | n/a | n/a | GM Spec(9) | yearly |
| geofilm/polymer treated(8) (% strength retained) | See § 5.6.4 | n/a | 85 | 80 | n/a | 85 | 80 | yearly |

n/a = non applicable with respect to this property

(1) (These values are maximum (all others are minimum)

(2) For both cap and carrier fabrics for nonwoven reinforced GCLs; one, or the other, must contain a scrim component of mass $\geq 100\,g/m^2$ for dimensional stability

(3) Calculated value obtained from difference of coated fabric to as-received fabric

(4) Value is both site-specific and product-specific and is currently being evaluated

(5) First value is for smooth geomembrane; second for textured geomembrane; third for geofilm

(6) Mass of the GCL and bentonite is measured after oven drying per the stated test method

(7) Value represents GCL permeability after permeation with a 0.1 M calcium chloride solution (11.1 g $CaCl_2$ in 1-liter water)

(8) Value represents the minimum percent strength retained from the as-manufactured value after oven aging at 60°C for 50 days

(9) Durability criteria should follow the appropriate specification for the geomembrane type used; i.e., GRI GM-13 for HDPE, GRI GM-17 for LLDPE or GRI GM-18 for fPP

Table 3. Differences between geosynthetic clay liners and compacted clay liners.

| Characteristic | GCLs | CCLs |
|---|---|---|
| Material | Bentonite clay, adhesives, geotextiles and/or geomembranes | Native soils or blends of soil and bentonite clay |
| Construction | Factory manufactured and then installed in the field | Constructed and/or amended in the field |
| Thickness | ≅8 mm | 300 to 900 mm |
| Permeability of clay | $10^{-10}$ to $10^{-12}$ m/s | $10^{-9}$ to $10^{-10}$ m/s |
| Speed and ease of construction | Rapid, simple installation | Slow, complicated construction |
| Installed Cost | \$0.05 to \$0.10 per $m^2$ | Highly variable (estimated range \$0.07 to \$0.30 per $m^2$) |
| Experience | CQC and CQA are critical | Has been used for decades |

presented. Also included is the minimum testing frequency of the various tests for proper manufacturing quality control purposes.

## 5 EQUIVALENCY ISSUES

Since CCLs (both natural soil and amended soil types) have been used historically as liquid barriers, it is only fitting that GCLs should have to compare favorably with, or be better than, CCLs in order to be used as replacement barrier materials. They may even have to be better than CCLs since GCLs are the replacement material, and concerns are often voiced when the use of new materials is contemplated. The obvious issues are due to fundamental differences listed in Table 3.

At first glance, we would assume that a technical equivalency argument could be based on the flow rate or flux through the competitive materials. Such a calculation is straightforward and is routinely used for such purposes. However, this particular calculation is only the beginning of a complete equivalency comparison since numerous hydraulic, physical/mechanical, and construction issues need evaluation. Within each issue there are specific questions that can be raised in order to arrive at a complete equivalency assessment. Furthermore, for waste containment systems, we can identify functional differences between a barrier material beneath a waste facility (e.g. landfills, surface impoundments, heap leach pads, and waste piles) and a barrier material placed above a waste facility (e.g. landfill and agricultural covers and closure situations). In addition, the comparison may differ depending on whether the GCL is compared to a CCL when each is used by itself (as with a single barrier) or when they are used in a composite barrier, as with a GM/GCL compared to a GM/CCL.

The aforementioned contrasts can be arranged via a comparison that includes the various issues for liners versus covers. See Tables 4 and 5 respectively, for a relatively complete set of equivalency issues that often require consideration. The tables can serve best as a guide or checklist for a site-specific comparison to be made by the user.

In both Table 4 (for liners) and Table 5 (for covers) it is seen that regarding the *hydraulic issues,* the chemical adsorptive capacity of a GCL compared to the typical CCL is generally not equivalent. It is site-specific just how dominant an issue this is. If it is significant, the use of a combined GCL/CCL composite is an alternative. Similarly the water and solute breakout times for the geotextile-covered GCLs are probably not equivalent to CCLs, but the geomembrane backed GCL probably is. Again the relevancy of breakout time must be assessed in light of site specific considerations. Intimate contact of geomembranes with both GCLs and CCLs is an area in need of appropriate CQC and CQA.

Regarding *physical/mechanical issues,* GCLs are generally equivalent to or better than CCLs, with the exception of squeezing, or bearing capacity, when the GCLs are of high moisture content and trafficked without sufficient soil cover. This issue must be avoided by proper specification values and follow-through in the CQC and CQA activities. The all-important issue of shear strength (internal and interface) is a site-specific and/or product specific issue.

Table 4. Generalized technical equivalency assessment for GCL **liners beneath** landfills and surface impoundments, Koerner & David (1995).

| Category | Criterion for evaluation | Probably superior | Probably equivalent | Probably not equivalent | Equivalency dependent on site or product |
|---|---|---|---|---|---|
| Hydraulic issues | Steady flux of water | | √ | | |
| | Steady solute flux | | √ | | |
| | Chemical adsorption capacity | | | √ | |
| | Breakout time | | | | |
| | Water | | | | √ |
| | Solute | | | | √ |
| | Horizontal flow in seams or lifts | | √ | | |
| | Horizontal flow beneath geomembrane | | √ | | |
| | Generation of consolidation water | √ | | | |
| Physical/ mechanical issues | Freeze-thaw behavior | √ | | | |
| | Total settlement | | √ | | |
| | Differential settlement | √ | | | |
| | Stability on slopes | | | | √ |
| | Bearing stability, or squeezing | | | √ | |
| Construction issues | Puncture resistance | | | √ | |
| | Subgrade conditions | | | √ | |
| | Ease of placement | √ | | | |
| | Speed of construction | √ | | | |
| | Availability of materials | √ | | | |
| | Requirements for water | √ | | | |
| | Air pollution concerns | √ | | | |
| | Weather constraints | | | | √ |
| | Quality assurance considerations | | √ | | |

Table 5. Generalized technical equivalency assessment for GCL **covers above** landfills and abandoned dumps, Koerner & Daniel (1995).

| Category | Criterion for evaluation | Probably superior | Probably equivalent | Probably not equivalent | Equivalency dependent on site or product |
|---|---|---|---|---|---|
| Hydraulic issues | Steady flux of water | | √ | | |
| | Breakout time of water | | | | √ |
| | Horizontal flow in seams or lifts | | √ | | |
| | Horizontal flow beneath geomembrane | | √ | | |
| | Generation of consolidation water | √ | | | |
| | Permeability to gases | | √ | | |
| Physical/ mechanical issues | Freeze-thaw behavior | √ | | | |
| | Shrink-swell behavior | √ | | | |
| | Total settlement | | √ | | |
| | Differential settlement | √ | | | |
| | Stability on slopes | | | | √ |
| | Vulnerability to erosion | | | | √ |
| | Bearing stability, or squeezing | | | √ | |
| Construction issues | Puncture resistance | | | √ | |
| | Subgrade conditions | | | √ | |
| | Ease of placement | √ | | | |
| | Speed of construction | √ | | | |
| | Availability of materials | √ | | | |
| | Requirements for water | √ | | | |
| | Air pollution concerns | √ | | | |
| | Weather constraints | | | | √ |
| | Quality assurance considerations | | √ | | |

Regarding *construction issues* it appears that only the puncture resistance and need for very careful subgrade preparation of GCLs are limiting issues. The self-healing characteristics of bentonite clay, however, must be considered in regard to puncture. Regarding subgrade conditions Scheu et al. (1990) describe GCLs placed over very rough subgrades. Even further, GCLs are sometimes used as protection mats in Germany placed *over* geomembranes and beneath coarse drainage stone in leachate collection layers. Lastly, a key issue, as with all geosynthetics and natural soil materials, is proper CQC and CQA insofar as their installation is concerned.

Thus it is felt that in most cases a GCL can replace a CCL on the basis of technical equivalency. One important issue not addressed in Tables 4 or 5 is cost. In areas where proper natural clay soils are plentiful, CCLs will be competitive to GCLs. In areas where they are not, and blending of native soils with admixed bentonite clay is necessary for a proper CCL, the GCLs will usually be very cost effective. This is obviously a site specific consideration.

## 6 SUMMARY

The purpose of this opening chapter in the book is to properly position GCLs in the context of liquid and gaseous barriers in environmental and hydraulic engineering applications. The subsequent chapters are of considerably more detail and are focused on the particular GCL issue being addressed. Bentonite, the core material of GCLs, is of paramount importance. It must be of the proper type and amount. The geosynthetic cap and carrier materials are also investigated. The all-important issue of field performance of GCLs is addressed in the book by Kavazanjian. Thus, from this first chapter to the final one, GCLs are presented as being bona-fide engineering materials which rightly are included in geosynthetics as a now established engineering discipline.

## ACKNOWLEDGEMENTS

The financial assistance of the member organizations of the Geosynthetic Institute and its related institutes for research, information, education, accreditation and certification is sincerely appreciated. Their identification and contact member information is available on the institute's web site at <www.geosynthetic-institute.org>.

## REFERENCES

Boardman, B. T., and Daniel, D. E. 1996. "Hydraulic Conductivity of Desiccated Geosynthetic Clay Liners," *J. Geotechnical and Geoenvironmental Eng., ASCE*, Vol. 122, No. 3, pp. 204–208.

Daniel, D. E., Shan, H.-Y., and Anderson, J. D. 1993. "Effects of Partial Wetting on the Performance of the Bentonite Component of a Geosynthetic Clay Liner," *Proceedings of Geosynthetics '93,* IFAI, pp. 1483–1496.

Daniel, D. E., Trautwein, S. J., and Goswami, P. K. 1997. "Measurement of Hydraulic Properties of Geosynthetic Clay Liners Using a Flow Box," in *Testing and Acceptance Criteria for Geosynthetic Clay Liners*, ASTM STP 1308, ed. Larry W. Well, ASTM, pp. 196–207.

Estornell, P., and Daniel, D. E. 1992. "Hydraulic Conductivity of Three Geosynthetic Clay Liners," *Jour. of Geotechnical Eng.,* ASCE, Vol. 118, No. 10, pp. 1592–1606.

Giroud, J.-P., Badu-Tweneboah, K., and Soderman, K. L. 1997. "Comparison of Leachate Flow through Compacted Clay Liners and Geosynthetic Clay Liners in Landfill Liner Systems," *Geosynthetics Int.*, Vol. 4, Nos. 3–4, pp. 391–431.

Heerten, G., von Maubeuge, K., Simpson, M. and Mills, C. 1993. "Manufacturing Quality Control of Geosynthetic Clay Liners – A Manufacturers Perspective," *Proceedings of 6th GRI Seminar, MQC/MQA and CQC/CQA of Geosynthetics*, IFAI, 1993, pp. 86–95.

Hsuan, Y. G. and Koerner, R. M. 2002. "Durability and Lifetime of Polymer Fibers With Respect to Reinforced Geosynthetic Clay Barriers," *Proc. Clay Geosynthetic Barriers,* ed. H. Zanzinger, R. M. Koerner, and, E. Gartung, A. A. Balkema, 73–86.

Jeon, H.-Y. 2009. "Clay Nanoparticle Formulated Geotextiles Used in Geoenvironmental Applications," *Proceedings of GRI-22 Conference,* Salt Lake City, GII, Folsom, PA, USA, (on CD).

Koerner, G. R. 2004. "Comparing GCL Performance via Different Permeameters," *Second Symposium on Geosynthetic Clay Liners*, STP1456, ASTM, pp. 110–120.

Koerner, R. M. 2005. *Designing With Geosynthetics*, 5th Edition , Prentice Hall Book Co., Upper Saddle River, NJ, 799 pgs.

Koerner, R. M. and Daniel, D. E. 1995. "A Suggested Methodology for Assessing the Technical Equivalency of GCLs to CCLs," in *Geosynthetic Clay Liners*, ed. R. M. Koerner, E. Gartung, and H. Zanzinger, A. A. Balkema, pp. 73–100.

Koerner, R. M., Koerner, G. R., and Eberlé, M. A. 1996. "Out-of-Plane Tensile Behavior of Geosynthetic Clay Liners," *Geosynthetics Int.*, Vol. 3, No. 2, pp. 277–296.

Koerner, R. M. and Narejo, D. 1995. "On the Bearing Capacity of Hydrated GCLs," Jour. of Geotechnical Engineering Division, ASCE, Vol. 121, No. 1, pp. 82–85.

Kolstad, D. C., Benson, C. H. and Edil, T. B. 2004. "Hydraulic Conductivity and Swell of Nonprehydrated Geosynthetic Clay Liners with Multispecies Inorganic Solutions," *Jour. Geotechnical and Geoenvironmental Engineering*, ASCE, Vol. 130, No. 12, December, pp. 1236–1249.

Kraus, J. B., Benson, C. H., Erickson, A. E., and Chamberlain, E. J. 1997. "Freeze-Thaw Cycling and Hydraulic Conductivity of Bentonite Barriers," *Jour. of Geotechnical and Geoenvironmental Eng.*, ASCE, Vol. 123, No. 3, pp. 229–238.

Rowe, R. K., Bostwick, L. and Thiel, R. 2009. "GCL Shrinkage and the Potential Benefits of Heat-Tacking GCL Seams," *Proceedings of Geosynthetics 2009,* Salt Lake City, IFAI, Roswell, MN, USA (on CD).

Scheu, C., Johannssen, K., and Soatloff, F. 1990. "Nonwoven Bentonite Fabrics – A New Fiber Reinforced Mineral Liner System," *4th Intl. Conf. on Geotextiles, Geomembranes and Related Products,* ed. Den Hoedt, A. A. Balkema, pp. 467–472.

Thiel, R., Daniel, R. E. Erickson, R. B. Kavazanjian, E. and Giroud, J. P. 2004. GSE Gundseal GCL Design Manual, GSE Lining Technology, Inc., Houston, Texas, 200 pgs.

von Maubeuge, K. P., and Lucas, S. N. 2002. "Peel and Shear Test Comparison and Geosynthetic Clay Liner Shear Strength Correlation," *Proc. Clay Geosynthetic Barriers*, H. Zanzinger, R. M. Koerner and E. Gartung, Eds., A. A. Balkema Publ., pp. 104–110.

Zanzinger, H., Koerner, R. M. and Gartung, E., Eds. 2002. *Clay Geosynthetics Barriers*, A. A. Balkema, 399 pgs.

Zornberg, J. G., McCartney, J. S. and Swan, R. J. Jr. 2005. "Analysis of a Large Database of GCL Internal Shear Strength Results," *Jour. of Geotechnical and Geoenvironmental Engineering*, ASCE, Vol. 131, No. 3, pp. 367–380.

# CHAPTER 2

# Durability and lifetime of the geotextile fibers of geosynthetic clay liners

Y.G. Hsuan
*Drexel University, Philadelphia, PA, USA*

R.M. Koerner
*Geosynthetic Institute, Folsom, USA*

## 1 INTRODUCTION AND BACKGROUND

Since their introduction in the early 1990's, reinforced geosynthetic clay liners (as opposed to the original nonreinforced products), have been used in many permanent applications involving sloping surfaces. Such situations are as follows;

- final covers of steeply sloped closed landfills,
- base liners in canyon-type landfills liners along the sloping sidewalls,
- liners along the sloping sides of canals and waterways,
- liners for secondary containment of storage tanks, and
- liners beneath steeply sloping roadways or shoulders of roadways

Geosynthetic clay liners (GCLs) are often used by themselves or are incorporated with additional geosynthetics in liner systems of all types. While most are in landfill applications, there are also many uses in geotechnical, transportation and hydraulic engineering. When the field situation calls for the GCL to be used on a sloped surfaces of approximately 7 to 10 degrees or more, one should consider a fiber reinforced GCL. Two different manufacturing types are currently available; needlepunched and stitch bonded as shown in Figures 1(a) and 1(b), respectively.

Both types of reinforced GCLs have been extensively studied in the laboratory using relatively large shear boxes to challenge the internal strength of reinforcement and an extensive database is available, Zornberg (2005). In all cases, however, shear rates were relatively rapid in the context of permanent applications that are designed for many decades of service lifetime. Service lifetimes are clearly site specific decisions, but probably vary from a minimum of 30-years for highway applications, to 100's of years for landfill liners and covers.

Long-term shear tests are rarely conducted in the laboratory. Some exceptions being reported by Heerten et al. (1995) and Koerner et al. (2000), both of which were on the as-manufactured products, i.e., the fibers were not aged, per se. Regarding full-scale field performance evaluation, the Cincinnati test plots have been evaluated for over 10-years and the 3(H)-to-1(V) test sections have been incorporated in the project's final cover, Daniel et al. (1998). At this point in time all of the slopes are stable. Time will tell as to the eventual behavior. In the interim, investigation into the anticipated behavior of GCL reinforcement fibers and yarns should be undertaken.

In addition to fiber durability, some GCLs are associated with geomembranes and geofilms. These are typically made from polyethylene resins and their durability must also be addressed. As will be seen in commentary at the end of this paper this issue is much more tractable than with fiber durability, the latter being the focus of the paper.

## 2 STRUCTURE AND DEGRADATION OF POLYMER FIBERS

It can be anticipated that numerous factors could affect the functioning of GCLs over their service lifetime. One of the factors is the durability of the reinforcement fibers. Note that the word "fibers" will be used throughout this paper, since it should be recognized that the yarns used in stitch

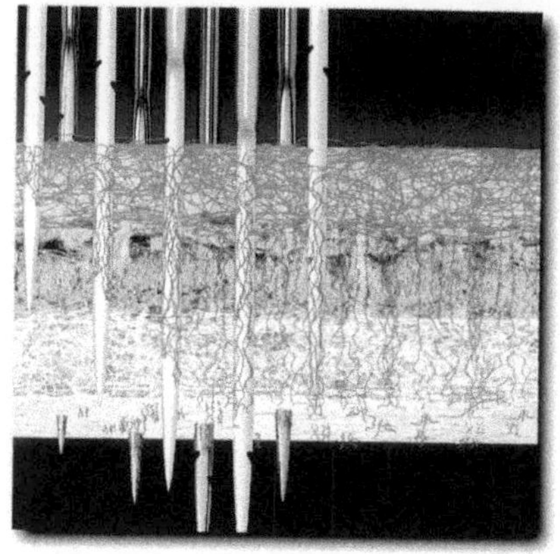

(a) Needle punched GCL (ref., Naue, GmbH)

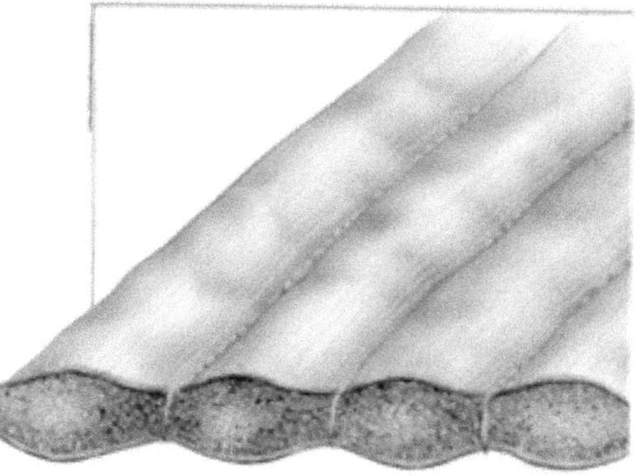

(b) Stitch bonded GCL (ref., Huesker Synthetic, GmbH)

Figure 1. Two different types of internally reinforced GCLs.

bonding consist of individual fibers. If the polymeric fibers in the GCL undergo degradation, they could eventually fail under the applied shear stresses at times earlier than the anticipated lifetime of the system.

### 2.1 *Polymers used for GCL fiber reinforcement*

The type of polymer that is commonly used for reinforced GCLs is polypropylene (PP). The exception is one GCL made from polyethylene (PE) geotextiles and fibers. However, both PP and PE are members of the polyolefin family, which is known to be susceptible to oxidative degradation. As a result, antioxidants are incorporated into the formulation of the resin to protect the polymer. In this paper, issues that are related to polyolefin degradation and stabilization are discussed.

### 2.2 *Fiber configuration of GCL reinforcement*

As mentioned previously, there are two types of reinforced GCLs. The first (and most commonly used) type of GCL is fiber reinforced and referred to as a needlepunched GCL, recall Figure 1(a). The bentonite is contained between the upper and lower geotextiles by needlepunching. One of the geotextiles must be nonwoven so that fibers from this geotextile can be punched through the bentonite layer and opposite geotextile thereby creating an interlocked system. The opposite geotextile can be either woven or nonwoven. It should be noted that a composite woven/nonwoven geotextile is recommended for double nonwoven related GCLs. The second type of reinforced GCL is referred to as stitch bonded in that bentonite is confined between two geotextiles via parallel rows of fiber yarns stitched through the entire structure, recall Figure 1(b). Both upper and lower geotextiles in this type of product are wovens, consisting of slit film yarns, monofilament yarns or monofilament fibers.

Fibers, in either short staple form in needlepunching or continuous yarn form in stitchbonding are manufactured via melt spinning extrusion and oriented to yield a diameter ranging from 40 to 80 μm. The degree of fiber orientation varies in different manufacturing processes and is clearly a product-specific decision of the manufacturer.

### 2.3 *Oxidative effects on polymer structure*

While there are many possible degradation mechanisms, oxidation of polyolefins is almost always present. Oxidative degradation is even present in a hydrated state, although in a reduced rate. Figure 2 illustrates three possible zones of fibers within a GCL where oxidation can occur. Zone A is often adjacent to an overlying geomembrane as is typical of a geomembrane/GCL composite liner. The amount of oxygen available is probably low. Zone B has the fibers encased within

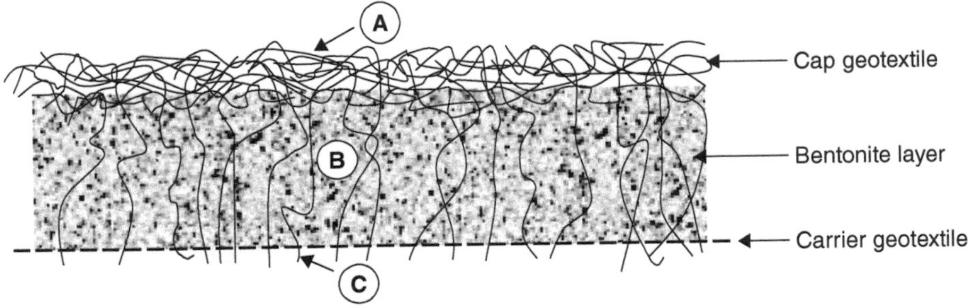


Figure 2. Various zones of a GCL where oxidation can occur to different degrees.

hydrated bentonite. The oxygen level is also probably low. Zone C has the GCL above a drainage geocomposite in a double liner system or adjacent to a soil subgrade in a single liner system. Its oxygen level is probably the highest of the three zones described.

Polypropylene and polyethylene have similar oxidation mechanisms. The oxidation is initiated by the formation of free radicals, which subsequently react with oxygen and promote a series of chain reactions (Grassie and Scott, 1985). The process leads to an auto-acceleration reaction. There are three material properties that can affect the oxidative degradation of polyolefin fibers.

### 2.3.1 *Tertiary carbon atoms*

Pure non-stabilized PP is susceptible to oxidation more than other polyolefins due to the number of tertiary carbon atoms. Tertiary carbon atoms are carbon atoms which bond with three other carbon atoms and one hydrogen atom. The C—H bond in this case has greater tendency to be disassociated forming a free radical in comparison to other carbon atoms (i.e., primary and secondary carbon atoms), as shown in Equation (1). In PP, every second carbon atom along the polymer chain is a tertiary carbon. This increases the probability of free radical formation compared to those having fewer tertiary carbon atoms, such as in linear polyethylene.

$$\mathrm{C{-}\underset{\displaystyle C}{\overset{\displaystyle H}{C}}{-}C} \longrightarrow \mathrm{C{-}\underset{\displaystyle C}{\dot{C}}{-}C} + \mathrm{H^{\bullet}} \qquad (1)$$

### 2.3.2 *Crystallinity*

In polymers, oxidation takes place almost entirely in the amorphous phase of the polymer. Oxygen diffuses into the amorphous phase easier than the crystalline phase (Michaels and Bixler, 1961). Rapoport et al. (1975) demonstrated that the molecular mobility within the amorphous phase plays a vital part in oxygen diffusion. The crystallinity of the fiber is dependent on its manufacturing processing and the degree of orientation.

### 2.3.3 *Orientation*

The microstructure of oriented polymers is very different from that of non-oriented polymers. The Peterlin model (1966) describes the polymer orientation mechanism under tensile stress. This elongation results in a much denser amorphous phase and higher crystallinity than that in the comparable isotropic material. This dense structure retards the diffusive mobility of oxygen; thus, the rate of oxidation is retarded.

## 3 STABILIZATION OF POLYPROPYLENE (AND POLYETHYLENE) FIBERS

Free radical oxidation reactions in polyolefins are prevented by incorporating antioxidants into the formulation. The purposes of the antioxidants are (i) to prevent degradation during high termperature processing, and (ii) to prevent oxidative reactions taking place during the long-term service

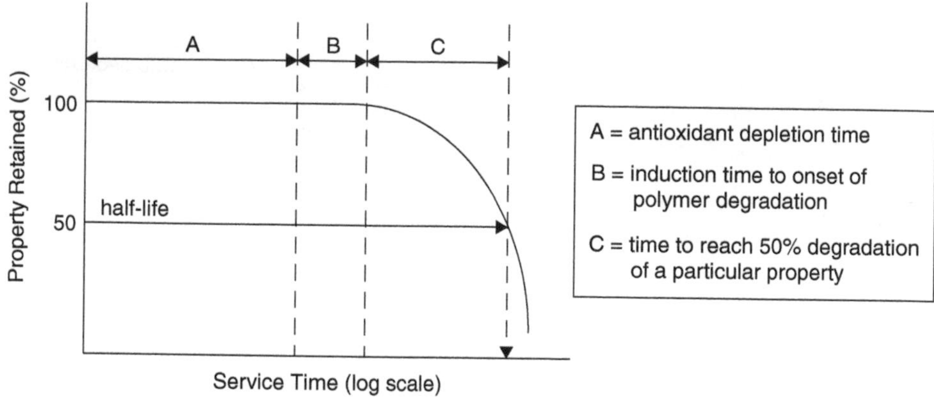


Figure 3. Various stages in degradation of polyolefin fibers.

life. Antioxidants are designed to react with free radicals and alkyperoxides that form in the polymer, converting them to stable molecules. Polymers that are formulated with appropriate amount and types of antioxidants can extend the lifetime of material significantly, as shown in Figure 3. The entire Stage A is contributed by the antioxidants formulated in the product. The lifetime of a non-stabilized polyolefin fiber is limited to stages B and C. As indicated in Figure 3, the "halflife" is often taken as the service life of the polymer under investigation and includes all three stages as illustrated.

Fay and King (1994) describe the types and functions of various antioxidants. For geosynthetic products, Hsuan and Koerner (1998) explain the function of antioxidants in high-density polyethylene. It is expected that polypropylene would have similar protection and degradation mechanisms as polyethylene.

## 4 ANTIOXIDANTS DEPLETION MECHANISMS

As indicated in Figure 3, the degradation insofar as a change in physical and mechanical properties starts after all antioxidants are depleted, i.e., at the end of Stage A. The depletion rate of antioxidants (i.e., the duration of Stage A) is critical to the overall oxidation degradation of stabilized polyolefins. The amount of antioxidants in a stabilized polymer decreases gradually as aging progresses. The depletion can be caused by two mechanisms: (i) chemical reactions of the antioxidant, and (ii) physical loss of antioxidants from the polymers by leaching. The two mechanisms can occur simultaneously. In addition, when one considers fibers (as opposed to geomembranes and geofilms), the large specific surface area has a significant effect on these two mechanisms.

### 4.1 *Chemical reactions*

As a chemical reaction, the antioxidants are consumed by free radicals and alkyperoxides present in the material. The rate of consumption, which progresses from the surface of the fiber inward depends on the concentration of these two species within the polymer internal material structure and external ambient factors.

### 4.2 *Physical loss*

The two major concerns with respect to the physical stability of antioxidants in the polymer are their volatility and extractability (Luston, 1986). Research has indicated that the distribution of antioxidants in semicrystalline polymers is not uniform, owing to the presence of crystalline and amorphous phases. It appears that a greater concentration of antioxidants is found in the amorphous region which is fortunate because the amorphous region is also the most sensitive to degradation. Hence the mobility of antioxidants in the amorphous phase controls these two physical processes.

The volatility of antioxidants is a thermally activated process and temperature change affects not only the evaporation of the stabilizers from the surface of the polymer but also their diffusion from the interior to the surface layer. For most applications, the typical temperature is below 40°C, see Koerner & Koerner (2005). Thus, volatility is probably not a concern. Because of this, however, one must avoid inducing such a mechanism in accelerated laboratory aging tests, i.e., the test temperature cannot be so high than volatilization takes place. The issue of laboratory incubation will be discussed later in the paper.

Extractability of antioxidants plays a part wherever the fiber comes into contact with liquids such as water or leachate. The rate of extraction is controlled by the dissolution of antioxidants from the surface and diffusion from the interior structure to the surface. Smith et al., (1992) performed an aging study on a medium density polyethylene pipe. They found that the consumption of antioxidants was three times faster in water than in air at temperatures of 80, 95 and 105°C. Hsuan and Koerner (1995) also found the same acceleration factor for depletion of antioxidants in a HDPE geomembrane at test temperatures of 55, 65, 75 and 85°C.

### 4.3 *Specific surface area*

Fibers with small diameters have a greater specific surface area than those with larger diameters and particularly so in comparison to flat sheet such as geomembrane and geofilm. Both depletion mechanisms stated above are highly affected by the specific surface area. For chemical reactions, the oxidation products (free radicals and alkyperoxides) result from the oxygen that diffuses through the surface of the fiber and then reacts with polymer chains. The greater the specific surface area, the more oxygen can diffuse into fibers producing more free radicals and alkyperoxides to react with antioxidants. The leaching of antioxidants is based on the chemical interaction between antioxidants and the surrounding liquid. Thus, fibers with high specific surface area provide more sites for these two substances to interact.

## 5 EXTERNAL ENVIRONMENTAL EFFECTS ON OXIDATION

The oxidation reaction in polyolefins is somewhat sensitive to the surrounding environment. An environment which can accelerate the formation of free radicals would increase the rate of antioxidant depletion via chemical reactions. An environment that can induce leaching of antioxidants also increases the depletion of antioxidants via physical loss.

### 5.1 *Energy level*

Energy is required to generate free radicals. Sunlight and heat are the two energy sources that could be encountered by the GCL. However, unless GCLs are utilized in an exposed application, sunlight will not be the source leading to oxidative degradation. Conversely, heat is always a source of energy. All other things being equal, polyolefins will degrade faster at higher temperatures as opposed to lower temperatures. Increase in temperature can enhance oxygen diffusion as well as the rate of reactions.

### 5.2 *Oxygen concentration*

The concentration of available oxygen is an essential component to any oxidation reaction. In a recent durability study (Elias, 1999), one part of the project focused on the evaluation of oxidation degradation of polyolefin geotextiles and geogrids at oxygen concentrations of 8 and 21%. Figure 4 shows the strength retained at 70 and 80°C of these two different oxygen concentrations after being incubated in forced air ovens. The contrast between these two incubation environments is obvious. Clearly, the availability of oxygen has major effect on the degradation of polyolefins.

The availability of oxygen is much higher for geosynthetics that are exposed to air than those that are in liquid or in bentonite as with a GCL, recall Figure 2. The reinforcing fibers in GCLs are surrounded by hydrated bentonite. The oxygen content in the liquid is lower than air. The oxygen content of water is typically 8% in comparison to air at 21%. In a landfill liner system, however, the cap geotextile of the GCL might be exposed to lower oxygen levels, whereas the carrier geotextile

Table 1. Estimated oxygen levels in different applications (Hsuan, 2000).

| Application | Condition | Estimated $O_2$ Content |
|---|---|---|
| Tank farms | Exposed or thin cover | 21% |
| Gas tank secondary liner | Water moist under cover | 5–8% |
| Canal liners | Water immersed | 8% |
| Highways and airfields | Moist to dry | 8–21% |
| Primary composite liner | Leachate saturated | 2–8% |
| Secondary composite liner | Water saturated | 2–8% |
| Cover composite liner | Water moist under soil cover | 5–8% |

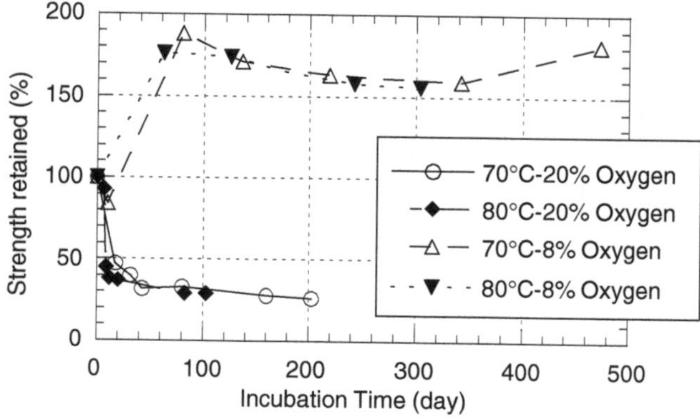


Figure 4. Tensile strength retained as 21% and 8% oxygen environments.

might be much higher. Unfortunately, data is not available to our knowledge. Table 1 shows the estimated oxygen levels that might occur in various applications.

## 5.3 *Liquid chemistry*

Since the fibers of GCLs are most likely surrounded by liquid, i.e., hydrated bentonite or leachate, the potential interactions between the liquid and antioxidants should be considered. Certain transition metals have a catalyzing effect on the oxidation. Wisse (1988) demonstrated the influence of iron oxide on PP geotextiles. Although the amount of transition metal ions that can be reached and sustained in the GCL is not known, the potential effect should be evaluated.

## 5.4 *Applied fiber stresses*

Other than an outright mechanical failure, applied tensile stresses can increase the oxidation of PP fibers. Horrocks and D'Souza (1990, 1992) found that applied stress increased the degradation rate of the polypropylene tapes that were tested. Obviously, GCLs on side slopes are subjected to shear stress; hence, sustained tensile stresses are generated in the reinforcement fibers.

## 5.5 *Annealing*

Annealing is a treatment of the finished fiber by heating. Rapoport et al., (1977) observed that annealing counteracts the oxidation retardation generated by stretching. This phenomenon is illustrated in Figure 5. The amorphous phase of an oriented material is in a non-equilibrium elongated configuration. Annealing brings the chains back to equilibrium, increasing chain mobility and reducing packing density. Thus, oxygen can diffuse into the amorphous phase easier, causing more rapid oxidation than in the non-annealed oriented material. However, for such changes to occur, the annealing temperature needs to be higher than the temperature at which the fiber stretching was conducted.

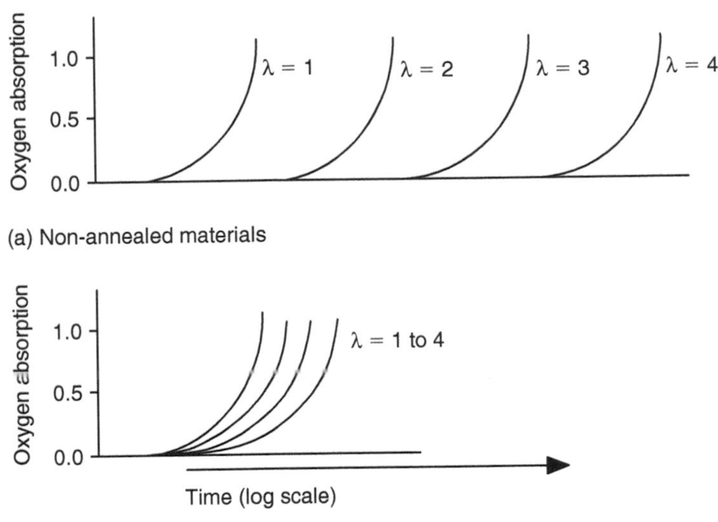


Figure 5. Effect of annealing on the induction period of oxidation.

## 6 DURABILITY EVALUATION AND SPECIFICATION

A durability evaluation of the fibers in a GCL obviously requires some type of laboratory incubation and subsequent testing. In general, this would be a controlled temperature long-term tensile test which assesses the retained strength or elongation of the fibers over time. Such a procedure would generally be prohibitively long. Alternatively, one could focus on the antioxidants and their depletion time due to oxidation. If this very conservative option is selected, the antioxidant depletion time, which is graphically shown as Stage A in Figure 3, would be the goal. Oxidative induction time (OIT) tests can be performed on incubated fibers using either ASTM D3895 (the standard OIT test), or ASTM D5885 (the high pressure OIT test). Both are described in detail by Hsuan and Koerner, 1999. Such testing (tensile strength, tensile elongation or OIT) can be done in a performance manner (which is largely research oriented) or by index tests (which can lead to a generic specification). Each procedure will be described along with a suggested generic specification.

### 6.1 *Performance testing*

Performance tests are designed to investigate the performance of the material under test conditions simulating field situations as closely as possible. However, to simulate the field situation of reinforcement fibers in GCLs is a challenging task, particularly with respect to the normal stress component. Figure 6 shows two possible incubation devices, in which the fibers in the GCL samples are subjected to both liquid and temperature effects. The difference is the applied normal stress on the incubating sample; one without and the other with normal stress.

The tensile load of the fibers is controlled by the applied shearing load. For consistent purpose of the tests, the liquid should use distilled water. However, other types of liquid could also be utilized. The incubation temperatures must not be above the fiber draw temperature to prevent an annealing effect. Temperatures at or below 70°C are probably acceptable for polyolefin fibers. There should be a minimum of three different temperatures, e.g., 50, 60 and 70°C.

The incubated samples are removed after predetermined periods of time. Peel strength of the GCL might be used to monitor changes in the specimens although the paper by Zornberg (2005) challenges its relationship to shear strength. The incubation time should be sufficiently long so that more than a 50% decrease in peel strength is achieved. The time to reach 50% ($t_{50}$) strength is the halflife of the specimen, as shown in Figure 7. The inverse of $t_{50}$ is the rate of reaction, which is then plotted against inverse incubation temperature to yield an Arrhenius graph, as shown in Figure 8.

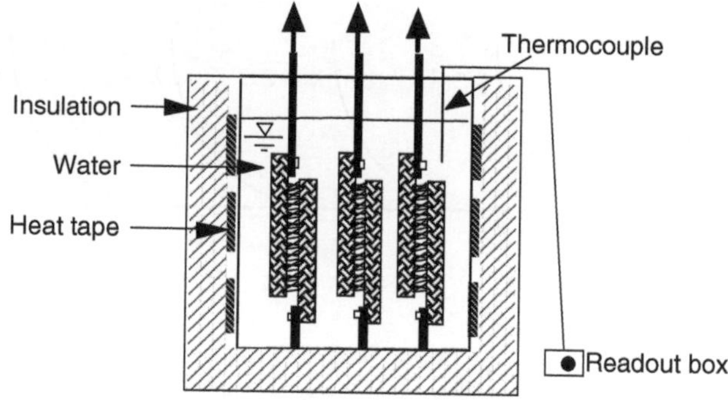


(a) No normal stress on samples

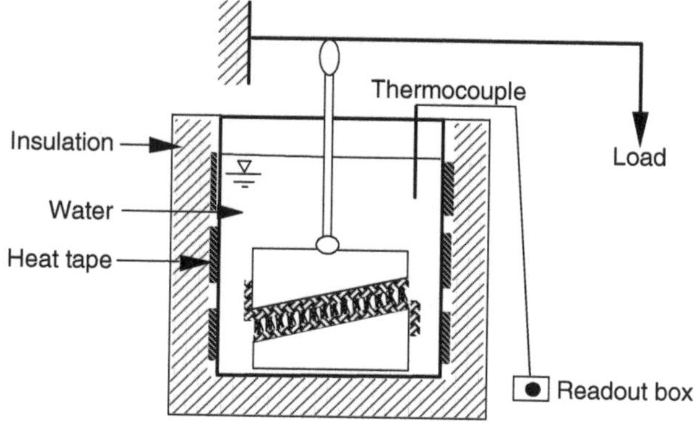


(b) Normal stress applied to sample

Figure 6. Two possible incubation setups for assessing the long-term durability of reinforced GCL specimens, after Hsuan (2000).

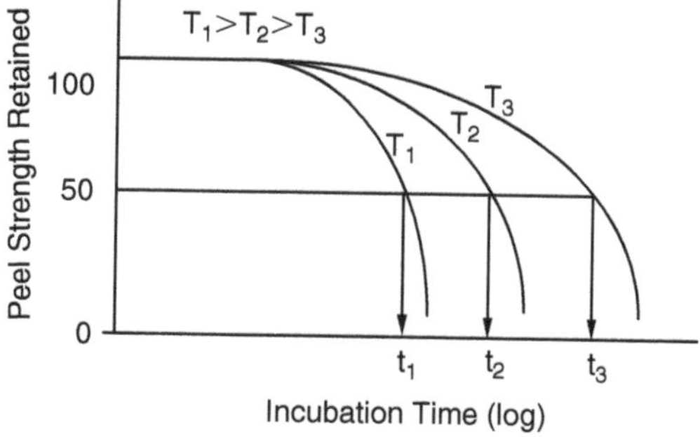


Figure 7. Determination of the halflife of the GCL specimen via peel strength.

The reaction rate ($1/t_{50}$) at the site specific temperature (e.g., at 20°C) can be determined using an extrapolation of the linear behavior, as shown in Figure 8. The halflife of the fibers at a given temperature, say 20°C or ($t_{20}$), can be subsequently determined using the following equation:

$$\ln\left(\frac{1}{t_{20}}\right) = \ln A + \left(\frac{-E}{R}\right)\left(\frac{1}{20+273}\right) \tag{2}$$

where $t_{20}$ = halflife of the fibers at 20°C; A = intercept on the y-axis; E = activation energy; R = gas constant; and E/R= slope of the Arrhenius curve.

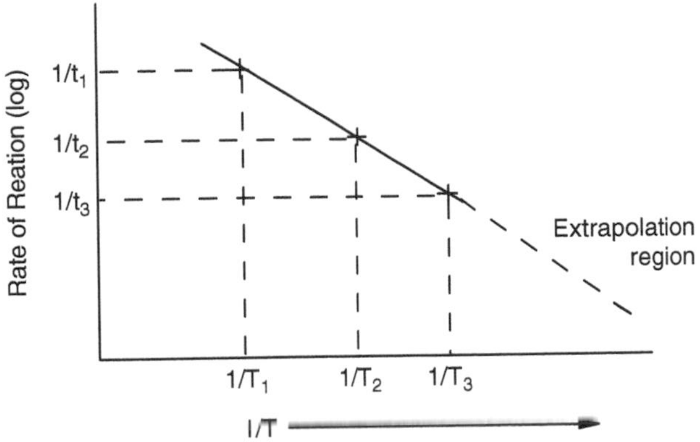


Figure 8. Arrhenius graph for PP fibers.

## 6.2 *Incubation methods for subsequent index testing*

The purpose of an incubation procedure is to accelerate a reaction in the material being evaluated. For subsequent index testing, this is typically done by air oven aging which is quite appropriate for polyolefin fibers and clearly represents a worst-case scenario.

The goal of such incubation is to achieve consistent results within a reasonable length of time and use the incubated samples for subsequent testing. In many situations, material properties might be changed under the condition of the incubation. As long as such changes are the same for all intended test materials, the tests are appropriate and the resulting data are valuable. For example, high temperatures are often used in oven aging to evaluate various stabilization packages in a polymer. The selected incubation temperature usually is just below the melting point of the polymer. Gray (1990) used a forced air oven at 150°C and 120°C to study different antioxidant packages of polyolefin materials. At 150°C, polypropylene fibers would be annealed and the orientation effect is completely removed as indicated in Figure 5. However, this negative impact would be the same for all PP fibers tested regardless the type of manufacturing processes. A proposed CEN test method entitled "Themo-oxidation Resistance of Polypropylene Split Film Yarns for Application as Filter Fabrics in Hydraulic Structures" requires 150°C for the oven test to determine the embrittlement time. This is to validate the amount and type of antioxidant package in the fiber. In the same CEN method, the extractability of antioxidants is also evaluated by heating the fiber at 90°C for 14 days and then subjecting the fiber to oven aging to determine the embrittlement time. Such oven aging procedures have been used by a number of researchers focusing on polyolefins used in fiber manufacturing. They will be described in the next section along with the particular test procedures which were used for the evaluation.

## 6.3 *Index test methods and procedures*

Some selected studies on polyolefin fibers used for incubated geosynthetic products (mainly geotextiles) are presented.

Salman, et al. (1998) used both forced air ovens and starved-air incubation, followed by strength testing on staple and continuous polypropylene fibers. They used different temperatures allowing for Arrhenius modeling for lifetime prediction. Their results indicated values of 51 years (at 21% $0_2$) and 240 years (at 8% $0_2$) for antioxidant depletion and induction time, and then a 50% strength loss at 20°C of 38 years. In the context of the three stages of Figure 3, the lifetime predicted was 89 years for full oxygen conditions, and 278 years under starved oxygen conditions for the particular fibers selected.

Müller & Jakob (2000) used forced air oven aging followed by OIT testing on polypropylene and polyethylene nonwoven geotextiles. They conclude that an antioxidant package along with carbon black or ultraviolet light stabilizers should be added to the formulation to produce high OIT-values. However, no specific numbers were provided.

Table 2. Lifetime prediction of a backfilled HDPE geomembrane as a function of in situ service temperature.

| In service Temperature (°C) | GSI* Stage A (yr.) Std OIT | HP-OIT | Stage B (yr.) | Stages A + B (yr.) | Gidde et al. (1994)** Total Lifetime (yr.) |
|---|---|---|---|---|---|
| 20 | 200 | 215 | 30 | 230 | 2124 |
| 25 | 135 | 144 | 25 | 160 | 1103 |
| 30 | 95 | 98 | 20 | 115 | 585 |
| 35 | 65 | 67 | 15 | 80 | 317 |
| 40 | 45 | 47 | 10 | 55 | 175 |

* Tests were performed on 1.5 mm thick HDPE geomembrane
** Tests were performed on 2.2 mm thick HDPE pipe

Schroeder, et al. (2000) aged a number of polypropylene geotextiles in auto-claves at a high pressure of 50 bar. They compared the results to forced air oven incubation at 110°C using the induction time and time to 80% residual strength. The data indicate that high pressure can significantly accelerate the oxidation; however, no predictive lifetime results.

Von Maubeuge and Ehrenberg (2000) incubated polypropylene and polyethylene fibers in forced air ovens per the CEN protocol mentioned earlier. Using tensile strength as their criterion, Arrhenius modeling resulted in halflife values of 25–50 years for polypropylene fibers and more than 100 years for polyethylene fibers for the fibers under investigation.

Thomas (2002) found that a 50% strength loss would occur after 30 years when exposed to air at 21% oxygen, whereas 100 years was attainable when exposed to starved-air conditions at 8%. Both situations assumed in-situ temperature of 20°C.

It should be mentioned that Müller & Jakob (2000) recommend that a PP geotextile should retain more than 50% of its strength after incubation for 1-year at 80°C.

## 7 GEOMEMBRANES AND GEOFILM

As with the fibers of geotextiles associated with GCLs, so must the geomembranes and geofilms be comparably durable to perform over their service lifetimes. In this regard, there is a considerable body of knowledge concerning polyethylene geomembrane durability and lifetime prediction. Table 2 shows the available lifetime data for commercially available 1.5 mm thick high density polyethylene (HDPE) geomembranes that are associated with some GCLs. For comparison, the lifetime of HDPE pipe is also included in the table.

This type of data strongly suggests that 1.5 mm thick HDPE geomembranes will be superior in anticipated lifetime to the PP fibers that were the focus of this paper.

The same cannot be said for the very thin ($\simeq$0.10 mm) geofilms that are used with some GCLs. In addition to the relative thinness, the polyethylene used in geofilms is of considerably lower density than HDPE. Thus, lifetime prediction of the geofilms is a topic in need of durability research in the future.

## 8 SUMMARY AND CONCLUSIONS

This paper has been focused on the degradation of polyolefin (polypropylene and polyethylene) fibers and yarns that are used to internally reinforce needlepunched and stitch bonded GCLs. Separate sections were presented on polymer structure, degradation and stabilization; laboratory performance incubation and testing; and index testing leading to lifetime prediction. Also commented upon was the lifetime of geomembranes and geofims which are associated with some GCLs.

### 8.1 *Polymer structure, degradation and stabilization*

The structure of polypropylene and polyethylene is well known. So much so that the major degradation mechanism, i.e., oxidation, is also well established. These aspects were described herein and elsewhere via a number of references. To retard this degradation, antioxidants are added to polymer formulations (along with carbon black or ultraviolet light inhibitors). Subsequently, three stages of degradation, leading to lifetime prediction, are established.

- Stage A – antioxidant depletion
- Stage B – induction time
- Stage C – degradation time to a specific limit (50% strength or elongation reduction is often used in this regard)

The analytic procedure to numerically quantify the above stages was also presented. It is based on test results generated from elevated temperature incubation and then are plotted on an Arrhenius graph to determine a predicted lifetime value.

### 8.2 *Laboratory performance incubation and testing*

The above procedure is most valid if the incubation simulates the application as close as possible. This is quite difficult for a composite material like GCLs (bentonite, textiles and fibers) in a complex structure (three-dimensional "truss-like" configuration). Nevertheless, two new incubation methods, as well as subsequent testing methods, are proposed. Both use the complete GCL, with or without compressive stress, and apply a shearing stress. By necessity, sample size is small which may be questionable for stitch bonded products due to possible scale effects. The methods appear to be reasonable for the needlepunched products. It is hoped that these concept ideas will generate dialogue as well as possible future activity.

### 8.3 *Index testing with a suggested specification*

The performance tests just noted are too research oriented and lengthy for use in a quality control context; thus, the need for shorter index testing which could be embodied in a generic specification. Almost certainly, this will take the form of fiber oven aging for the incubation, followed by either tensile testing (for strength and/or elongation), or oxidation induction time (OIT) testing. Both of these approaches have been recently used and the literature was reviewed and presented in this regard. Perhaps the most provocative study was that of Salman, et al. (1998) who identified the three stages of a particular polypropylene fiber as having a lifetime of 89 years in full oxygen conditions (i.e., exposed to ambient environment) and 278 years under starved oxygen condition (i.e., buried beneath soil or solid waste). Of course, this is for one particular type of fiber and its unique formulation.

Using this procedure, however, it is sufficient (and in the authors opinion necessary) to embody the results in a generic specification. In it the endurance section should challenge not only the reinforcement fibers, but the geotextiles and the stability of the bentonite, as well. Such a generic specification was presented in this book in the paper by Koerner and Koerner.

### 8.4 *Conclusion and recommendation*

Some concluding comments along with selected recommendations to the paper are as follows:

(i) When GCLs are subjected to long-term shear stresses, fiber durability is obviously important, particularly with respect to sloping surfaces and canyon-type landfill liners.
(ii) Factors involved in fiber durability are polymer formulation, stress level, and environmental conditions (e.g., oxygen level).
(iii) The key to the polymer formulation is the process manufacturing of the fibers, and the type and amount of antioxidants.
(iv) Lifetime assessment testing was suggested using two different performance testing schemes.
(v) Results should produce a three-stage degradation mechanism; (i) depletion of antioxidants, (ii) induction time, and (iii) halflife of a targeted property. For fibers, the latter would probably be strength or elongation. One would be quite conservative to predict lifetimes on the basis of antioxidant depletion alone.
(vi) The suggested performance test setup will be most difficult. The two incubation schemes offered are conceptual and promise to be troublesome particularly the containment of the

bentonite over long time periods. Also, the peel test that was suggested to monitor degradation behavior has large statistical scatter and may not be indicative of the materials shear strength which is the primary focus. Fiber testing is an alternative, but is also very sensitive and not without its own difficulties.

(vii) Oven incubation and index testing were reviewed in the context of past research with the goal of lifetime prediction. When such procedures are developed they should be embodied in a generic specification.

(viii) It should be mentioned in closing that fiber durability is not only an issue in reinforced GCLs. Many long-term applications of polyolefin geotextiles, geogrids, geonets and geomembranes utilize the same technology as presented in this paper.

(ix) As a result, there is some activity in developing appropriate antioxidant packages consisting of different types and amounts of chemical compounds. Furthermore, modifications in the polymeric fiber microstructure could also alter the oxidation mechanisms.

(x) A short section was devoted to degradation and lifetime of geomembranes and geofilms which are associated with some GCLs. In general, it is felt that properly formulated geomembranes will have service lifetime for exceeding geotextile fibers. Conversely, very thin geofilms are essentially unknown as to their durability and lifetime prediction. A study on these materials appears to be warranted.

## ACKNOWLEDGEMENT

The financial assistance of the member organizations of the Geosynthetic Institute and its related institutes for research, information, education, accreditation and certification is sincerely appreciated. Their identification and contact member information is available on the institute's web site at «geosynthetic-institute.org».

## REFERENCES

Daniel, D.E., Koerner, R.M., Bonaparte, R., Landreth, R.E., Carson, D.A., & Scranton, H.B. 1998. "Slope Stability of Geosynthetic Clay Liner Test Plots," Journal Geotechnical and Geoenvironmental Engineering, ASCE, Vol. 124, No. 7, pp. 628–637.

Elias, V., Salman, A., Juran, I., Pearce, E., & Lu, S. 1999. *Testing Protocols for Oxidation and Hydrolysis of Geosynthetics*, Federal Highway Administration, Publication No. FHWA-RD-97-144, 186 pgs.

Fay, J.J., & King, R.E. 1994. "Antioxidants for Geosynthetic Resins and Applications", Geosynthetic Resins, Formulations and Manufacturing, edited by Hsuan, Y.G., and Koerner, R.M., GRI-8, GRI Conference Series, pp. 77–96.

Gedde, U.W., Viebke, J., Leijstrom, H., and Ifwarson, M., 1994. "Long-Term Properties of Hot-Water Polyolefin Pipes – A Review" Polymer Engineering and Science, Vol. 34, No. 24, pp. 1773–1787.

Grassie, N. & Scott, G. 1985. *Polymer Degradation and Stabilization*, Cambridge University Press, New York, U.S.A., 221 pgs.

Gray, R.L. 1990. "Accelerated Testing Methods for Evaluating Polyolefin Stability", *Geosynthetic Testing for Waste Containment Applications, ASTM STP 1081*, R. M. Koerner, Ed., American Society for Testing and Materials, pp. 57–71.

Heerten, G., Saathoff, F., Scheu, C. & von Maubeuge, K. P. 1995. "On the Long-Term Shear Behavior of Geosynthetic Clay Liners (GCLs) in Capping Sealing Systems," Proc. International Symposium on Geosynthetic Clay Liners at Nurnberg, Germany, R. M. Koerner, E. Gartung and H. Zanzinger, Eds., A. A. Balkema Publishers Ltd., Rotterdam, pp. 141–150.

Horrocks, A.R. & D'Souza, Z.A. 1990. "Physiochemical Changes in Laboratory-aged Oriented Polypropylene Tapes – The Effects of Stress and Humidity", Geotextiles, Geomembranes and Related products, Den Hoedt, Ed., Balkema, Rotterdam, pp. 709–714.

Horrocks, A.R. & D'Souza, Z.A. 1992. "Degradation of Polymers in Geomembranes and Geotextiles", *Handbook of Polymer Degradation*, S. H. Halim, M. B. Amin, and A. G. Maadhah, Eds., Marcel Dekker, Inc., New York, pp. 433–506.

Hsuan, Y.G. 2000. "The Durability of Reinforcement Fibers and Yarns in Geosynthetic Clay Liners," Proceedings of the GRI-14 Conference on "Hot Topics in Geosynthetics – I," R.M. Koerner, G.R. Koerner, Y.G. Hsuan and M. V. Ashley, Eds., GII Publications, Folsom, PA, pp. 211–225.

Hsuan, Y.G. & Koerner, R.M. 1999. "Rational and Background for the GRI-GM13 Specification for HDPE Geomembranes," Proc. Geosynthetics '99, IFAI, Roseville, MN, pp. 385–222.

Hsuan, Y.G. & Koerner, R.M., 1995. "Long-Term Durability of High Density Polyethylene Geomembranes: Part I – Depletion of Antioxidants," Geosynthetic Research Report #16, Published by Geosynthetic Institute, Folsom, PA, 37 pgs.

Hsuan, Y.G. & Koerner, R.M. 1998. "Antioxidant Depletion Lifetime in High Density Polyethylene Geomembranes", Journal of Geotechnical and Geoenvironmental Engineering, Vol. 124, No. 6, pp. 532–541.

Koerner, G.R. & Koerner, R.M. 2005. "In-Situ Temperature Monitoring of Geomembranes," Proc. GRI-18 Conference on Geosynthetics Research and Development In-Progress, GeoFrontiers, ASCE, pp. 318–323.

Koerner, R. M., Soong, T.-Y., Koerner, G. R. & Gontar, A. 2000. "Creep Testing and Data Extrapolation of Reinforced GCLs," Proceedings of the GRI-14 Conference on "Hot Topics in Geosynthetics – I," R.M. Koerner, G.R. Koerner, Y.G. Hsuan and M. V. Ashley, Eds., GII Publications, Folsom, PA, pp. 189–210.

Luston, J. 1986. "Physical Loss of Stabilizers from Polymers," *Developments in Polymer Stabilization – 2*, Chapter 5, Scott, G., Ed., Published by Applied Science Publishers, Ltd, London, pp. 185–240

Michaels, A.S. & Bixler, H.J. 1961. "Solubility of Gases in Polyethylene," Journal of Polymer Science, 50, pp. 393–412.

Müller, W. & Jakob, I. 2000. "Comparison of Oxidation Stability of Various Geosynthetics," Proceedings of EuroGeo 2, Bologna, Italy, pp. 449–454.

Peterlin, A. 1966. "Molecular Mechanism of Plastic Deformation of Polyethylene," *The Meaning of Crystallinity of Polymers*, Price, F.P., Ed., Journal of Polymer Science, Part C, Polymer Symposia, No. 8, pp. 123–132.

Rapoport N. Ya., Berulava, S. I., Kovarskii, A.L., Musayelyan, I. N., Yershow, Yu.A. & Miller, V.B. 1975. "The Kinetics of Thermo-oxidative Degradation of Oriented Polypropylene in Relation to the Structure and Molecular Mobility of the Polymer," Vysokomol. Soyed, A17:11, pp. 2521–2527. (Translated in Polymer Sci. U.S.S.R., 17:11, pp. 2901–2909.)

Rapoport, N. Ya., Livanova, N. M. & Miller, V. B. 1977. "On the Influence of Internal Stress on the Kinetics of Oxidation of Oriented Polypropylene," Vysokomol. Soyed. A18 : 9, 1976, pp. 2045–2049. (Translated in Polymer Science U.S.S.R., 18, p. 2336–2341).

Salman, A., Elias, V. & DiMello, A. 1998. "The Effect of Oxygen Pressure, Temperature and Manufacturing Processes on Laboratory Degradation of Polypropylene Geosynthetics," Proceedings 6 ICG, IFAI, Roseville, MN, pp. 683–690.

Schroeder, H. F., et al. 2000. "Durability of Polyolefin Geosynthetics Under Elevated Oxygen Pressure," Proceedings of EuroGeo 2, Bologna, Italy, pp. 459–464.

Smith, G.D., Karlsoon, K. & Gedde, U.W. 1992. "Modeling of Antioxidant Loss From Polyolefins in Hot-Water Applications, I: Model and Application to Medium Density Polyethylene Pipes," Polymer Engineering and Science, Vol. 32, No. 10, pp. 658–667.

Thomas, R. W. 2002. "Thermal Oxidation of Polypropylene Geotextile Used in a Geosynthetic Clay Liner," Proc. on Clay Geosynthetic Barriers, A. A. Balkema, The Netherlands, pp. 87–96.

Von Maubeuge, K. & Ehrenberg, H. 2000. "Long-Term Resistance to Oxidation of PP and PE Geotextiles", Proceedings of EuroGeo 2, Bologna, Italy, pp. 465–468.

Wisse, J.D.M. 1988. "The Role of Thermo-Oxidative Aging in the Long-Term Behavior of Geotextiles", *Durability of Geotextiles*, RILEM, Chapmen and Hall, London, UK, pp. 207–216.

Zornberg, J. G., McCartney, J. S. and Swan, R. Jr. 2005. "Analysis of a Large Database of GCL Internal Shear Strength Results," Jour. of Geotechnical and Geoenvironmental Engineering, Vol. 131, No. 3, pp. 367–380.

# CHAPTER 3

# Mineralogy and engineering properties of Bentonite

W.J. Likos & J.J. Bowders
*University of Missouri, Columbia, MO, USA*

W.P. Gates
*Monash University, Melbourne, Australia*

## 1 INTRODUCTION

The term "bentonite" is used somewhat loosely and without universal agreement on its precise definition, origin, or scope. Strictly speaking, bentonites are rocks composed of the swelling clay mineral smectite and variable amounts of other minerals (Grim and Güven, 1978). The term "bentonite" in and of itself, however, does not directly describe the mineralogy of the material (i.e. it is not a mineral name) nor does it imply a specific location or geologic formation within which it may be found. Rather, bentonite has frequently been used to describe any naturally occurring deposit (or mined ore) composed primarily of the clay mineral montmorillonite that has been derived from alteration of volcanic ash. Unfortunately, this usage implies that either other smectites are unimportant components of bentonites or that other mechanisms of formation are non-existent, both of which are false.

Smectite is a class of hydrated 2:1 layer silicate minerals that form stable colloidal suspensions and have an expandable volume due to the retention of hydrated cations. These attributes arise from the unique structural chemistry of the mineral. High quality bentonites contain greater than 70% smectite by mass. Minor constituents may include quartz, volcanic glass, opaline silica, feldspar, zeolite, and carbonate, as well as discrete kaolin, mica, or illite or mixed-layer clay minerals such as illite-smectite and kaolin-smectite. It is the smectite component, however, with its large specific surface area (as high as 850 $m^2/g$), cation exchange capacity (80–150 meq/100 g), and capability for interlayer swelling, that is responsible for the desirable physical and chemical attributes that make the use of bentonite ubiquitous in many industrial, commercial, and engineering applications (Murray, 1999; Harvey and Lagaly, 2006).

## 2 MINERALOGY OF SMECTITE

### 2.1 *Crystalline structure*

As illustrated in Figure 1, the basic structure of minerals in the smectite group is characterized by having one octahedral sheet joined on either side by two silica tetrahedral sheets (thus the term 2:1 layer silicate) forming a unit layer (Brindley and Brown, 1980; Newman, 1987; Moore and Reynolds, 1997). In an ideal 2:1 layer, aluminum (gibbsite-like) or magnesium (brucite-like) octahedra are coordinated with four oxygen atoms and two hydroxyl ions. The tetrahedra are comprised of silicon atoms coordinated with four oxygen atoms; one of these, the apical oxygen, is also bonded to the octahedra.

If the central cations in the octahedral sheet are predominantly trivalent (e.g. $Al^{3+}$) then only two thirds of the cation sites (2 of 3) need to be occupied for electroneutrality and the structure is termed "dioctahedral." Dioctahedral smectites include the montmorillonite, beidellite, and nontronite species. If the cations in the octahedral sheet are pre-dominantly divalent (e.g. $Mg^{2+}$) then all of the cation sites (3 of 3) need to be occupied and the structure is termed "trioctahedral." Trioctahedral smectites include the saponite, hectorite, and sauconite species. Montmorillonite is the most common naturally occurring species in the smectite group, and the following discussion will be limited to montmorillonite unless other minerals are specifically mentioned. The structures

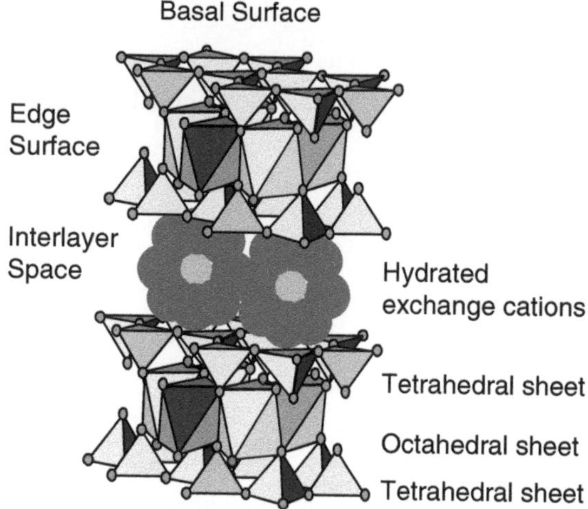


Figure 1. Structure of montmorillonite, a hydrated 2:1 layer silicate or smectite. Cations commonly occupying tetrahedral sheets are $Si^{4+}$ (yellow tetrahedra) and $Al^{3+}$ (green) and those occupying the octahedral sheet are $Al^{3+}$ (yellow octahedra), $Mg^{2+}$ (blue), $Fe^{3+}$ (pink) and $Fe^{2+}$ (not shown). Oxygen (grey circles) and hydroxyl ions (red) constitute the anionic lattice. Hydrated exchange cations (green circles surrounded by blue circles), which neutralize charge due to isomorphic substitutions occupy the basal or interlayer surface and enable montmorillonite to swell in water.

of other phyllosilicates are more thoroughly discussed in relevant chapters of the Handbook of Clay Science (Bergaya et al., 2006).

Oxygen anions are shared between the central octahedral sheet and the two tetrahedral sheets, thus forming a strong bond that preserves the 2:1 unit layer. The thickness of one unit layer is ~9.6 Å (~0.96 nm). The thickness of the interlayer plus one unit layer (i.e. the structural "repeat" distance) is variable and is referred to as basal spacing, indexed as d(*001*). Several tens or even hundreds of unit layers constitute a crystallite. Many million layers and thousands of crystallites form montmorillonite particles with sizes on the order of 2 μm effective spherical diameter.

Adjacent montmorillonite layers are separated by an interlayer space that spans the distance between opposing basal or interlayer surfaces (see Fig. 1). These surfaces are elongated flat plates and constitute the largest surface area in smectites. The interlayer surfaces are also where most of the layer charge of the mineral is expressed (see Section 2.3) and where exchangeable cations required to balance the layer charge reside. The high hydration energies of the exchange cations enable montmorillonite to absorb large amounts of water. While bonding between unit layers arising from electrostatic and short-range van der Waals attractive forces is relatively weak, the interaction of water with exchangeable cations within the interlayer space can be strong enough to allow partial or complete disassociation of smectite unit layers in aqueous solution, a phenomenon referred to as "interlayer swelling." Thus a unit cell – the smallest replicating average chemical formula – of a smectite clay mineral undergoes volume change as a function of hydration.

Depending on the extent of interlayer disassociation, smectite particles range in thickness from individual unit layers, upward to thicknesses corresponding to several tens, hundreds, or thousands of stacked unit layers. Detecting changes in basal spacing associated with interlayer sorption of water or other polar sorbates is one of the key processes by which the presence of smectite in a mineral mixture is identified using X-ray diffraction (XRD) (see Section 4.1).

The specific gravity of smectite ranges from about 2.35 to 2.70. Bulk density is on the order of 12 kN/m$^3$ (~75 lbs/ft$^3$). External specific surface area (the surface area of the interparticle space per unit mass) ranges from about 30 to 120 m$^2$/g. The internal specific surface area (the surface area related to the interlayer space) ranges from about 700 to 850 m$^2$/g, which is considerably larger than the total surface area of other clay minerals such as kaolin ($S_s$ ~ 10 to 20 m$^2$/g) or illite ($S_s$ ~ 60 to 100 m$^2$/g). The large difference in the external and internal specific surfaces of smectite reflects the fact that sorption may occur on various levels of scale within the particle fabric and is dependent on the properties of the sorbate. Non-polar sorbate molecules, such as $N_2$, generally do

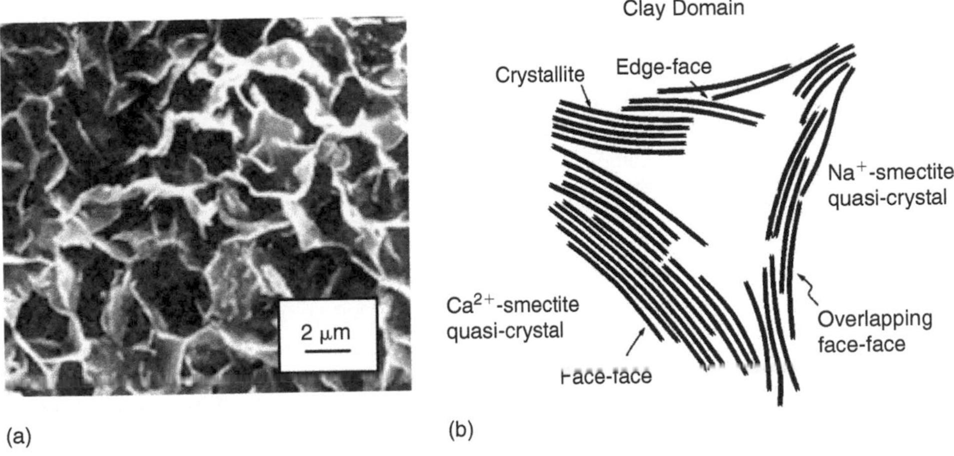


Figure 2. (a) Microstructure of $Na^+$-montmorillonite (modified from Tessier, 1990); (b) Layer interactions and quasi-crystal formation.

not enter the interlayer pore space, and may therefore be used to probe the external surface area. Polar sorbate molecules, such as $H_2O$ or ethylene glycol, do enter the interlayer pore space and may therefore be used to probe the internal surface area.

## 2.2 *Morphology and fabric*

The morphology of smectite often resembles a thin, crumpled film or flake and it can be difficult to differentiate one "particle" from another. Multiple levels of scale and multiple types of mineral interactions must be considered. The volume fraction, arrangement, and orientation of the solids and pore spaces on these multiple levels of scale define the fabric of the system. This fabric (also referred to as texture or microstructure) influences bulk physical properties of the bentonite such as swelling, sealing, and hydraulic conductivity, and is critically important in governing the corresponding performance of bentonite as a barrier material.

As illustrated on Figure 2, individual crystallites comprised of multiple unit layers interact in face-face, overlapping face-face, edge-face and edge-edge associations to form quasi-crystals (Aylmore and Quirk, 1960; Tessier, 1990, Hetzel et al., 1994). A quasi-crystal is an equilibrium structure smectite gels adopt that reduces the net potential energy of particle interaction. Attractive van der Waals force interactions between quasi-crystals enable the formation of larger, interconnected clay domains. Clay domains can have various size and durability, depending on the surrounding environment (Lagaly, 2006).

The number of layers that constitute a quasi-crystal, its size, shape and interaction with other quasi-crystals is generally dependent on the layer charge characteristics of the mineral, electrolyte concentration, and identity of the exchange cation. The average thickness of $Na^+$-smectite quasi-crystals (right-hand side of Fig. 2b) is small; typically $<100$ individual layers, but they generally have a large aspect ratio (Tessier, 1990). The resulting gel fabric or micro-structure has a distribution of inter-particle pore sizes, which optimizes water retention, inhibits water flow (Egloffstein, 2001), and influences how interlayer swelling translates to corresponding bulk volume changes (Likos and Lu, 2006). On the other hand, $Ca^{2+}$-smectite gels (left-hand side of Fig. 2b) are typically composed of larger crystallites made up of as many as several hundred individual layers (Tessier, 1990). This results in a fabric with a pore size distribution more conductive to water (Egloffstein, 2001) and which more efficiently translates interlayer volume change to bulk volume change (Likos and Lu, 2006).

Swelling clays can absorb $>200\%$ their mass in water and increase their original volume many fold. The clay fabric at high water content is instrumental in translating properties from the individual mineral layer or crystallite level to the macro-scale level. When under confinement, as is the usual condition in geotechnical applications, a $Na^+$ clay fabric will seal because domains of

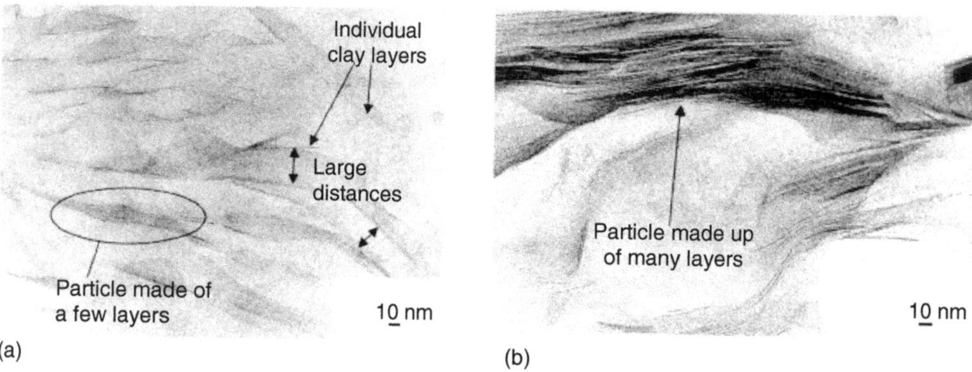


Figure 3. Transmission electron micrographs of bentonite from a GCL in the (a) gelled phase and (b) hydrated solid phase [note difference in scale]. (modified from Guyonnet et al., 2005).

quasi-crystals will necessarily have to re-arrange and expand into the available inter-particle pore space during wetting. This results in a reduction in void sizes and inter-connectivity of pores with an associated increase in tortuosity. The highly flexible nature of clay platelets enables overlapping face-face associations to dominate the fabric of $Na^+$ saturated smectite gels (right hand side of Fig. 2b). These particle interactions are relatively unimportant in $Ca^{2+}$-smectite gels, where quasi-crystals are generally larger and more rigid (left hand side of Fig. 2b).

Figure 3 shows transmission electron microscope (TEM) images for two GCL bentonites. Figure 3a shows the occurrence of a gelled phase obtained after hydration and percolation through a natural sodium bentonite with low ionic strength NaCl solution ($10^{-3}$ mol/L). Individual unit layers or crystallites made up of just a few unit layers are separated by large distances. The crystallites have largely disassociated due to long-range interlayer swelling caused by the predominance of sodium in the exchange complex, effectively eliminating the quasi-crystal structure (Section 3.1). Corresponding hydraulic conductivity of the material imaged in Figure 3a is relatively low ($k = 1.0 \times 10^{-9}$ cm/s; Norotte et al., 2004). Figure 3b, on the other hand, shows the occurrence of a hydrated sodium-activated calcium bentonite comprising thick crystallites made up of multiple unit layers, a microstructure caused by percolation with aqueous NaCl ($10^{-3}$ mol/L). Hydraulic conductivity is about one order of magnitude greater than the previous case ($k = 1.7 \times 10^{-8}$ cm/s). The limited interlayer swelling observed illustrates the potentially important effects of ion exchange from internal cation sources in activated bentonites, corresponding effects on fabric, and ultimately on barrier performance. Such changes are also characteristic of the effect on clay fabric of polyvalent cations in the exchange complex (Fig. 2b and Section 3.1).

## 2.3 *Layer charge*

Smectite is characterized by extensive isomorphous substitution that results in charge deficiencies within the crystalline structure. Substitution may occur either in the octahedral layer (e.g. $Mg^{2+}$ for $Al^{3+}$) or in the tetrahedral layer (e.g. $Al^{3+}$ for $Si^{4+}$), such that the overall structure carries a net negative layer charge. The negative layer charge must be neutralized by cations residing in the interlayer, which in the case of smectites, are hydrated and exchangeable.

A generalized structural formula for a unit cell of montmorillonite is

$$M_x^+[Si_{8-a}Al_a](Al_{4-b-c-c'}Mg_bFe_c^{3+}Fe_{c'}^{2+}) \cdot O_{20}(OH)_4 \cdot nH_2O$$

where $M^+$ represents a hydrated monovalent exchange cation; subscript $x$ is the number of atoms of exchange cation in the unit cell; subscript $a$ is the number of $Al^{3+}$ atoms substituting in the tetrahedral sheet; subscripts $b$, $c$, and $c'$ the number of other atoms ($Mg^{2+}$, $Fe^{3+}$, and $Fe^{2+}$ respectively) substituting within the octahedral sheet and $n$ is the number of $H_2O$ molecules associated with the exchange cation and mineral surfaces. Cations enclosed within brackets, [ ], have tetrahedral coordination, and those within parentheses, ( ), have octahedral coordination.

Table 1. Structural formulae and layer charge properties of the montmorillonite fractions from several bentonites calculated from chemical analysis (Gates, 2005).

| | Occupancies | | | | | Layer Charge X | | |
|---|---|---|---|---|---|---|---|---|
| | Tetrahedral | | Octahedral | | | Total | Tet. | Oct. |
| | $Si^{4+}$ | $Al^{3+}$ | $Al^{3+}$ | $Fe^{3+}$ | $Mg^{2+}$ | | $(e^-)$ | |
| Bentonite source | per $O_{20}(OH)_4 \cdot nH_2O$ | | | | | | | |
| Jelŝovy Poltok, Slovakia | 7.92 | 0.08 | 3.01 | 0.19 | 0.79 | 0.93 | 0.08 | 0.85 |
| Otay, CA, USA | 7.99 | 0.01 | 2.68 | 0.11 | 1.29 | 1.05 | 0.01 | 1.04 |
| SWy, WY, USA | 7.77 | 0.23 | 3.08 | 0.44 | 0.45 | 0.77 | 0.23 | 0.54 |
| Ossean, South Africa | 7.78 | 0.22 | 2.92 | 0.27 | 0.83 | 0.94 | 0.22 | 0.72 |
| Arumpo, NSW, Australia | 7.92 | 0.08 | 2.80 | 0.37 | 0.85 | 0.87 | 0.08 | 0.79 |
| JC Lane, WY, USA | 7.94 | 0.06 | 3.07 | 0.41 | 0.43 | 0.74 | 0.06 | 0.67 |
| Miles, QLD, Australia$^{+}$ | 7.74 | 0.26 | 3.02 | 0.52 | 0.45 | 0.72 | 0.23 | 0.46 |
| Redhill, Surrey UK | 7.87 | 0.13 | 2.50 | 0.77 | 0.74 | 0.83 | 0.12 | 0.70 |

$^{+}$after accounting for $SiO_2$ as opal-CT (Gates et al., 2002).

Table 1 illustrates the variability observed in the chemistries of different montmorillonites separated from bentonites. It is important to note that the chemistries suitable for calculation of structural formulae can only be determined on the bentonite fraction containing a known amount of montmorillonite and minimal amount of impurities. Montmorillonites thus isolated display a range in $Al^{3+}$ for $Si^{4+}$ substitution in the tetrahedral sheets as well as $Fe^{2+}$ and $Mg^{2+}$ for $Al^{3+}$ substitution in the octahedral sheet.

The exchange cation (e.g. $Ca^{2+}$) is neglected in Table 1, and instead the total layer charge for each montmorillonite, and its distribution between octahedral and tetrahedral sheets is shown. The montmorillonites have permanent negative layer charge arising primarily in the octahedral sheet, due to partial substitution of $Mg^{2+}$ for $Al^{3+}$, which is expressed mostly at the basal or interlayer surfaces and neutralized by hydrated exchange cations. The permanent layer charge, X, of montmorillonite may be calculated from its structural formula by $X = 12 - ((8 - a) - (4 - b - c))$; such measure is the total layer charge.

Determined in this way, the total layer charge and the distribution of layer charge between the octahedral and tetrahedral sheets are informative on how the montmorillonite can be expected to behave. Generally, lower total layer charge, e.g. X less than ~0.85 e$^-$ per unit cell, is more conducive to swelling than higher layer charges ($X > 0.85\,e^-$ per unit cell). Note that several montmorillonites displayed in Table 1 have total layer charges <0.85 e$^-$ per unit cell. This charge deficiency is relatively low among the 2:1 layer silicates, which is partially responsible for the limited extent of electrostatic attractive forces available to counteract interlayer swelling (Brindley and MacEwen, 1953). This is because a higher surface charge results in stronger electrostatic interactions between the exchange cation and the smectite surface. Thus, these clays demonstrate considerably less swelling than those having lower total layer charge. To a lesser degree, clays having a greater proportion of layer charge within the octahedral sheet can be expected to have greater swelling and water retention properties than those with high levels of tetrahedral layer charge.

Some variable charge (also called pH-dependent charge or amphoteric charge) may be present on crystallite edges due to broken bonds at the crystalline discontinuity. The crystallite edges may contribute to the total negative layer charge at high solution pH, or detract from the total negative layer charge when the solution pH is low. Variable charge can also occur at the boundaries of clay domains. Variable layer charge is negligible for a 1 $\mu$m smectite particle, but may contribute significantly to the total layer charge for particles less than 0.2 $\mu$m.

## 2.4 *Exchangeable cations*

The net negative charge arising from isomorphous substitution is offset by exchangeable interlayer cations. These cations are predominantly located between the unit layers within the interlayer space, but can also exist between particles or crystallites within the interparticle pore space. The cations are referred to as exchangeable because they may be replaced by other cations from solution under energetically favorable conditions. Exchangeable cations in naturally occurring clay minerals are primarily $Ca^{2+}$, $Mg^{2+}$, $Na^{+}$, and $K^{+}$, usually in that decreasing order of abundance (Mitchell, 1993).

The predominant type of exchangeable cation may be used as a qualifier for differentiating one form of the same mineral from another. We refer, for example, to $Na^{+}$-bentonite (or the $Na^{+}$ form of bentonite) if $Na^{+}$ is the predominant type of exchangeable cation. $Ca^{2+}$ forms of bentonite are common as well. The predominant type of exchangeable cation often reflects the environment within which mineral formation took place, as well as the chemistry of the precursor materials. Bentonites altered from volcanic ash in a seawater environment, for example, are predominantly sodium form, whereas bentonites formed within a freshwater environment are predominantly calcium form. Extensive deposits of high-quality $Na^{+}$-bentonite are located in the Wyoming/Montana region of the United States. Although deposits of $Na^{+}$-bentonite exist worldwide, $Ca^{2+}$-bentonites are much more predominant on a global scale. In some cases, magnesium may exceed calcium and sodium (e.g. Churchman et al., 2002).

Bentonites used in GCLs commonly have approximately equal fractions of $Ca^{2+}$ and $Na^{+}$ cations on the exchange complex when delivered from the GCL factory (Shackelford et al., 2000). Naturally predominant $Na^{+}$-bentonites or activated ($Na^{+}$ exchanged) $Ca^{2+}$-bentonites are preferred for barrier applications due to the beneficial swelling, sealing, and hydrologic characteristics associated with sodium in the exchange complex.

The exchangeability of exchange cations is governed by their valence, size, hydration energies and concentration (Teppen and Miller, 2005). For equivalent concentrations, the propensity for one type of cation to replace another follows the sequence:

$$Li^{+} < Na^{+} < K^{+} < Mg^{2+} < Ca^{2+} < Al^{3+} < Fe^{3+}$$

As noted previously, the higher affinity for polyvalent cations (e.g. $Ca^{2+}$ over $Na^{+}$) leads to important considerations regarding ion exchange and the corresponding changes in engineering behavior (e.g. increases in hydraulic conductivity) that may take place over the service life of a GCL (e.g. Petrov and Rowe, 1997; Shackelford et al., 2000; Jo et al., 2001; Meer and Benson, 2007).

The magnitude and equivalency of charge in montmorillonite is usually quantified by cation exchange capacity (CEC) rather than by layer charge (as in Table 1). In an ideal case, these measures should be identical, but since the CEC is an operationally defined measure (it depends on the method used; Section 4.2) there is usually some disparity between CEC determined by wet-chemical methods and layer charge determined by transitional mineralogical methods.

The CEC is expressed in milliequivalents per 100 grams of dry clay (meq/100 g) or as centimoles of charge per kilogram (cmol/kg) [1 meq/100 g = 1 cmol/kg]. A higher CEC reflects a greater layer charge and a consequently larger surface activity for exchange. A related property, surface charge density, or $\sigma$, is defined as CEC divided by specific surface area, and thus has units of meq/100 $m^2$. It must be stressed that high CEC (and therefore high layer charges or surface charge densities) are not necessarily indicative of bentonite quality for optimal barrier performance. This will be discussed in more detail later.

## 2.5 *Sodium activation of bentonites*

Bentonites often must be processed to improve their suitability for various industrial purposes. The simplest form of processing is drying and crushing to increase the reactive surface area, but other, more complex processes may be required to achieve the desired characteristics. For example, natural $Ca^{2+}$-bentonites must generally be "beneficiated" for these materials to have the swelling and fluid loss properties associated with $Na^{+}$-bentonites. Termed soda or sodium activation, this process (first realized industrially in Germany during the 1930's) is based on the dry-mixing of $Ca^{2+}$ or $Mg^{2+}$ bentonite forms with a few percent soda ash (sodium carbonate or sometimes sodium bicarbonate).

Traditionally, beneficiation has served the purpose of improving the viscosity and suspension density of bentonite slurries used as drilling muds (Odom, 1984), but has become standard practice to improve bentonite properties used in GCLs (Landis and von Mauge, 2004). Recent evidence suggests however, that traditional beneficiation procedures may not be suitable for all bentonites, in particular those earmarked for use as environmental barriers. For example Volzone and Garrido (2001) reported differences in the behavior of soda-activated montmorillonites: montmorillonites having high total layer charges predominantly located within the octahedral sheet (Otay type montmorillonite) were found to exhibit lower viscosities and swelling than montmorillonites having lower (i.e. $<0.9\,e^-$ per unit cell) total layer charge (Wyoming type montmorillonites). Guyonnet et al. (2005) observed that calcium carbonate present in bentonites used in GCLs might be detrimental to barrier performance, as re-wetting of the activated bentonite dissolved the carbonate and resulted in $Ca^{2+}$ for $Na^+$ exchange.

Gates et al. (2008 – submitted) considered that the calcium carbonates formed during sodium activation of $Ca^{2+}$-bentonites were highly soluble. While sodium activation resulted in complete replacement of $Ca^{2+}$ by $Na^+$ of a reference smectite (high tetrahedral layer charge), a re-distribution of $Ca^{2+}$ within the beneficiated clay occurred during aging under ambient (RT, ~30% RH) conditions. Upon re-wetting dissolution of the carbonates resulted in $Ca^{2+}$ re-exchanging for $Na^+$ on the clay. In light of field exhumations of landfill cover applications of GCLs (Meer and Benson, 2007), it would appear that desiccation and rewetting of $Na^+$-bentonites in the presence of soluble $Ca^{2+}$ and carbonate (both common soil constituents) can have adverse impact on barrier performance.

## 3 ADSORPTION AND SWELLING BEHAVIOR

The primary function of the bentonite layer in a GCL is to impede the flow of migrating liquids and dissolved chemical species. This is achieved by its low hydraulic conductivity, which is on the order of $10^{-8}$ cm/s to $10^{-10}$ cm/s. This low hydraulic conductivity results primarily from the smectite component's small particle size, large surface area, swelling potential, and capability to adsorb and effectively immobilize pore water through a variety of short-range and long-range hydration mechanisms. The adsorption and swelling behavior of bentonite is affected by many factors that are inherent to the bentonite in its natural state. These include: layer charge (CEC), the location of layer charge, exchange cation identity, the clay fabric in its natural state, and impurities present.

### 3.1 *Short-range hydration mechanisms*

Colloidal clay minerals in general, and smectite in particular, are capable of significant interactions with water. Interactions occur on the external particle surfaces and, for swelling clay minerals, on the interlayer surfaces as well. The surface area and the surface charge properties of the mineral dictate the extent and range of the interactions. The large specific surface area of smectite is clear indication of the importance of solid-liquid interaction in governing its behavior.

The initial hydration of a clay mineral surface involves four basic short-range interaction mechanisms: (i) hydrogen bonding, (ii) dipole-charged surface attraction, (iii) van der Waals attraction, and (iv) hydration of exchangeable cations (Mitchell, 1993).

Hydrogen bonding occurs between water dipoles and oxygens along the surface of the interlayer, coordination under-saturated oxygen at crystallite edges and most importantly between water molecules. Interlayer surfaces have a permanent, but low residual charge, thus water interacts with these surface via the proton. At high pH, water interacts with the edge surfaces via the proton, but at low pH, water interacts with the protonated edge surfaces via the oxygen. Charged surface-dipole attraction occurs as water dipoles become attracted to and oriented with the negative electrical field emanating from the mineral surface. van der Waals attractive fields arise from instantaneous atomic interactions between the atoms comprising the mineral surface and the atoms comprising the pore water.

Most importantly for montmorillonite, however, hydration of exchangeable cations occurs as the positively charged ions attract water dipoles, which form a hydration shell surrounding the cation. Because the cations are restrained by the layer charge field, their water of hydration shell is restrained as well. The energy associated with cation hydration is a function primarily of the cation size and valence, where small size, yet low valence provides favorable energetics for hydration (Bohn et al., 1985). Secondary layers of water can H-bond with the primary layer of hydration water, forming multiple hydration shells. These H-bonds decrease in strength rapidly with the number of layers. Beyond two or three layers, the water of these shells behaves as bulk water.

Table 2. Properties of common exchange cations (Kielland, 1937; Conway, 1981).

| Element | Symbol & valence | Ionic radius (pm) | Hydrated radius (pm) | Δradius/ valence ratio | Number of hydration shells |
|---|---|---|---|---|---|
| Lithium | $Li^+$ | 94 | 600 | 506 | 3+ |
| Sodium | $Na^+$ | 117 | 450 | 333 | 3+ |
| Potassium | $K^+$ | 149 | 300 | 151 | 1 |
| Rubidium | $Rb^+$ | 163 | 250 | 87 | 1 |
| Cesium | $Cs^+$ | 186 | 250 | 64 | 1 |
| Beryllium | $Be^{2+}$ | 31 | 800 | 384 | 2+? |
| Magnesium | $Mg^{2+}$ | 72 | 800 | 364 | 2 |
| Calcium | $Ca^{2+}$ | 100 | 600 | 250 | 2 |
| Strontium | $Sr^+$ | 126 | 500 | 187 | 1 |
| Barium | $Ba^{2+}$ | 142 | 500 | 179 | 1 |
| Lanthanum | $La^{3+}$ | 116 | 900 | 261 | 1 |

Table 2 lists the ionic and hydrated radii of common exchange cations. Ions with small ionic radii have a high charge potential emanating from their "surface." They interact strongly with water (Teppen and Miller, 2005) and therefore attract more water within the interlayer of clays than larger ions of similar valence. While the size of hydrated $Ca^{2+}$ is larger than $Na^+$, its higher valency enables a greater attraction to the clay surface. Greater attraction energy enables cations like $Ca^{2+}$ to interact strongly with both surfaces of adjacent clay layers, therefore $Ca^{2+}$ is retained more strongly than $Na^+$. The ratio of the difference in ionic and hydrated radii to the cation valence (Δradius/valence) provides a measure of this interaction.

The physical consequence of these short-range solid-liquid interaction mechanisms is to attract, align, and impart some order into the molecular arrangement of the vicinal (near surface) pore water. The potential of the vicinal water, therefore, is reduced relative to the potential of free (bulk) pore water. This local gradient in potential serves to drive additional water into the interlayer, but is highly dependent on the properties of the interlayer cation, as described above. For some cations, for example $Na^+$ and $Ca^{2+}$, the attraction of additional water results in volume change phenomena known as "crystalline swelling" at relatively low water contents and "osmotic," swelling at higher water contents. Crystalline swelling can occur for most cations, but osmotic swelling is limited to $Na^+$ and $Li^+$ (and possibly $Be^{2+}$) where the Δradius to valence ratio is greater than 300, and more than 3 hydration shells can occur.

### 3.2 *Crystalline and osmotic swelling regimes*

Norrish (1954) applied XRD techniques to observe basal spacing up to about 130 Å in $Na^+$-montmorillonite within the "gelled" phase at high water content. A synthesis of these results is illustrated in Figure 4, where it can be observed that interlayer swelling is generally divided into two stages. In the first stage, or within the *crystalline swelling regime*, basal spacing increases in a step-wise fashion as molecular layers of $H_2O$ are sequentially adsorbed in the interlayer around the interlayer cations. Total water content corresponding to the upper limit of crystalline swelling (three or four water layers) is less than about 25% by mass. Transition into an *osmotic swelling regime* is manifest as a jump in basal spacing from ~20 Å to ~40 Å, whereupon a significant amount of water is adsorbed, disassociation of the crystallites continues, and the system enters a gelled phase.

The transition between the crystalline swelling regime and osmotic swelling regime is not well understood, but insight has been gained by molecular simulation of hydrating mineral systems (e.g. Meleshyn and Bunnenberg, 2005). Understanding interlayer swelling within each regime provides a basis to understand how, where, and to what extent water is adsorbed by clay mineral systems, and thus provides insight into the corresponding effects on their overall hydrologic and engineering behavior.

### 3.3 *Crystalline swelling*

Crystalline swelling is a process whereby expandable 2:1 phyllosilicates sequentially intercalate one, two, three, or four discrete layers of $H_2O$ molecules between the mineral interlayers

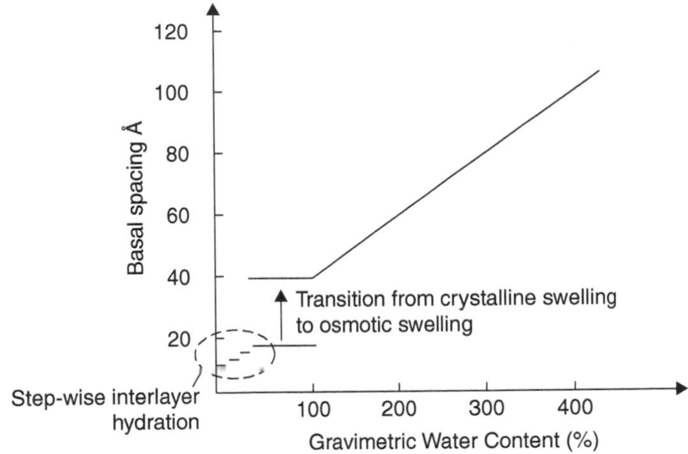


Figure 4. Basal spacing of $Na^+$-montmorillonite over range of water content (after Norrish, 1954).

(Norrish, 1954). This process, which has also been referred to as Type I swelling (e.g. Newman, 1987; Van Olphen, 1991), occurs prior to osmotic (Type II) swelling associated with longer-range electrical diffuse double layer effects.

Interlayer hydration and dehydration in the crystalline swelling regime has been explored extensively using a variety of approaches. Studies have taken the form, for example, of basal spacing measurements using XRD under controlled relative humidity (RH) conditions (Chipera et al., 1997), XRD of saturated pastes equilibrated in different ionic strength solutions (Norrish, 1954; Norrish and Quirk, 1954; Slade and Quirk, 1990; Slade et al., 1991; Laird et al., 1995; Laird and Shang, 1997; Gates, 2004), mathematical considerations of interlayer geometry (Barshad, 1949), measurements of water vapor sorption isotherms (Collis-George, 1955; Keren and Shainberg, 1975; Berend et al., 1995; Likos, 2004), investigations of water film formation using infrared spectroscopy (Sposito and Prost 1982), neutron diffraction and neutron scattering (Powell et al. 1997), and NMR spectroscopy (Weiss and Gerazimowicz 1996), as well as modeling using molecular dynamics and Monte Carlo simulations (Karborni et al., 1997; Sposito et al., 1999).

Figure 5 illustrates four consecutive stages of hydration for a conceptual smectite particle composed of two unit layers (only two unit layers are shown for clarity). Initially (Fig. 5a), the particle system is completely dry. Exchangeable cations (e.g. $Na^+$ or $Ca^{2+}$) occupy the interlayer pore space. As water of hydration is adsorbed by the particle/cation system (Fig. 5b through 5d), the basal spacing transitions from one stable hydrate state to the next, occurring generally in a discrete or step-wise fashion. The basal spacing of smectite proceeds, for example, from ~9.7 Å to 12 Å upon transition from the zero-layer hydrate state to the one-layer hydrate state, from 12 Å to 15.5 Å during the one-layer to two-layer transition, and from 15.5 Å to 18.3 Å during the two-layer to three-layer transition. Gravimetric water content of pure $Na^+$-smectite at the one-layer, two-layer, and three-layer hydrate states is approximately 10%, 19%, and 22%, respectively (Karaborni et al., 1997.)

Crystalline swelling is the result of the balance between forces of attraction and repulsion operating between adjacent interlayer surfaces (Norrish, 1954; van Olphen, 1965; Kittrick, 1969). The net potential energy of interaction (Laird, 1996; 2006) is dominated by electrostatic attraction between the exchange cations and the basal surfaces of the clay. The positively charged cations act as charge bridges or links between adjacent negatively charged clay layers. The net potential energy of repulsion, on the other hand, is dominated by the hydration energy of the exchange cations. The exchange cations can exist within the interlayer in partially hydrated, thermodynamically stable phases that interact strongly with their surroundings: they will attract or release water of hydration until the net potential energies of attraction and repulsion are balanced. Unsaturated conditions or saturated conditions with high electrolyte concentrations favor the dominance of net forces of attraction, while fully saturated conditions of low electrolyte concentration favor the dominance of net forces of repulsion.

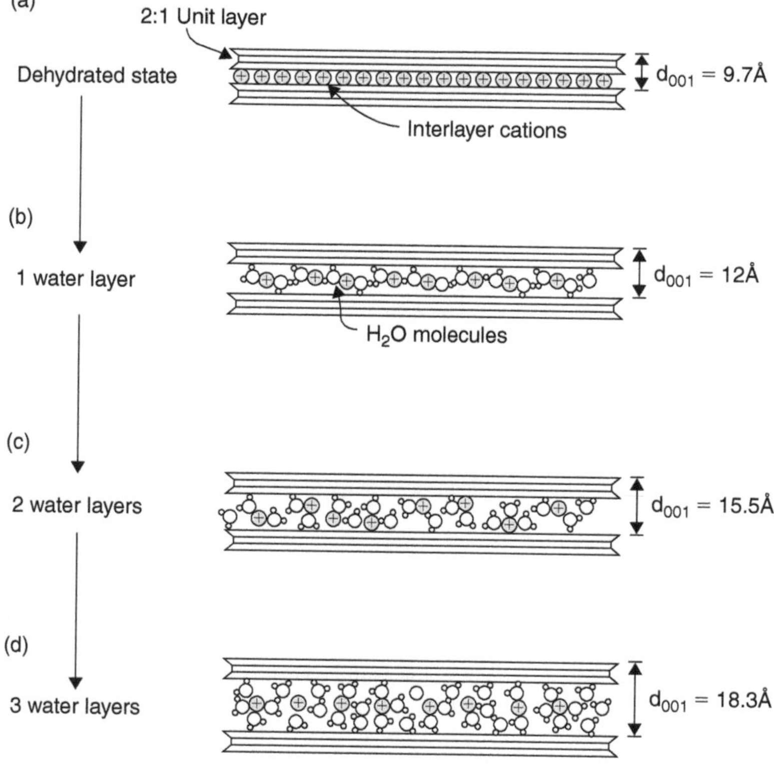


Figure 5. Step-wise interlayer hydration in the crystalline swelling regime (after Likos, 2004).

Interlayer attractive energy can be semi-quantitatively assessed by the proportionality

$$E \propto \frac{\sigma ve}{2d\varepsilon} \tag{1}$$

where $\sigma$ is the surface charge density of the mineral, $v$ is the interlayer cation valence, $e$ is the elementary charge, $d$ is the half distance between the unit layers, and $\varepsilon$ is the permittivity of the intervening fluid. Thus, high surface charge density, high cation valency, and small separation distance produce stronger attractive forces that must be overcome by cation hydration for the interlayer to separate. The relatively low surface charge density of montmorillonite (Section 2.3) makes its structure particularly amenable to interlayer disassociation through hydration of interlayer cations.

For divalent mineral forms where relatively large interlayer attractive forces must be overcome (e.g. $Ca^{2+}$-smectite), crystalline swelling ceases at three or possibly four molecular water layers (Slade and Quirk, 1990; Slade et al., 1991), thus forming stable quasi-crystals such as those observed previously in Figure 3b (Quirk and Aylmore, 1971). Because interlayer sorption is limited, corresponding bulk volume changes of $Ca^{2+}$-smectite are relatively small. Bulk volume change upon inundation with water under free swell conditions may be on the order of 125% (Grim, 1962). Limited interlayer sorption in $Ca^{2+}$-smectite is also reflected in engineering indices such as Atterberg limits (Table 3), which although high by comparison with other minerals, are small by comparison with smectite having sodium as the predominant type of interlayer cation.

For monovalent mineral forms such as $Na^{+}$-smectite, where the interlayer attractive forces are relatively weak, interlayer sorption may continue beyond the three or four layers of water associated with crystalline swelling. Additional interlayer separation continues under an osmotic mechanism driven by a gradient in chemical potential between the interlayer water and bulk pore water residing in the interparticle pore space. It is this transition into the osmotic swelling regime that allows for more complete disassociation of the mineral interlayers, large macroscopic volume changes (e.g. up

Table 3. Atterberg limits for $Na^+$ and $Ca^{2+}$ smectite.

| Source | Liquid limit, LL (%) | | Plasticity index, PI (%) | |
|---|---|---|---|---|
| | $Na^+$ | $Ca^{2+}$ | $Na^+$ | $Ca^{2+}$ |
| Cornell (1951) | 710 | 510 | 656 | 429 |
| Grim (1968) | 700 | 124 | 603 | 52 |
| Mesri and Olson (1970) | 880 | 200 | NA | 167 |
| Eykholt (1988) | 590 | 123 | 553 | 85 |
| Gleason (1993) | 603 | 124 | 567 | 98 |

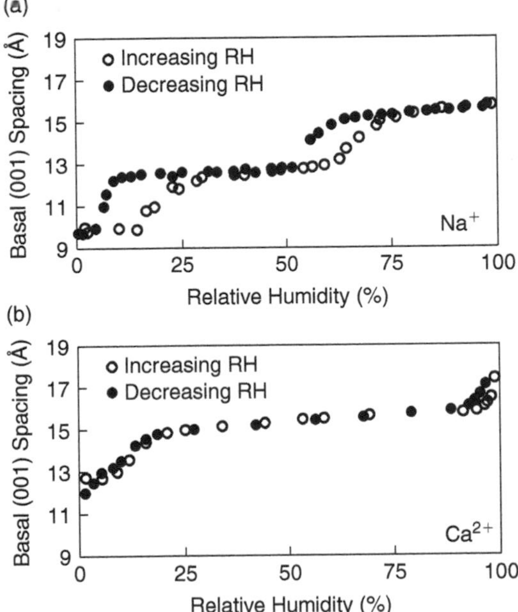


Figure 6. Basal spacing measured as a function of relative humidity from humidity-controlled XRD tests: (a) $Na^+$-smectite, and (b) $Ca^{2+}$-smectite (after Chipera et al. 1997).

to 1,600% for free swell), high Atterberg indices (Table 3), and gelling behavior partly responsible for the self-healing and low hydraulic conductivity characteristics required for barrier applications.

The onset and progression of crystalline swelling fundamentally depends on the availability of water to the system, which may be quantified in terms of relative humidity (RH) or water activity. Specific regimes of RH correspond to stable hydrate states (e.g. one water layer, two water layers, etc) that are reflected in the step-wise interlayer adsorption and desorption behavior.

The step-wise nature of crystalline interlayer swelling and corresponding RH stability regimes are demonstrated in Figure 6, which summarizes a series of humidity-controlled XRD tests conducted for $Na^+$-smectite (Fig. 6a) and $Ca^{2+}$-smectite (Fig. 6b) (Chipera et al., 1997). Results are presented in the form of equilibrium basal spacing measured along paths of increasing and decreasing RH. Relatively flat portions of the trends indicate humidity regimes corresponding to stable hydrate states. Relatively steep portions indicate humidity regimes corresponding to interstratifications of two hydrate states during transition from one state to another. The $Na^+$-smectite transitions through the zero-layer (9.7 Å), one-layer (12 Å), and two-layer (15.5 Å) states over the 0% RH to 100% RH range. The $Ca^{2+}$-smectite remains stable at the two-layer state over a much wider range of RH. Both clays exhibit hysteresis between the wetting and drying cycles.

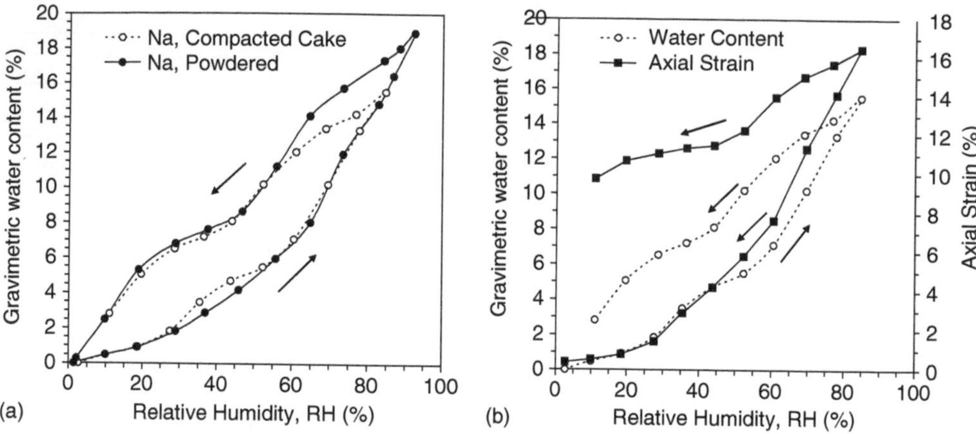


Figure 7. (a) Water vapor sorption isotherm for $Na^+$-bentonite in the crystalline swelling regime, and (b) corresponding bulk volume change (Likos and Lu, 2006).

Laird (2006) discussed the dependence of crystalline swelling of hydrated smectites on their surface charge characteristics. At equal-molar electrolyte concentrations, low charge smectites ($X < 0.85$ e- per unit cell) tend to swell more than high charge smectites ($X > 0.85$ e-) (Slade and Quirk, 1990; Laird et al., 1995). For a given layer charge, dioctahedral smectites with most of the charge originating in the octahedral sheet (i.e. montmorillonites) have greater swelling than those with most of the layer charge originating in the tetrahedral sheet (i.e. beidellites) (Slade et al., 1991). Because swelling underpins smectite performance in GCLs, it follows that those GCLs containing bentonites with moderate to low CEC, such as the Wyoming bentonites, will most likely have greater swelling indices than those with higher CEC.

Figure 7a shows water vapor sorption isotherms obtained for compacted and powdered specimens of Wyoming $Na^+$-bentonite. The isotherms are characterized by steep portions where a relatively large amount of water is gained or lost with a change in RH as well as relatively flat portions where a lesser amount of water is gained or lost. This characteristic type of sorption behavior is described more generally by Rouquerol (1999) as "wavy, with an ill-defined double step" and has been extensively documented for swelling clay minerals including smectite and vermiculite (e.g. Mooney et al., 1952, Collis-George, 1955; Keren and Shainberg, 1975; Cases et al., 1992, Berend et al., 1995). The observation may be interpreted as the bulk manifestation of crystalline interlayer swelling, where the relatively steep portions of the isotherm indicate transitions from one hydrate state to another and the relatively flat portions correspond to stable RH regimes. Considering the XRD results noted above, the steep portion noted on the wetting loop located between about 30% RH and 50% RH, indicates transition from the zero-layer hydrate state to the one-layer state. The steep portion between about 60% RH and 85% RH indicates transition from the one-layer state to the two-layer state. The fact that the steps are ill-defined reflects the occurrence of concurrent water adsorption in the interparticle pore space (in addition to interlayer adsorption) and interstratification among interlayers (i.e. crystalline swelling is not a homogenous process).

Figure 7b shows the sorption data for the compacted $Na^+$-bentonite specimen now superimposed with corresponding bulk volume change measured during the sorption-desorption process. Bulk volume change is reported as axial strain (%) measured for a cylindrical specimen under zero axial and radial confining pressure (Likos and Lu, 2006). Superimposing the sorption and volume change results in this manner illustrates correlation between bulk volume change and corresponding transitions in the interlayer hydrate states, as well as the hysteretic behavior observed in both. This volume change originates on the sub-particle scale as a change in basal spacing and ultimately translates to a macroscopic volume change of the bulk multiparticle system. Corresponding bulk volume change and swelling pressure are complex functions of fabric-related variables that are necessarily introduced by scaling from the sub-particle scale to the multiparticle scale. Among others, these include particle size, particle size distribution, void ratio, pore size distribution, and confining pressure.

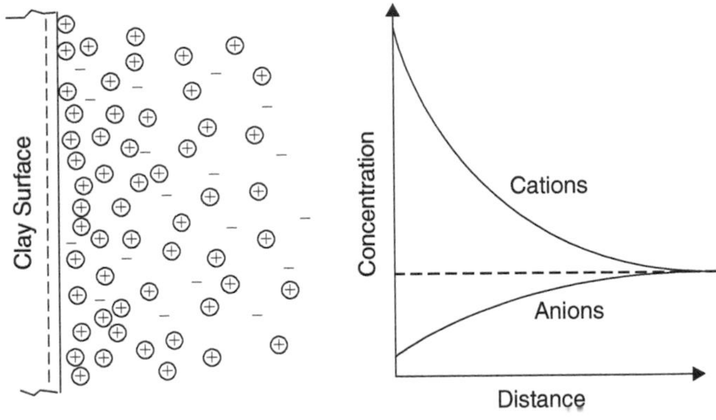


Figure 8. Schematic of the clay-water-ion double layer system (after Mitchell, 1993).

### 3.4 *Osmotic swelling*

Osmotic swelling occurs as the result of pore water flow driven by a gradient in dissolved solute concentration between the interparticle pore fluid and the interlayer pore fluid. Figure 8 shows a schematic for the distribution of ions relative to a charged surface in aqueous solution. The charge field emanating from charge deficiencies in the clay mineral is greatest at the surface and decays with increasing distance. As a result, cations are concentrated near the surface and tend to be diffuse at distance.

A common model for this distribution is the Guoy-Chapman theory of a diffuse double layer (DDL), which models the decay exponentially (e.g. Van Olphen, 1991; Mitchell, 1993). The thickness (1/K) of the diffuse double layer is:

$$\frac{1}{K} = \sqrt{\frac{\varepsilon_0 DkT}{2n_0 e^2 v^2}} \tag{2}$$

where $\varepsilon_0$ is the permittivity of vacuum ($8.8542 \times 10^{-12}\ C^2J^{-1}m^{-1}$), $D$ is the dielectric constant of the pore fluid ($D_{H2O} = 80$), $k$ is the Boltzmann constant ($1.38 \times 10^{-23}\ JK^{-1}$), $T$ is absolute temperature (K), $n_0$ is the far-field (bulk pore fluid) ion concentration (ions/m$^3$), $e$ is the electronic charge ($1.602 \times 10^{-19}$ C), and $v$ is the ion valence. Equation 2 is instructive because it allows one to appreciate the impact of changes in the system variables on the thickness of the double layer. For example, a double layer will generally decrease in thickness, or "collapse," if the ion concentration is increased, if the ion valence is increased (e.g. if $Na^+$ is replaced with $Ca^{2+}$), or if the dielectric constant of the pore fluid is decreased. This affects the net interlayer and interparticle interaction force, the corresponding particle and pore fabric, and ultimately, the engineering behavior of the clay.

When two charged surfaces in an electrolyte solution approach one another, their electrical double layers at some point begin to interact. If the energetics of the two surfaces favor attraction, the double layers will overlap. Overlapping double layers are the normal condition governing the behavior of clay-water systems of geotechnical importance, such as hydrated GCLs. Overlapping double layers in clay-water-ion systems may occur on both the particle scale (i.e. between adjacent parallel or non-parallel particles in the interparticle pore space) and on the sub-particle scale (i.e. between quasi-crystals, crystallites, and ultimately unit layers in the interlayer pore space). The region of overlapping double layers is marked by a relatively high concentration of cations because they are effectively restrained from diffusing to other regions of the pore fluid. This concentration gradient results in osmotic pressure that acts to drive additional water into the interlayer, resulting in disassociation of the unit layers, large bulk volume change, and the formation of a gelled fabric.

Osmotic swelling can be better understood by considering the simple experiment illustrated in Figure 9. Free water is shown in contact with a solute solution through a semi-permeable membrane that is permeable to water molecules but not to the solute. The dissolved solutes decrease the potential of the solution in the left-hand reservoir by an amount dependent on the solute concentration, thus creating a gradient that acts to drive water through the membrane from the

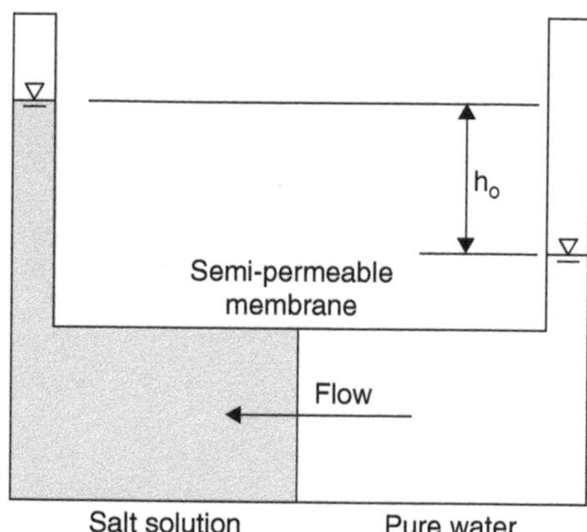


Figure 9. Generation of osmotic flow and pressure through a semi-permeable membrane.

right-hand reservoir. Equilibrium is reached when the pressure head in the left reservoir ($h_0$) balances the osmotic forces tending to cause inflow. This pressure head is the osmotic pressure, $\pi$, which, if the solute solution is ideal and dilute, may be approximated by the van't Hoff equation:

$$\pi = \rho_s g h_0 = RTC_s \tag{3}$$

where $\rho_s$ is the density of the solute solution, $g$ is gravitational acceleration, $R$ is the universal gas constant, $T$ is absolute temperature, and $C_s$ is the molar concentration of the solution. Osmotic pressure, therefore, is directly proportional to the concentration difference between the reservoirs.

In clays, the effect of an imperfect semi-permeable membrane is introduced by the restraint in ion mobility that the negatively charged mineral surfaces impose on the interlayer cations in overlapping double layers. Once sufficient separation distances between unit layers are achieved during crystalline swelling (for $Na^+$-smectite, this is ~20 Å), attractive forces no longer dominate, and the interlayer can be considered a region of high electrolyte concentration compared to the external bulk solution. If cations are retained in the interlayer space in concentrations greater than in the bulk pore fluid, osmotic pressure arises from the difference in ion concentration, thus causing water to be drawn into the zone of relatively high concentration. The volume change associated with this process is referred to as osmotic swelling (e.g. Bolt, 1956, Warkentin et al., 1957; Quirk and Marčelja 1997).

During osmotic swelling, smectites can form a stable gel phase. A gel phase is a borderline equilibrium state where both crystalline and osmotic swelling as well as dispersion may occur during further uptake of water on the one hand, and shrinking, flocculation and fabric rearrangement occur during water loss on the other hand. Stated another way, a gel phase is a transitional hydrous (saturated) phase existing between the condensed (plastic-like) phase and the dispersed (fluid-like) phase. Typically clay gels are materials having reduced void sizes and minimal inter-connectivity of pores, and therefore exhibit increased tortuosity of flow path. In order to effectively seal under confinement, domains of quasi-crystals must be able to re-arrange and expand into the available inter-particle pore space during wetting. Typically wetting with low ionic strength solutions forms optimal bentonite gels. Low solution concentrations of counter ions fail to fully shield repulsive forces operating between adjacent clay layers, thus enabling repulsive forces to override the attractive forces. A stable gel phase then results. Conversely, higher counter ions tend to shield the repulsive forces and result in attractive forces within the gel becoming dominant. This results in collapse of the gel phase to a plastic phase.

## 4 ENGINEERING PROPERTIES AND THEIR MEASUREMENT

A variety of techniques may be employed for evaluating the composition, properties, and quality of bentonite. Many of these techniques are applicable for evaluating bentonite in general, while others are applicable for more specifically evaluating the quality of bentonite for use in GCL applications.

The remainder of this chapter describes several physical, chemical, and engineering properties of bentonite considered relevant to the quality control and performance of GCLs. Experimental techniques commonly employed for their measurement are summarized. Key engineering properties not described here, such as hydraulic conductivity, shear strength, and compressibility, are described in detail in subsequent chapters of this volume.

## 4.1 *Mineralogical analysis*

The mineral composition of clayey materials is most commonly determined using X-ray diffraction (XRD). XRD relies on the interaction of X-rays with atomic planes in the mineral crystalline structure. Reflections defining atomic plane spacings, most notably the $d_{001}$, or basal spacing, in combination with their integrated areas, become the primary criteria for mineral identification. Kaolin and illite, for example, have a basal spacing of 7.15 Å and 10 Å, respectively. The basal spacing of smectite varies depending on the extent of interlayer separation, which can be controlled for analysis by equilibrating samples with ethylene glycol or glycerol vapor, which expands the basal spacing to about 17 Å.

Figure 10 shows XRD traces obtained for orientated films of a natural Wyoming bentonite (Fig. 10a) and Colorado bentonite (Fig. 10b). Results are presented in the form of radiation intensity (counts per second) measured from an incident X-ray beam at an angle °θ, or more conveniently °2θ. Atomic plane spacing corresponding to the incident angle is computed from the wavelength of the X-ray radiation and using Bragg's Law (Moore and Reynolds, 1997).

The XRD patterns in Figure 10 were obtained for evaporated sample mounts using an automated Scintag XDS-2000 XRD system employing CuKα radiation. Analysis indicates that the Wyoming bentonite is comprised of discrete smectite with a minor quartz component. The presence of smectite is indicated by the transition of the 001 peak from 7.1°2θ (12.4 Å) in the air dried condition to 5.6°2θ (15.8 Å) after glycolation. The sharp reflection in the air-dry pattern at 9.7°2θ and to a lesser extent at 15.8°2θ indicates zeolite. The basal (001) spacing of 12.4 Å in the air-dried condition (ambient relative humidity) indicates the presence of ~1 molecular layer of interlayer water and reflects the predominance of $Na^+$ in the exchange complex (see Fig. 6). The Colorado bentonite is comprised of smectite with trace amounts of kaolinite, quartz, calcite, and feldspar. The presence of kaolinite is verified by the destruction of its 001 peak (12.35°2θ, 7.16 Å) upon heating to 550°C. The air-dried basal spacing at 6.0°2θ (14.7 Å) reflects ~2 layers of interlayer water and the predominance of exchangeable $Ca^{2+}$ and $Mg^{2+}$.

Mean thicknesses of smectite crystallites may be estimated using the Fourier decomposition technique of Bertaut-Warren-Averbach applied to the XRD patterns for the glycolated clays (Eberl et al., 1996). Mean crystallite thicknesses determined from the patterns in Figure 10 are 12.7 nm and 11.1 nm for the Wyoming and Colorado clays, respectively, which correspond to 7.6 and 6.6 unit layers (m) per crystallite. These values are near the upper limit reported by Mystkowski et al. (2000) for a wide range of natural smectites ($3.4 \leq m \leq 7.3$).

Quantitative mineralogical analysis requires identification of mineral components in a bentonite by random powder XRD and estimation of their proportions by Reitfeld refinement (Taylor and Hinczak, 2004) using powder diffraction patterns of known international reference minerals. This is generally performed on bulk bentonite, showing montmorillonite and other constituent contents as a percentage of bulk. Detailed (and more expensive) analysis can be performed on size-fractionated bentonites (Gates et al., 2002), showing montmorillonite and other constituents contents as a percentage of size fraction (i.e. ≤0.5 μm). Such information is useful in determining the degree of fineness of the smectite component, as generally non smectite minerals (impurities) are found in the coarser fractions (i.e. >0.5 μm). Further analyses, such as a quantitative estimate of any illite-smectite interstratifications can be performed (e.g. Moore and Reynolds, 1997), but these are generally not required for most bentonite specifications.

## 4.2 *Cation exchange capacity*

CEC is most commonly determined by replacing the cations in the natural exchange complex with a solution of a known or "index" cation species. The quantity of index cations required to satisfy the exchange sites is determined analytically. The composition of the native exchange complex may also be determined from analysis of the extract.

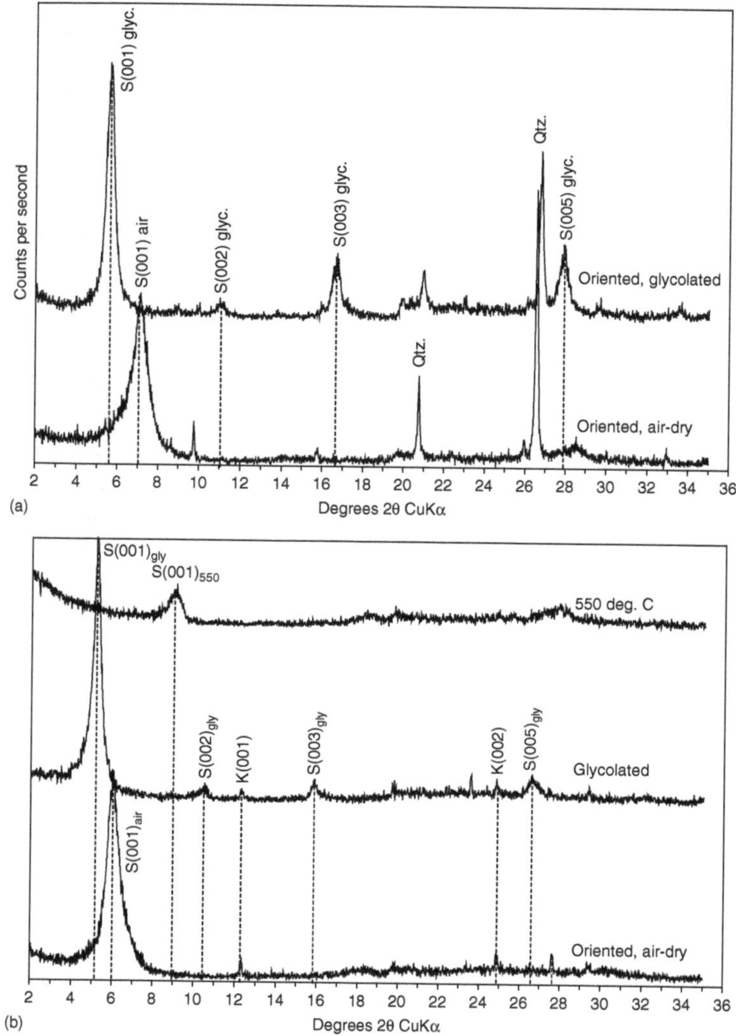


Figure 10. XRD traces of oriented samples of (a) Wyoming and (b) Colorado bentonites.

CEC is generally determined using the ammonium displacement method. Alternative methods for determining CEC include Methylene Blue titration methods (e.g. Cokca and Birand, 1993), surface tension methods (e.g. Burrafato and Miano, 1993) and methods based on mapping of spectral adsorption features (e.g. Kariuki et al., 2003). CEC can also be determined from measurement of the changes in interlayer displacement upon the adsorption of alkylammonium cations (Mermut and Lagaly, 2001). Another method uses barium ($Ba^{2+}$) saturation (Battaglia et al., 2006) and quantification directly on the sample using X-ray fluorescence. This method is robust in that it can be performed on bulk material or size-fractionated material, thereby getting information specific to the bentonite or the montmorillonite. An additional strength is that the entire chemistry of the clay can be determined on the same material enabling structural formulae and layer charge distribution to be determined.

The structural chemistry and layer charge characteristics of the clay can be used to calculate a CEC. For montomorillonites, CEC determined in this way should give a true representation of the CEC, but high layer charge clays or illite:smectite interstratifications may retain cations more effectively, thus CEC measured by other methods may be displaced to lower values. CEC determined in any of the above methods usually always differ (Table 4) due to differences in errors

Table 4. CEC (cmol/kg) estimated for two reference bentonites using various methods.

| Bentonite | Structural formulae | $Ba^{2+}$ displacement | $NH_4^+$ displacement | Alkylammonium method |
|---|---|---|---|---|
| SWy (Wyoming, USA) | 103 | 101* | 76 | 89** |
| Miles (QLD, Australia) | 96 | 98 | 78 | not determined |

*Bataglia et al. (2006); **Mermut and Lagaly (2001); all other measures this study.

associated with each technique and assumptions made. The caveats associated with the various techniques should be understood, and only rarely should one measure be relied on.

### 4.3 *Surface area*

The specific surface area is defined as the combined area of all surfaces as determined by some experimental technique or combination of techniques. A specific surface area measurement rarely represents a true surface area and each experimental method used will provide a different value because (i) the solid surface can be altered during preparation for analysis (e.g. drying exposes more surface) and (ii) if the measurement requires a specific reaction (i.e. a surface complexation reaction), only the surfaces involved in the reaction will be measured.

A variety of methods exist for estimating specific surface areas of bentonites. These include physical methods, such as X-ray diffraction to determine crystallographic information, or electron microscopic measurements to determine shape, and dimension of particles. These methods tend to provide information on total specific surface area. Other methods include positive adsorption, where the accumulation of some substance at the surface is measured, or negative adsorption, where the loss of some substance from the surface is measured. These latter methods tend to provide information on particular surfaces specific to the reactions involved [thus the term specific surface areas, coined by Sposito (1984)].

Specific surface area is most commonly determined by sorption analysis using either non-polar (e.g. nitrogen) or polar [e.g. ethylene glycol monoethyl ether (EGME), water vapor, methylene blue (MB) dye] sorbate molecules. Multilayer or monolayer sorption models [e.g. Brunauer-Emmett-Teller (BET); Brunauer et al., 1938], are considered to calculate surface area from the measured sorption isotherms. These procedures should be used with caution for application to bentonites because the sorption mechanisms are complex and depend upon the layer charge characteristics of the montmorillonite.

Knowledge of the structural formula as well as unit cell information determined by X-ray powder diffraction, enables estimation of the crystallographic surface area ($S_o$), as well as surfaces associated with the interlayer ($S_I$) and edge ($S_e$) surfaces (see Table 5). While such determinations are not routinely made they can provide baseline information to which other measures can be compared.

### 4.4 *Particle and granule size distribution*

Non-hydrated GCLs contain bentonite in either granular (aggregated) or powdered form. Gradation is often reported in terms of the distribution of the dry, aggregated granules and the distribution of the dispersed particles comprising those granules (e.g. Fig. 11). The former is obtained by mechanical sieving. The latter may be obtained by hydrometer or pipette sedimentation analyses. Results are improved by limiting the amount of bentonite in the sedimentation cylinder to about 10 grams, allowing longer presoaking times (48 hours), and by aggressively dispersing the suspension with ultrasonic processing. High quality bentonite is comprised of 70% to 90% particles less than 2 $\mu$m.

The aggregate particle size of a bentonite product may impact its initial hydration. Vangpaisal and Bouazza (2004) observed that powdered bentonites generally hydrate uniformly from the outer surfaces of the GCL toward the center (Fig. 12), resulting in rapid development of an effective seal against further advective water movement. On the other hand, the outer surfaces of each individual granule within granulated bentonites wets first, and particles within aggregates are then wet more slowly. The initial advective water movement is higher in a granulated than a powdered bentonite GCL.

Table 5. Crystallographic specific surface areas of montmorillonites purified from reference bentonites.

| Bentonite | m.u.c† (g) | $S_o$ ($m^2/g$) | $S_I$ ($m^2/g$) | $S_e$ ($m^2/g$) |
|---|---|---|---|---|
| SWy (Wyoming, USA) | 748 | 865 | 828 | 37.1 |
| Miles (QLD, Australia) | 751 | 829 | 792 | 37.0 |

† mu.c. = mass of the unit cell considering $Ca^{2+}$ as the exchange cation.

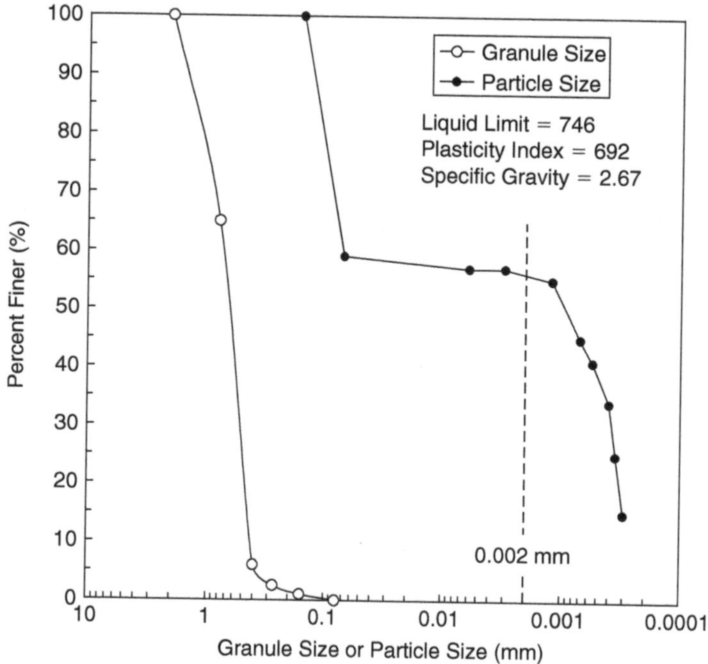


Figure 11. Granule and particle size distribution of a GCL bentonite (after Jo et al., 2001).

## 4.5 *Water content*

The water content of GCL bentonite is an important quality control parameter because of its role in governing mass per unit area. Water content is expressed gravimetrically ($w\% = m_w/m_s \times 100\%$, where $m_w$ = mass of water and $m_s$ = mass of soil solids) and may be determined by oven drying to constant mass following ASTM D2216. Post (1989) points out that drying may take significantly longer than 24 hours for highly plastic soils such as bentonite. The presence of adhesives in the GCL may also complicate determination of moisture content in many cases (Koerner, 1998). The water content of GCLs in the "dry" or non-prehydrated state typically range from about 10%–18% (Koerner, 1998). This suggests that non-prehydrated GCLs generally fall within the stable one-layer or two-layer hydrate states of the crystalline swelling regime. Subsequent hydration promotes transition into the osmotic swelling regime.

## 4.6 *Swell index*

ASTM Standard D5890 describes procedures for conducting the free swell test for bentonite quality control. The result of a free swell test is an index value (Swell Index) that may be used to assess bentonite swell potential as an indirect measure of its effectiveness as a hydraulic barrier. Testing involves grinding the bentonite to pass a #200 (75 μm) sieve, filling a 100 mL graduated cylinder

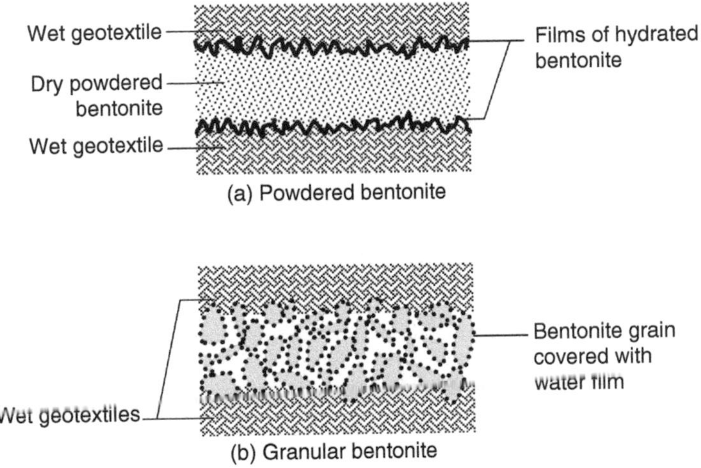


Figure 12. Hydration mechanisms of different forms of bentonite (from Vangpaisal & Bouazza, 2004).

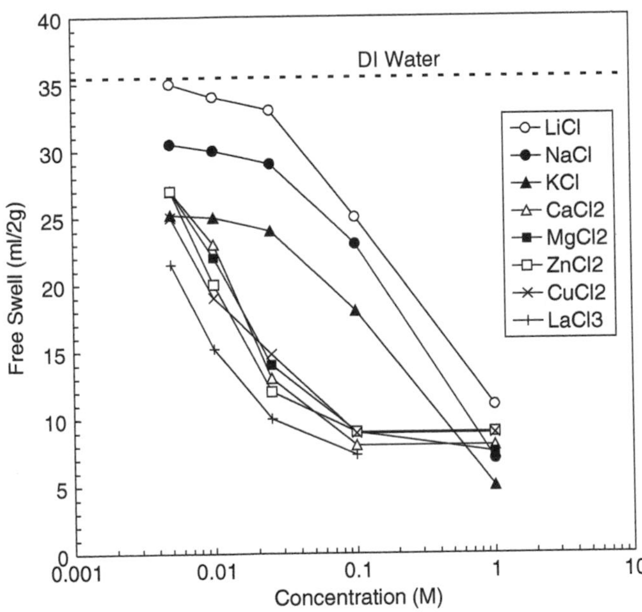


Figure 13. Effect of concentration on free swell of $Na^+$-bentonite from GCL (after Jo et al., 2001).

with ~90 mL of distilled, de-ionized water, adding 2 grams of oven-dried, powdered bentonite to the cylinder in 0.1 g increments, filling the cylinder to the 100 mL mark, and recording the equilibrium volume (in mL/2 g) of the hydrated clay mass that has settled to the bottom of the cylinder after 16 to 48 h.

$Ca^{2+}$-bentonites typically exhibit swell indices on the order of 5 to 10 mL/2g. $Na^+$-bentonites may exhibit swell indices of 25–35 mL/2 g. Most GCLs contain bentonite with swell indices of at least 24 mL/2 g (Trauger, 1994). Many bentonites require additives to achieve acceptable free swelling values. Common additives include the nonionic poly(acrylamide), anionic poly(acrylate) and low cationicity poly(dimethyldiallylammonium) salts. These water soluble polymers serve to enhance dispersion and the formation of stable gels.

Figure 13 presents free swell results reported by Jo et al. (2001) for a GCL bentonite in de-ionized (DI) water and various salt solutions. The greater tendency for particle dispersion and unit layer disassociation is reflected by the large swell indices measured for DI water (35.5 mL/2 g), while inhibited interlayer swelling is evident for the tests conducted using salt solutions. Swell is most effectively repressed in the high valence, high concentration solutions.

### 4.7 *pH, electrical conductivity and carbonate content*

Soil pH plays an important role in ion mobility, precipitation and dissolution phenomena, and oxidation-reduction equilibria. Soil pH is typically reported as that for a 1:1 (mass:volume) suspension ratio of soil in distilled water, usually measured with a glass electrode (Thomas, 1996). The pH of a $Na^+$-bentonite suspension usually ranges from 8.5 to 10.5, whereas the pH of $Ca^{2+}$-bentonite will usually be lower than 7.0 to 8.5 (Trauger, 1994).

The electrical conductivity (mS/m) of the soil solution reflects the abundance of ions and may be measured for either a soil extract or soil suspension. Abu-Hassanein et al. (1996) explored correlations between bentonite content and electrical conductivity and proposed procedures for estimating bentonite content using electrical conductivity for mixtures in slurry form.

Soluble calcium compounds present in GCLs can pose long-term problems associated with the preferential exchange of $Ca^{2+}$ for $Na^+$ in the exchange complex (Guyonnet et al., 2005; Gates et al., 2008). Carbonate content can be determined quantitatively by XRD, semi-quantitatively using infrared spectrometrometry or using gasometric techniques. Total carbonate content of commercial bentonites is approximately nil to 8% by weight (Egloffstein, 1994). For bentonites used in GCL applications, carbonate content should ideally be negligible.

### 4.8 *Fluid loss*

ASTM D 5891 describes test procedures for determining the fluid or filtrate loss behavior of the clay component of GCLs. Results are reported as the volume of fluid (mL) lost through a filter cake formed from an equilibrated 6%-solids clay slurry pressurized to 100 psi (690 kPa) against a porous surface (porous stone and filter paper). Fluid loss indirectly indicates the extent of particle disassociation and dispersion in suspension and the corresponding ability of the bentonite to form stable gels and a low-permeability, self-sealing barrier. The lower the fluid loss volume (reported in mL/30 min @ 100 psi), the more effective the bentonite at forming a barrier. Additives, such as polymers, can enhance fluid loss characteristics of bentonites.

## 5 CLOSING REMARKS

The primary functional component of a geosynthetic clay liner is its internal bentonite layer. The engineering properties of bentonite are largely controlled by its mineralogy, and most notably by that of its smectite component. It is essential, therefore, to understand smectite mineralogy, how this mineralogy influences layer charge, surface area, morphology, and fabric, and in turn, how these surface and fabric properties influence water sorption, swelling, and corresponding engineering behavior. This chapter has intended to serve as an introduction to these topics, and thus provide a foundation for the more detailed treatments of bentonite behavior in the context of GCL applications that are presented elsewhere in this volume.

## REFERENCES

Abu-Hassanein, Z.S., Benson, C.H., Wang, X., Blotz, L.R. 1996. Determining bentonite content in soil-bentonite mixtures using electrical conductivity. *Geotechnical Testing Journal* 19(1): 51–57.

Aylmore, L.A.G. and Quirk, J.P. 1971. Domains and quasicrystalline regions in clay systems. *Soil Science Soc. Amer. Proc.* 35, 652–654.

Barshad, I. 1949. The nature of lattice expansion and its relation to hydration in montmorillonite and vermiculite. *American Mineralogist* 34: 675–684.

Battaglia, S., Leoni, L., Sarori, F. 2006. A method for determining the CEC and chemical composition of clays via XRF. *Clay Minerals*, 41:717–725.

Bergaya, F., Lagaly, G., Vayer, M. 2006. Cation and anion exchange. In. Bergaya, F., Theng, B.K.G., Lagaly, G. (eds.) *Handbook of Clay Science. Developments in Clay Science* Vol. 1. pp.979–1002.

Berend, I., Cases, J., Francois, M., Uriot, J., Michot, L., Maison, A., and Thomas, F. 1995. Mechanism of adsorption and desorption of water vapor by homoionic montmorillonites. *Clays and Clay Minerals* 43(3): 324–336.

Bohn, H.L., McNeal, B.L., and O'Connor, G.A. 1985. *Soil Chemistry*, 2nd Edition, John Wiley and Sons, New York.

Bolt, G.H. 1956. Physico-chemical analysis of the compressibility of pure clays, *Geotechnique* 6(2): 86–93.

Brunauer, S., Emmett, P.H., and Teller, E. 1938. Adsorption of gases in multimolecular layers. *Journal of the American Chemical Society* 60: 309.

Brindley, G.W., and Brown, G. 1980. *Crystal structures of clay minerals and their X-Ray identification.* Mineralogical Society Monograph No. 5.

Brindley, G.W. and MacEwan, D. 1953. Structural aspects of mineralogy of clays and related silicates, in *Ceramics*, British Ceramic Society: 15–19.

Burrafato, G. and Miano, F. 1993. Determination of the cation exchange capacity of clays by surface tension measurements. *Clay Minerals* 28: 475–481.

Cases, J.M., Berend, I., Besson, G., Francois, M., Uriot, J.P., Thomas, F., and Poirier, J.E. 1992. Mechanism of adsorption and desorption of water vapor by homoionic montmorillonite. I. The sodium exchanged form. *Langmuir*, 8, 2730–2739.

Chipera, S.J, Carey, J.W., and Bish, D.L. 1997. Controlled-humidity XRD analyses: Application to the study of smectite expansion/contraction. In *Advances in X-Ray Analysis* 39, J.V. Gilfrich et al. ed: 713–721.

Churchman, G.J., Askary, M., Peter, P., Wright, M., Raven, M.D., Self, P.D. 2002. Geotechnical properties indicating environmental uses for an unusual Australian bentonite. *Applied Clay Science*, 20: 199–209.

Collis-George, N. 1955. The hydration and dehydration of Na-montmorillonite (Belle Fourche). *Journal of Soil Science* 6(1).

Conway. 1981. *Ionic hydration in chemistry and biophysics*, Elsevier, New York.

Cornell University. 1951. Final report on soil solidification research, Ithaca, NY.

Eberl, D.D., Drits, V.A., Środoń, J. 1996. MUDMASTER; a program for calculating crystallite size distributions and strain from the shapes of X-ray diffraction peaks, *US Geological Survey Open File Report* 96–171.

Egloffstein, T. 1994. Properties and test methods to assess bentonite used in geosynthetic clay liners. *Proc. Int. Symp. GCLs*, Koerner et al. (eds.), Nürnberg.

Egloffstein, T.A. 2001. Natural bentonite – influence of the ion exchange and partial desiccation on permeability and self-healing capacity of bentonites used in GCLs. *Geotextiles and Geomembranes*, 19, 427–444.

Gates, W.P. 2004. Crystalline swelling of organo-modified bentonite in ethanol-water solutions. *Applied Clay Science*, 27, 1–12.

Gates, W.P. 2005. Infrared spectroscopy and the chemistry of dioctahedral smectites. In Kloprogge, J.T. (ed.) The Application of Vibrational Spectroscopy to Clay Minerals and Layered Double Hydroxides. CMS Workshop Vol. 13, The Clay Minerals Society, Aurora, CO. pp. 125–168.

Gates, W.P., Anderson, J., Raven., M.D., Churchman, G.J. 2002. Mineralogy of a bentonite from Miles, Queensland, and characterization of its acid-activation products. *Applied Clay Science*, 20, 189–197.

Gates, W.P., Hitchcock, A.P., Dynes J.J. (2008 – submitted) Scanning transmission X-ray microscopy study of Sodium beneficiation reactions of $Ca^{2+}$-smectite. *Clays and Clay Minerals.*

Grim, R. 1968. *Clay Mineralogy*, 2nd Ed., McGraw-Hill, New York.

Grim, R.E., Güven, N. 1978. *Bentonites: Geology, Mineralogy, Properties and Uses.* Chapter 4. Elsevier, Amsterdam.

Guyonnet, D., Gaucher, E., Gaboriau, H., Pons, C-H., Clinard, C., Norotte, V., and Didier, G. 2005. Geosynthetic clay liner interaction with leachate: correlation between permeability, microstructure, and surface chemistry. *Journal of Geotechnical and Geoenvironmental Engineering* 131(6): 740–749.

Harvey, C.C., Lagaly, G. 2006. Conventional Applications. Chapter 10.1 In: Bergaya, F., Theng, B.K.G., Lagaly, G. (eds.) *Handbook of Clay Science. Developments in Clay Science* Vol.1, Elsevier, Amsterdam, pp. 501–540.

Hetzel, F., Tessier, D., Jaunet, A.-M., Doner, H. 1994. The microstructure of three Na+ smectites: The importance of particle geometry on dehydration and rehydration. *Clays Clay Minerals*, 42, 242–248.

Jo, H.Y., Katsumi, T., Benson, C.H., and Edil, T.B. 2001. Hydraulic conductivity and swelling of nonprehydrated GCLs permeated with single-species salt solutions. *Journal of Geotechnical and Geoenvironmental Engineering* 127(7): 557–567.

Jo, H.Y., Benson, C.H., Shackelford, C.D., Lee, J-M., and Edil, T.B. 2005. Long-term hydraulic conductivity of a geosynthetic clay liner permeated with inorganic salt solutions. *Journal of Geotechnical and Geoenvironmental Engineering* 131(4): 405–417.

Kariuki, P.C., Van Der Meer, F., and Verhoef, P.N.W. 2003. Cation exchange capacity (CEC) determination from spectroscopy. *International Journal of Remote Sensing* 24(1): 161–167.

Karaborni, S., Smit, B., Heidug, W., and van Oort, E. 1996. The swelling of clays: Molecular Simulations of the hydration of montmorillonite. *Science* 271: 1102.

Keren, R., and Shainberg, I. 1975. Water vapor isotherms and heat of immersion of Na/Ca-montmorillonite systems-I: Homoionic clay. *Clays and Clay Minerals* 23: 193–200.

Kielland, J. 1937. Individual activity coefficients of ions in aqueous solutions, *J. Am. Chem. Soc.* 59:1675–1678.

Kittrick, J.A. 1969. Interlayer forces in montmorillonite and vermiculite. Soil Science Soci*ety of America Proceedings* 33: 217–222.

Koerner, R.M. 1998. *Designing with Geosynthetics*, 4th Edition, Prentice Hall, Upper Saddle River, New Jersey.

Lagaly, G., 2006. Colloid clay science. Chapter 5 In: Bergaya, F., Theng, B.K.G., Lagaly, G. (eds.) *Handbook of Clay Science. Developments in Clay Science*, Vol. 1. Elsevier. pp. 141–246.

Laird, D.A. 1996. Model for crystalline swelling of 2:1 phyllosilicates. *Clays and Clay Minerals*, 44, 553–559.

Laird, D.A. 2006. Influence of layer charge on swelling of smectites. *Applied Clay Science*. 34, 74–87.

Laird, D.A. Shang, C. 1997. Relationship between cation exchange selectivity and crystalline swelling in expanding 2:1 phyllosilicates. *Clays and Clay Minerals*. 45, 681–689.

Laird, D.A., Shang, C., Thomson, M.L. 1995. Hysteresis in crystalline swelling of smectites. *Journal of colloid and interface science*. 171, 240–245.

Landis, C.R., von Mauge, K., 2004. Activated and natural sodium bentonites and their markets. *Mining Engineering*, 56, 17–22.

Likos, W.J. 2004. Measurement of crystalline swelling in expansive clay. *Geotechnical Testing Journal* 27(6): 540–546.

Likos, W.J. and Lu, N. 2006. Pore-scale analysis of bulk volume change from crystalline interlayer swelling in Na and Ca-smectite, *Clays and Clay Minerals* 54(4): 516–529.

Meer, S. R., and Benson, C. G. 2007. "Hydraulic conductivity of geosynthetic clay liners exhumed from landfill final covers." *J. Geotechnical and Geoenvironmental Engineering*, 133(5), 550–563.

Meleshyn, A. and Bunnenberg, C. 2005. The gap between crystalline and osmotic swelling of Na-montmorillonite: A Monte Carlo study, *J. Chem. Phys*. 122, 034705.

Mermut, A.R., Lagaly, G. 2001. Baseline studies of the clay minerals society source clays: Layer charge determination and characteristics of those minerals containing 2:1 layers. *Clays and Clay Minerals*, 49, 393–397.

Mitchell, J.K. 1993. *Fundamentals of Soil Behavior*, John Wiley, New York.

Mooney, R.W, Keenan, A.G., and Wood, L.A. 1952. Adsorption of water vapor by montmorillonite. II. Effect of exchangeable ions and lattice swelling as measured by X-ray diffraction. *Journal of American Chemical Society*, 74, 1371–1374.

Moore, D.M., and Reynolds, R.C. 1997. *X-Ray Diffraction and the Identification and Analysis of Clay Minerals*, Oxford University Press.

Murray, H.H., 1999. Clays for our future. P. 3-11 In: Kodoma, H., Mermut, A.R., Torrence, J.K. (Eds.) Clays for Our Future. 11th International Clays Conference, Ottowa. ICC97 Organizing Committee, Ottawa.

Mystkowski, K., Środoń, J., and Elsass, F. 2000. Mean thickness and thickness distribution of smectite crystallites. *Clay Minerals*, 35, 545–557.

Newman, A.C.D. 1987. The interaction of water with clay mineral surfaces. In *Chemistry of Clays and Clay Minerals*, Mineralogical Society Monograph No. 6, A.C.D. Newman: 237–274.

Norrish, K. 1954. The swelling of montmorillonite. *Transactions Faraday Society* 18: 120–134.

Norrish, K., Quirk, J.P. 1954. Crystalline swelling of montmorillonite. Use of electrolytes to control swelling. *Nature*, 173, 255.

Norotte, V., Didier, G., Guyonnet, D., and Gaucher, E. 2004. Evolution of GCL hydraulic performance during contact with landfill leachate. *Advances in Geosynthetic Clay Liner Technology: 2nd Symp. Proc. ASTM STP 1456*, Mackey and Maubeuge, eds., ASTM International, West Conshohocken, PA, 41–52.

Odom, I.E. 1984. Smectite clay minerals: properties and uses. *Philosophical Transactions*, Royal Society. London, A., 311, 391–409.

Petrov, R. and Rowe, R.K. 1997. Geosynthetic clay liner (GCL) chemical compatibility by hydraulic conductivity testing and factors impacting its performance. *Canadian Geotechnical Journal* 34: 863–865.

Post, J.L. 1989. Moisture content and density of smectites. *Geotechnical Testing Journal* 12(3): 217–221.

Powell, D.H., Tongkhao, K., Kennedy, S.J., and Slade, P.G. 1997. A neutron diffraction study of interlayer water in sodium Wyoming montmorillonite using a novel difference method. *Clays and Clay Minerals* 45: 290–294.

Quirk, J.P. and Alymore, L.A.G. 1971. Domains and quasi-crystalline regions in clay systems. *Journal of the Soil Science Society of America* 35: 652–654.

Quirk, J.P., Marcelja, S. 1997. Application of double-layer theory to the extensive crystalline swelling of Li-montmorillonite. *Langmuir*, 13, 6241–6248.

Rouquerol, F. 1999. *Adsorption by powders and porous solids: principles, methodology, and applications*. Academic Press, San Diego, California.

Shackelford, C.D., Benson, C.H., Katsumi, T., Edil, T.B., and Lin, L. 2000. Evaluating the hydraulic conductivity of GCLs permeated with non-standard liquids. *Geotextiles and Geomembranes* 18(2–4): 133–162.

Slade, P.G., Quirk, J.P. 1990. The limited swelling of smectites in CaCl2, MgCl2, and LaCl3 solutions. *J. Colloid and Interface Science*, 144, 18–26.

Slade, P.G., Quirk, J.P., Norrish, K. 1991. Crystalline swelling of smectite samples in concentrated NaCl solutions in relation to layer charge. *Clays Clay Minerals*, 39, 234–238.

Sposito, G., 1984. *The Surface Chemistry of Soils*. Oxford University Press., New York. 234p.

Sposito, G. and Prost, R. 1982. Structure of water adsorbed on smectites. *Chemical Reviews* 82: 553–573.

Sposito, G., Park, S-H., and Sutton, R. 1999. Monte Carlo simulation of the total radial distribution function for interlayer water in sodium and potassium montmorillonites. *Clays and Clay Minerals* 47: 192–200.

Taylor J.C., Hinczak, I. 2004. *Reitveldt Made Easy*. Sietronics Pty Ltd. Belconnen, Australia, 201p.

Teppen, B.J., Miller, D.M. 2005. Hydration energy determines isovalent cation exchange selectivity by clay minerals. *Soil Science Society America J.*, 70, 31–40.

Tessier, D. 1990. Behavior and microstructure of clay minerals. In *Soil Colloids and Their Associations in Aggregates*, M.F. DeBoodt et al. eds., Plenum Press, New York.

Thomas, G.W. 1996. Soil pH and soil acidity. *In* D.L. Sparks (ed.), *Methods of Soil Analysis, Part 3, Chemical Methods*, Soil Science Society of America, Madison, WI.

Trauger, R.J. 1994. The structure, properties, and analysis of bentonite in geosynthetic clay liners. *Proceedings of the 8th GRI Conference*, Philadelphia, PA: 185–197.

Vangpaisal, T. and Bouazza, A. 2004. Gas permeability of partially hydrated geosynthetic clay liners. *Journal of Geotechnical and Geoenvironmental Engineering*, 130 (1), pp.93–102.

van Olphen, H., 1963. Compaction of clay sediments in the range of molecular particle distances. *Clays and Clay Minerals* 11: 178–187.

van Olphen, K. 1965. Thermodynamics of interlayer adsorption of water in clays. *Journal of Colloid Science* 20: 822–837.

van Olphen, H. 1991. *Clay Colloid Chemistry*, 2nd Edition, Krieger.

Volzone, C., Garrido, L.B. 2001. Changes in suspension properties of structural modified montmorillonites (Mundanças em propriedades de suspensões de montmorilonitas modifacadas estruturalmente). *Cerâmica*, 47, 4–8.

Wang, M.K., Wang, S.L., and Wang, W.M. 1996. Rapid estimation of cation exchange capacities of soils and clays with Methylene Blue exchange. *Soil Science Society of America* 60: 138–141.

Warkentin, B.P., Bolt, G.H., and Miller, R.D. 1957. Swelling pressure of montmorillonite, *Soil Science Society of America Proc.* 21(5): 495–497.

Weiss Jr, C.A. and Gerasimowicz, W.V. 1996. Interaction of water with clay minerals as studied by 2H Nuclear Magnetic Resonance spectroscopy. *Geochimica Cosmochimica Acta* 60: 265–271.

# CHAPTER 4

# Hydraulic conductivity of geosynthetic clay liners

T. Katsumi
*Kyoto University, Kyoto, Japan*

## 1 GEOSYNTHETIC CLAY LINERS IN BOTTOM LINERS AND TOP COVERS

Liners are the essential components of waste containment facilities since they prevent waste leachates from leaking and causing an adverse impact on the surrounding soil and groundwater. Liners make up the bottom layer system which consists of the layer with the barrier function. In addition, cover systems usually contain a layer which is expected to function as a barrier (Figure 1).

The bottom liner is the final defense against the leakage of waste leachates. Thus, the barrier performance of the bottom liner is crucial. The final cover functions to minimize the infiltration of rainfall water into the waste layer so that the generation of waste leachates can be minimized. In the U.S. and most European countries, the barrier function has been recognized as being essential in final covers. Recently, however, bio-reactor landfills have been proposed to promote the degradation and the stabilization of waste layers in which the final cover is not necessarily expected to have a perfect barrier function. Some amount of infiltration may promote degradation and prevent the formation of a "dry tomb" which is the result of the final cover having an "excellent barrier" function. Actually, this concept for waste landfills has already been put to practice in Japan (The Landfill System Technologies Research Association 2000). Japanese landfills have only a soil layer for the final cover, and it does not have a specific barrier function. Thus, rainfall water can be induced into the waste layer so that the degradation of the waste layer can be promoted. In any case, geosynthetic clay liners (GCLs) have been used for these bottom liners and final covers due to their hydraulic conductivity and additional properties/advantages which will be discussed in other chapters.

## 2 DEFINITIONS

### 2.1 *Hydraulic conductivity*

Among the several functions required of the GCLs applied to containment systems in landfills or liquid impoundments, the barrier function is a very important function. Thus, an index that can evaluate "how difficult it will be for liquid to pass through a GCL" is necessary. For GCLs in which granular or powdered bentonites are sandwiched between the geotextiles, bentonites clearly play a vital role in the barrier function. For GCLs in which bentonites are glued to the geomembranes, both the bentonites and the geomembranes are layers which resist leakage. In this chapter, the hydraulic conductivity of the bentonite portion is mainly discussed.

Bentonites are clay particles which have a high composition of smectic clay minerals. The bentonite layer in a GCL forms a "porous media," because it consists of solids (bentonite particles)

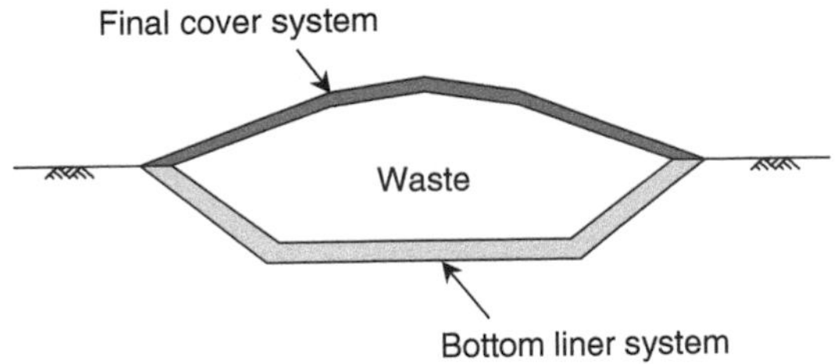


Figure 1. Bottom liner and final cover for waste landfills.

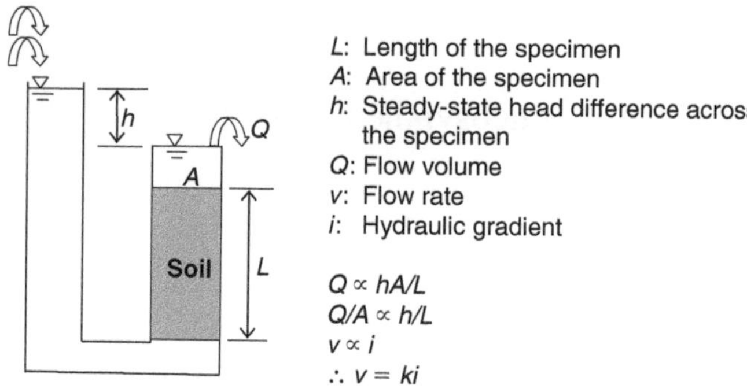


Figure 2. Concept of the hydraulic conductivity tests and the definition of hydraulic conductivity ($k$).

and voids (filled with air and/or liquids). The factors affecting "how difficult it will be for liquid to pass through a GCL" are summarized into the following three categories:

- **Media properties or volumes and the sizes of the voids in the porous media:** Larger sized voids and larger total volumes of voids result in more liquid passing through the GCL. The voids discussed here are effective void spaces, but not total spaces. Therefore, the volume of the water molecules which are strongly attracted to the clay surface should be excluded, and the development of a diffuse double layer should be taken into account.
- **Liquid properties, in particular, viscosity:** High viscosity decreases the potential for liquid flow. The friction and the cohesion between the liquid and the soil particles are included as factors affecting the viscosity.
- **Boundary conditions or the applied liquid pressure:** If the boundary conditions provide a higher liquid pressure across the element of porous media, a larger liquid flow will occur.

To evaluate the first two factors, hydraulic conductivity (or the coefficient of permeability) is used as an index. It is based on Darcy's law.

Darcy (1856) found that the flow volume through saturated soil correlates linearly to the difference in water (or liquid) head applied across the soil specimen, as shown in Figure 2. The flow volume is also the function of the length and the area of the specimen, and it increases linearly with an increase in the cross area and a decrease in the length of the specimen. Thus, flow volume $Q$, ($m^3/s$ or $cm^3/s$) can be expressed by the length of the specimen, $L$, (m or cm), the cross area of the specimen, $A$, ($m^2$ or $cm^2$), and the difference in water head across the soil specimen, $h$, (m or cm) as

$$Q \propto \frac{Ah}{L} \tag{1}$$

$$\frac{Q}{A} \propto \frac{h}{L}, \quad \text{or} \quad v \propto i \tag{2}$$

where $v$ is the flow rate (m/s or cm/s) and $i$ is the hydraulic gradient (non-dimensional). If it is expressed more strictly, $i = \partial h/\partial L$. Using $k$ (m/s or cm/s) as a constant, Equation (2) can be

$$v = k \times i \tag{3}$$

Constant $k$ is defined as the "hydraulic conductivity" (or the "coefficient of permeability," and it is the index which results from all the factors attributed to the media (volumes and sizes of the voids in porous media) and the liquids (e.g., viscosity), except for the factors governed by the boundary conditions.

The typical ranges of hydraulic conductivity values for soils are summarized in Table 1. In general, coarser particle sizes result in larger volumes of individual voids (but not the total volume of the voids). In contrast, the surface effect is more dominant for finer particles, and the repulsion works between the particles such that the total volume of the voids becomes larger. The void ratio ($e$ is defined as the volume of the void spaces over the volume of the occupying solid) for (coarse)

Table 1. Typical range of hydraulic conductivity values for soils.

| Type of soil | Range of hydraulic conductivity |
|---|---|
| Gravelly soil | $>1 \times 10^0$ cm/s |
| Sandy soil | $1 \times 10^0$ cm/s – $1 \times 10^{-2}$ cm/s |
| Silt | $1 \times 10^{-2}$ cm/s – $1 \times 10^{-6}$ cm/s |
| Clayey soil | $<1 \times 10^{-5}$ cm/s |
| Bentonite layer in GCLs | $1 \times 10^{-9}$ cm/s – $5 \times 10^{-9}$ cm/s |

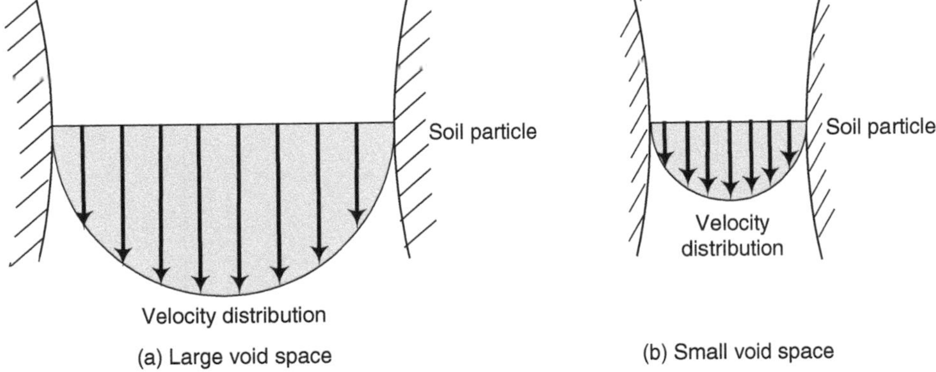


Figure 3. Viscosity of water permeating through the porous media (small void spaces limit the velocity).

sandy soil is typically 0.6 to 0.8, while $e$ for (fine) clayey soil is always larger than 1 and in some cases larger than 2 or even 3. Although clayey soils contain larger void spaces, clayey soils exhibit much lower levels of hydraulic conductivity than sandy soils. This is attributed to the viscosity of the water permeating through the soil and to the surface effect of the fine clay particles. Clayey soils provide small volumes for individual voids, so that the water cannot permeate as freely as it can in sandy soil due to the viscosity, as shown in Figure 3. This results in the low hydraulic conductivity of clayey soils. As for the surface effect of clay particles, since they are usually negatively charged, the polar molecules of liquids such as water are attracted to the clay surface. This results in the formation of a layer of liquid molecules that is strongly attracted to the clay surface. This layer, called a diffuse double layer (DDL), does not contribute to the space for liquid to permeate since liquid molecules are strongly attracted to the clay. Thus, as more DDLs develop, lower levels of hydraulic conductivity are obtained.

The hydraulic conductivity of the bentonite portions in GCLs permeated with pure water depends on the confining stress. From the reported publications, it typically ranges from $1 \times 10^{-9}$ to $5 \times 10^{-9}$ cm/s.

## 2.2 *Unsaturated hydraulic conductivity*

The hydraulic conductivity mentioned in Section 2.1 is "saturated" hydraulic conductivity. This is the hydraulic conductivity when the soil pores are completely filled (saturated) with water (or permeant liquids), and do not contain any air. Most GCLs subjected to liquid flow are thought to be saturated with liquids. However, GCLs employed in cover systems are sometimes considered to be "unsaturated."

In general, hydraulic conductivity is the function of the degree of saturation, as shown in Figure 4. This is because the flow rate is dependent on the volume occupied by water, so that a lower degree of saturation results in low hydraulic conductivity. Thus, evaluating the performance of such traditional barrier materials as compacted clay liners (CCLs) using saturated hydraulic conductivity, even under unsaturated conditions, will provide a conservative design. For GCLs, on the other hand, low hydraulic conductivity is achieved due to the swelling of bentonites. Therefore, a relatively large

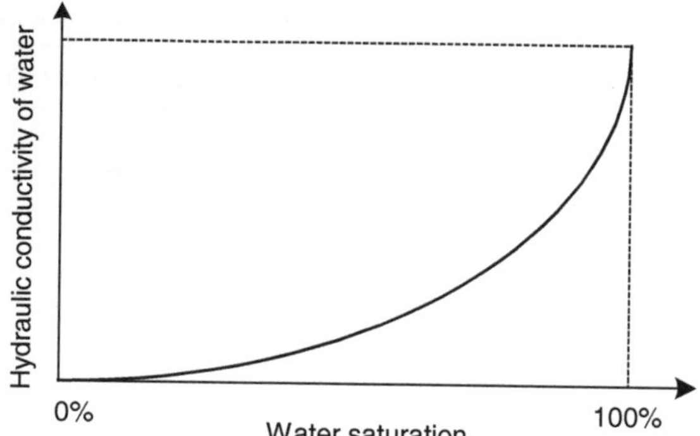


Figure 4. Typical relation between unsaturated hydraulic conductivity and the degree of saturation.

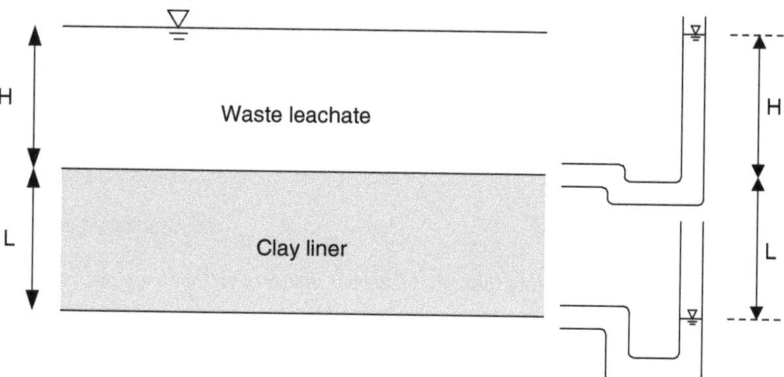


Figure 5. Cross section (used as an example for calculating the flow rate).

amount of permeation is observed for the early stage of the hydraulic conductivity tests, until the saturation of the bentonites by water is achieved and a sufficient amount of swelling occurs. Very few data have been reported on the unsaturated hydraulic conductivity of GCLs.

## 2.3 *Permittivity*

GCLs are thin materials and their thickness varies with the state of swelling and the in-situ conditions, in particular, the confining stress. When the barrier layer is thin, the flow rate is sensitive to changes in thickness. As an example, consider the conditions shown in Figure 5. Assuming that the clay layer is saturated, waste leachates accumulate with depth $H$ (cm or m) above the top of the clay layer, the water head at the bottom of the clay layer is equal to its elevation, and flow rate $v$ can be calculated as

$$v = k \times \frac{H+L}{L} \tag{4}$$

Firstly, consider the case for compacted clay liners (CCLs). When $k = 1 \times 10^{-7}$ cm/s and $H = 30$ cm, $v$ for $L = 30$ cm and 33 cm can be obtained as

$$v = 1 \times 10^{-7}(\text{cm/s}) \times \frac{30(\text{cm}) + 30(\text{cm})}{30(\text{cm})} = 2.0 \times 10^{-7}(\text{cm/s}) \tag{5-1}$$

$$v = 1 \times 10^{-7}(\text{cm/s}) \times \frac{30(\text{cm}) + 33(\text{cm})}{33(\text{cm})} \cong 1.91 \times 10^{-7}(\text{cm/s}) \tag{5-2}$$

An increase in thickness of five centimeters, which might be unrealistic for CCLs, will result in a decrease in the flow rate by approximately 4.5%.

For GCLs, assuming that $k = 1 \times 10^{-9}$ cm/s and $H = 30$ cm, $v$ for $L = 1.0$ cm and 1.1 cm can be obtained as

$$v = 1 \times 10^{-9}(\text{cm/s}) \times \frac{30(\text{cm}) + 1.0(\text{cm})}{1.0(\text{cm})} = 3.1 \times 10^{-8}(\text{cm/s}) \tag{6-1}$$

$$v = 1 \times 10^{-9}(\text{cm/s}) \times \frac{30(\text{cm}) + 1.1(\text{cm})}{1.1(\text{cm})} \cong 2.83 \times 10^{-8}(\text{cm/s}) \tag{6-2}$$

The flow rate decreased by approximately 8.8% when thickness $L$ of the GCL was increased by 10% (1.0 cm to 1.1 cm). This is similar to the conditions of the CCLs. This variation (or even more) in L is possible, however, for GCLs. Thus, only the hydraulic conductivity is insufficient when the barrier performance of the GCL is evaluated and a certain thickness is required.

Permittivity (cross-plane hydraulic conductivity) may be used as an index to combine the hydraulic conductivity and the thickness, and some publications have used this index (Giroud et al. 1997). The definition of permittivity is explained as follows:

$$v = k \times i = k \times \frac{\Delta h}{L} = \frac{k}{L} \times \Delta h \tag{7}$$

where $\Delta h$ is the water head difference across the clay layer. If $k/L$ is defined as the permittivity, $\psi$ (1/s) and flow rate $v$ can be expressed as

$$v = \psi \times \Delta h \tag{8}$$

If $\psi$ is known, the flow rate can be obtained for given conditions of the hydraulic head of the waste leachates above the clay liner.

### 2.4 *Intrinsic permeability*

Since several types of liquids are used as permeants in the experiment in order to evaluate the GCLs, and because several types of waste leachates may permeate through the GCLs in the field, the properties of the permeant liquids should be taken into consideration. Among the properties of the liquids, the viscosity is a very important factor. For example, Petrov et al. (1997) conducted hydraulic conductivity tests on needle-punched GCLs permeated with ethanol solutions to assess the chemical compatibility. While significant increases in hydraulic conductivity were observed for ethanol solutions of 75% and 100%, ethanol solutions of 25% and 50% resulted in slightly lower levels of hydraulic conductivity than pure water, as shown in Figure 6(a). This decrease in hydraulic conductivity for ethanol solutions of 25% and 50% is due to the viscosity effect of the permeants. To evaluate the conductivity while avoiding the effect of the liquid viscosity, intrinsic permeability is used. It can be defined as

$$k_l = K_l \frac{\gamma_l}{\mu_l} \tag{9}$$

where $\gamma_l$ is the unit weight of the permeant liquid, $\mu_l$ is the absolute viscosity of the permeant liquid, $K_l$ is the intrinsic permeability of the porous medium to the permeant liquid, and $k_l$ is the hydraulic conductivity of the porous medium to the permeant liquid. Equation (9) means that, for the same value of intrinsic permeability ($K_l$), a larger viscosity ($\mu_l$) will result in a lower level of hydraulic conductivity ($k_l$). When the hydraulic conductivity tests are conducted with chemical solutions and waste leachates, the interaction between the clay and the liquid, such as the development of a diffused double layer, sometimes cannot be clearly evaluated without excluding the effect of the viscosity control. As shown in Figure 6(b), the intrinsic permeability values ($K_l$) for ethanol solutions of 25% and 50% are almost the same as those for pure water. This means that ethanol concentrations lower than 50% had no effect on the pore structure of the bentonite. The intrinsic permeability ($K_l$) is an index indicating the pore structure effective to the liquid permeation, such as a diffused double layer.

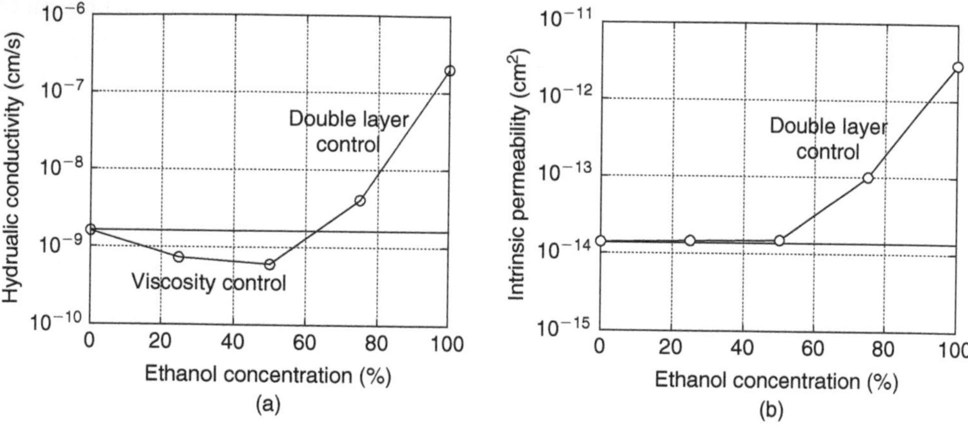


Figure 6. (a) Hydraulic conductivity and (b) intrinsic permeability of a needle-punched GCL permeated with water-ethanol mixtures (data from Petrov et al. 1997).

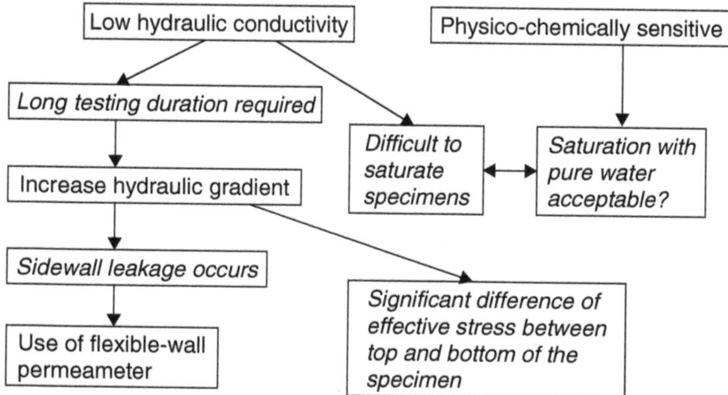


Figure 7. Concerns and measures involved with hydraulic conductivity testing for barrier materials such as GCLs (italicized words show the problems which arise).

## 3 HYDRAULIC CONDUCTIVITY TESTING

Generally, it is difficult to measure the hydraulic conductivity of barrier materials. This is simply because the level of hydraulic conductivity is extremely low. Therefore, hydraulic conductivity testing requires some special care and technical skills. Figure 7 shows the concerns as well as the measures involved with hydraulic conductivity testing. Conducting hydraulic conductivity tests for materials with low levels of hydraulic conductivity takes an extremely long time. A large hydraulic gradient is applied to shorten the testing period; however, this large hydraulic gradient may cause sidewall leakage. Using a flexible-wall permeameter, which will be discussed later, enables the minimization of this sidewall leakage. A large hydraulic gradient would also result in the problem of a significant difference in the effective stress levels between the top and the bottom of the specimens. Other problems with hydraulic conductivity testing are attributed to the physico-chemically sensitiveness of clays. In particular, the type of liquids used in the first step of saturation will affect consequent hydraulic conductivity testing. This effect is called the "first exposure effect," and it will be discussed in Chapter 6.

The important factors affecting hydraulic conductivity values are the type of permeameter employed, the effective stress, the hydraulic gradient, the size of the specimen, the type and the chemistry of the permeant, and the termination criteria. The effects of these factors on the hydraulic conductivity were recently summarized by Benson et al. (1994), Daniel (1994), Shackelford (1994), Kamon and Katsumi (2001) for clay liners, and by Daniel et al. (1997a), Petrov et al. (1997b), and Shackelford et al. (2000) particularly for GCLs.

### 3.1 *Types of permeameters*

One critical issue in hydraulic conductivity testing is the selection of the right type of permeameter. The permeameters typically used are (1) rigid-wall permeameters, (2) flexible-wall permeameters, and (2) consolidation-cells. Consolidation-cells are rarely used to determine the hydraulic conductivity of barrier materials. When materials with low levels of hydraulic conductivity are used for the tests, a high hydraulic gradient has to be induced into the specimen in order to reduce the testing period. However, such a high hydraulic gradient results in a large seepage force, which might in turn induce sidewall leakage (leakage between the specimen and the ring) if a rigid-wall permeameter is used. Furthermore, if the specimen is likely to shrink (due to the effect of permeant chemicals), sidewall leakage is unavoidable. Unlike rigid-wall permeameters, flexible-wall permeameters are effective in minimizing sidewall leakage, even for test specimens with rough sidewalls. A typical schematic diagram of a flexible-wall permeameter is shown in Figure 8. The test specimen is confined with end caps having porous disks on the top and the bottom and by a latex membrane on the sides. The cell is filled with water and is pressurized to press the membrane, and therefore, sidewall leakage is eliminated. There are several other advantages to flexible-permeameters, namely, back-pressure usage to saturate the specimen, control of the principal stress levels, and a reduction in the testing period (Daniel 1994). The disadvantages of flexible-wall permeameters are their relatively high cost and their more complicated operation. Flexible-wall permeameters usually require three pressure positions (for cell, influent, and effluent pressures), which cost more. If only the gravity force is used for these pressures, the permeameter system can be simplified.

Although flexible-wall permeameters are most commonly used to determine the hydraulic conductivity of materials with low levels of hydraulic conductivity, such as GCLs, other apparatuses are sometimes used. This is probably due to (1) chemical compatibility and (2) accurate measurements of the thickness.

For CCLs, Bowders and Daniel (1987) conducted a comprehensive testing program permeated with diluted organic chemicals. They found that the permeation of some organic chemical solutions provided lower hydraulic conductivity values when a flexible-wall permeameter was used, but higher hydraulic conductivity values when a rigid-wall permeameter was used, compared with the permeation of pure water. The chemical solutions had an effect on decreasing the diffuse double layer thickness, and clay specimens were likely to shrink. Micro-cracks were generated due to this shrinkage, so that higher hydraulic conductivity values were obtained with rigid-wall permeameters. For flexible-wall permeameters, however, these cracks closed up when cell pressure was applied. In the field, this will cause the overestimation of the hydraulic performance of clay liners. If sidewall leakage occurs due to the insufficient swelling of the bentonites in the GCLs, but not simply due to the hydraulic pressure applied, and if insufficient swelling is also considerable in the field, laboratory tests should be conducted with a rigid-wall permeameter or a flexible-wall permeameter making sure to select appropriate cell pressures which do not close up the cracks attributed to the chemical effects.

In testing GCLs, one of the key factors in obtaining an accurate hydraulic conductivity is to determine the thickness of the GCLs. This is because GCLs are thin, and their thickness values may easily vary with the testing conditions. Therefore, accurate measurements of the thickness are critical to obtaining accurate levels of hydraulic conductivity. Although flexible-wall permermeters have several advantages, as listed above, determining the thickness of the GCLs is rather troublesome. The thickness is usually determined by measuring the level of marks with a cathetometer from the outside of the cell. To determine the thickness more accurately and more easily, a triaxial compression test apparatus can be used instead of a flexible-wall permeameter, but this results in an extremely high cost. Petrov et al. (1997a) compared three different types of permeameters, namely, (1) flexible-wall permeameters, (2) fixed-ring cell permeameters, and (3) double-ring cell permeameters. Fixed-ring cell permeameters are rigid-wall cell permeameters which are able to directly measure the thickness with a dial gauge while simultaneously applying vertical pressure. A schematic drawing of a fixed-ring permeameter is presented in Figure 9. Double-ring cell permeameters are also rigid-wall permeameters, and they are facilitated to divide the inner and the

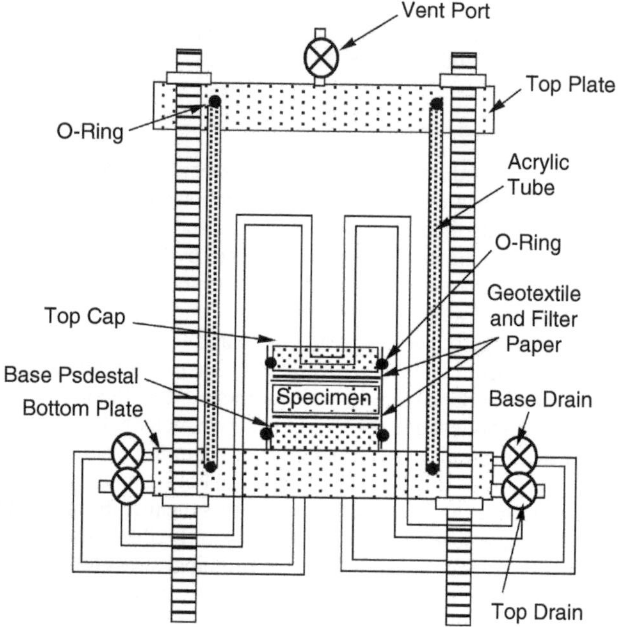


Figure 8. Schematic view of a flexible-wall permeameter.

outer effluents so that sidewall leakage can be determined. Unlike CCLs, sidewall leakage is not likely to occur with GCLs because bentonites swell. Even for a low confining stress (3–35 kPa) and for NaCl solutions (0.6 and 2.0 mol/L), this conclusion was found to be valid. Petrov et al. (1997a) concluded that these three types of permeameters provide similar hydraulic conductivity values, and fixed-ring permeameters, rather than flexible-wall permeameters, have been used for research since that time (e.g., Petrov and Rowe 1997; Rowe et al. 2004).

A flow box has been used to measure the hydraulic conductivity for special situations such as the overlapping of GCLs or GCLs which are underlain by a geomembrane with a hole (Daniel et al. 1997b). Examples of the hydraulic conductivity determined with this apparatus are addressed in Chapter 5.

### 3.2 *Critical issues*

Daniel et al. (1997a) addressed the key variables in measuring the hydraulic conductivity of GCLs as (1) the trimming of the GCL specimen, (2) the determination of the thickness of the specimen, (3) the selection of the effective stress, (4) the selection of the hydraulic gradient, and (5) the selection of the first wetting liquid. Among the above five variables, thickness was discussed in the previous section. The effective stress and the hydraulic gradient will be discussed in Chapter 4, and the first wetting liquid will be discussed in Chapter 5.

The loss of bentonites along the edges is crucial when the specimens of geotextile-encased GCLs are prepared (trimmed), as shown in Figure 10. To avoid the loss of bentonites, water is applied before cutting the GCLs. A mark is scribed on one geotextile to show what area should be cut out, and then a small amount of water is applied along the marked circle. After cutting the GCLs, the moistened bentonites do not fall out from the edges. In addition, the uncut group of fibers should not be left as shown in Figure 10 because they will create a preferential flow.

Daniel et al. (1997a) also performed a round-robin testing program for which eighteen laboratories from various academia and industries conducted independent hydraulic conductivity tests. They concluded that there was less variability than might be expected, considering the difficulty in accurately measuring the hydraulic conductivity of relatively impermeable materials such as GCLs.

Special care is required for chemical compatibility testing when hydraulic conductivity tests are conducted using chemical solutions and/or waste leachates. Details of this issue are discussed in Chapter 6.

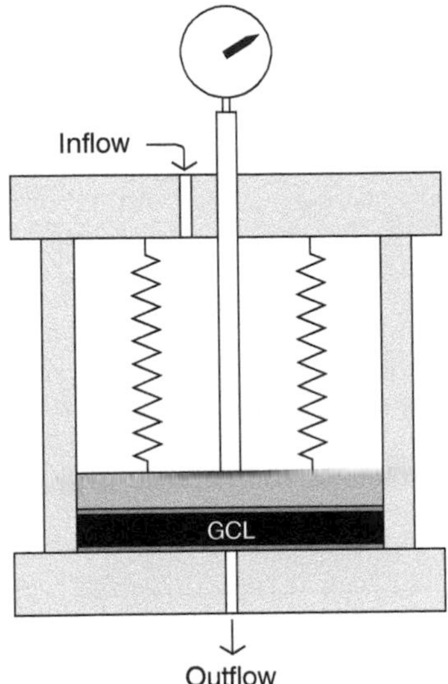


Figure 9. Schematic drawing of a fixed-ring cell permeameter (modified from Petrov et al. 1997a).

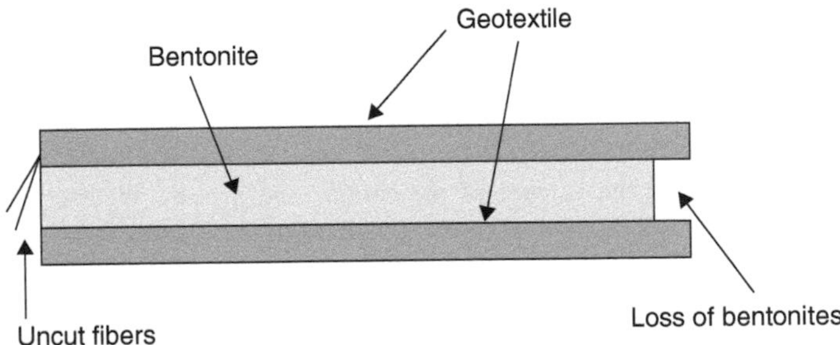


Figure 10. Loss of bentonites from the edge of the GCL (modified from Daniel et al. 1997a).

## 4 INTACT GCLS

### 4.1 *Why do GCLs provide extremely low levels of hydraulic conductivity?*

The most dominant reasons why GCLs provide low levels of hydraulic conductivity are the diffuse double layer (DDL) effect and osmotic swelling. The diffuse double layer effect generally occurs with clay particles, while osmotic phase swelling is a specific phenomenon of montmorillonites.

Diffused double layers (DDLs) are generally expressed by the Stern-Guoy model (Mitchell 1993). According to the Stern-Guoy model, oriented water dipoles and fixed hydrated cations are directly bound to the clay surface where a thin film called the "Stern layer" is formed. Adjacent to the Stern layer, a diffuse layer of hydrated cations, attracted to the clay surface, resides (Shang et al. 1994). The concentration of cations in the diffuse layer is the function of the electrical potential of the negatively charged clay surface, which decreases with distance from the clay surface. The Debye length, $\lambda$, corresponds approximately to the centroid of the diffuse layer, and is given by the following expression (e.g., Mitchell 1993):

$$\lambda = \sqrt{\frac{\varepsilon \varepsilon_0 RT}{2\nu^2 F_\eta^2}} \tag{10}$$

where $\varepsilon$ is the dielectric constant of the pore water (also referred to as the relative permittivity), $\varepsilon_0$ is the permittivity in a vacuum, $R$ is the universal gas constant, $F$ is Faraday's constant, $T$ is

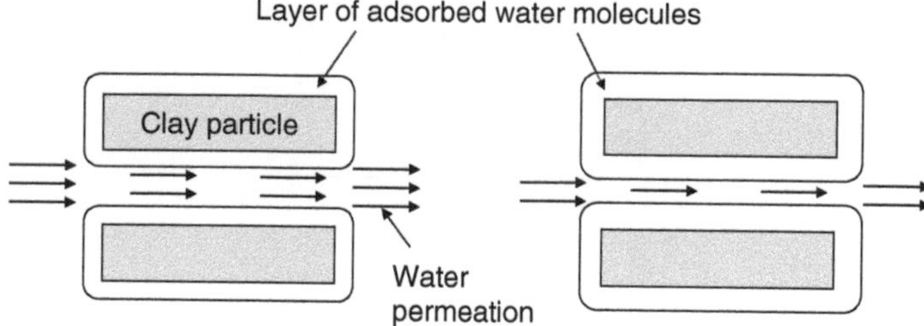


Figure 11. Effect of the thickness of the layer of adsorbed water molecules. (A thicker layer of adsorbed water molecules provides less effective pore space for water permeation and a lower hydraulic conductivity (right).)

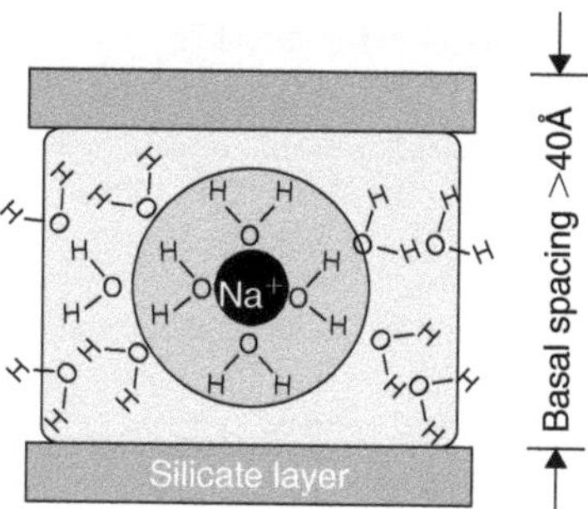


Figure 12. Osmotic swelling of the smectite clay mineral.

the absolute temperature, $\nu$ is the valence of the cation, and $\eta$ is the electrolyte concentration. The Debye length, $\lambda$, is usually thought to be the thickness of the diffuse double layer, although a distinct boundary does not exist between the diffuse layer and the bulk pore fluid. Equation (10) indicates that the electrolyte concentration, cation valence, and dielectric constant have an effect on the development of the DDL, which means they have an effect on the hydraulic conductivity and the swelling of the bentonites. This is because water molecules in the DDL are strongly bound to the clay surface; they do not contribute to providing space for water permeation. The effect of the adsorbed water molecules on the hydraulic conductivity is drawn in Figure 11.

In contrast to the DDL which develops on the external clay surfaces, osmotic swelling is attributed to the osmotic phase in which water molecules are attracted into the internal clay surfaces. The osmotic phase occurs only when the exchange sites of the internal surfaces contain monovalent cations, such as $Na^+$ (Norrish and Quirk 1954; McBride 1994). The interlayer region may retain numerous layers of water molecules, as shown in Figure 12. The number of layers of water molecules at equilibrium is proportional to the cation concentration in the bulk water (Onikata et al. 1999). Accordingly, when the bulk water contains a low concentration of monovalent cations, and the monovalent cations occupy the exchange sites, a larger fraction of the total water is bound so that less mobile water is available for flow, the swell volume is large, and the hydraulic conductivity is low. This condition is commonly observed when sodium bentonites are hydrated and/or permeated with de-ionized water.

In addition to the DDL development and osmotic swelling, the confining pressure has an effect on the hydraulic conductivity. A high confining pressure will provide a low void ratio, and will result in a low level of hydraulic conductivity. However, the effect of the confining pressure is minor compared to the DDL and osmotic swelling.

## 4.2 *Hydraulic conductivity of different types of GCLs*

Since there are several different types of GCLs, it is important to know the differences in hydraulic properties. Factors dominantly affecting the hydraulic conductivity are considered to be (1) the type of bentonite, (2) the bentonite mass per unit area, and (3) the state of the bentonite, namely, whether

Table 2. Reported hydraulic conductivity values for GCLs (data from Ruhl and Daniel 1997).

| Type of GCL | Hydraulic conductivity |
|---|---|
| Bentofix | $7 \times 10^{-10}$ cm/s |
| Bentomat | $9 \times 10^{-10}$ cm/s |
| Claymax 200R | $1 \times 10^{-9}$ cm/s |
| Contaminant-resistant Gundseal (only bentonite portion) | $1 \times 10^{-9}$ cm/s |
| Regular Gundseal (only bentonite portion) | $1 \times 10^{-9}$ cm/s |

Table 3. Hydraulic conductivity for thin clay layers simulating GCLs (data from Gleason et al. 1997).

| | Hydraulic conductivity (cm/s) | |
|---|---|---|
| Material tested | Permeation with water | Direct permeation with 0.25 mol/L $CaCl_2$ |
| Calcium bentonite (granular) | $6 \times 10^{-9}$ | $3 \times 10^{-8}$ |
| Sodium bentonite (granular) | $6 \times 10^{-10}$ | $6 \times 10^{-7}$ |
| Sodium bentonite (powdered) | $6 \times 10^{-10}$ | $9 \times 10^{-9}$ |

the bentonite is in a powdered or a granular form. The structure for containing the bentonites, such as needle-punching, is also important. Several publications reported hydraulic conductivity data for different GCLs. For example, Ruhl and Daniel (1997) conducted a comprehensive experimental program on the hydraulic conductivity testing of five different GCLs permeated with chemical solutions and waste leachates under a cell pressure of 35 kPa using a flexible-wall permeameter. For comparison, they obtained the hydraulic conductivity values for tap water, as listed in Table 2. Bentofix, Bentomat, and Claymax are geotextile-encased GCLs, and two types of Gundseal, both contaminant-resistant and regular, are geomembrane-glued GCLs. Bentofix and Bentomat are the needle-punched GCLs. Bentofix contained regular bentonite and Bentomat contained contaminant-resistant bentonite. Claymax contained a sodium-rich, water-soluble adhesive. As shown in Table 2, the hydraulic conductivity values were within the range of $7 \times 10^{-10}$ cm/s – $1 \times 10^{-9}$ cm/s. There was little difference among the hydraulic conductivity values for the five GCLs used in the research, which is probably because they contained sodium bentonites and chemical-resistant bentonites.

The type of bentonite used results in a substantial difference in the hydraulic performance. Table 3 lists the results of the hydraulic conductivity values for sodium and calcium bentonites reported by Gleason et al. (1997). Gleason et al. (1997) also conducted hydraulic conductivity tests on powdered sodium bentonites. Granular bentonites are rather common for GCLs. For water permeation, the hydraulic conductivity of calcium bentonite was larger than that of sodium bentonite, because osmotic swelling does not occur with calcium bentonite. Granular versus powdered did not result in a difference in the hydraulic conductivity. The permeation of a $CaCl_2$ solution provided different results in hydraulic conductivity from water permeation. Granular sodium bentonites permeated with a $CaCl_2$ solution exhibited a significantly (three orders of magnitude) larger hydraulic conductivity than that with pure water, which is attributed to the restriction of osmotic swelling. In contrast, powdered sodium bentonites exhibited only a one order increase in hydraulic conductivity from the water permeation to the permeation of the $CaCl_2$ solution. The difference between powdered and granular bentonitesw permeated with chemical solutions has been reported by Katsumi et al. (2002) and will be discussed in Section 6.3.

As shown in Table 3, an increase in hydraulic conductivity of only one order of calcium bentonite was observed from the water permeation to the permeation of the $CaCl_2$ solution. This is because calcium bentonites are less sensitive to chemical solutions. Similar results were obtained by Egloffstein (2002), who reported the hydraulic conductivity values of GCLs containing sodium bentonite and calcium bentonite. His results indicated that sodium bentonite GCLs provided a permittivity approximately equal to or lower than $1 \times 10^{-9}$ (1/s) permeated with pure water. However, the permeation of $CaCl_2$ solutions after the water permeation resulted in an increase in permittivity of one order (approximately $1 \times 10^{-8}$ (1/s)). In contrast, permittivity values for calcium bentonite GCLs were stable and fell within the range of $1.5 \times 10^{-8}$ to $4 \times 10^{-8}$ (1/s) even when the $CaCl_2$

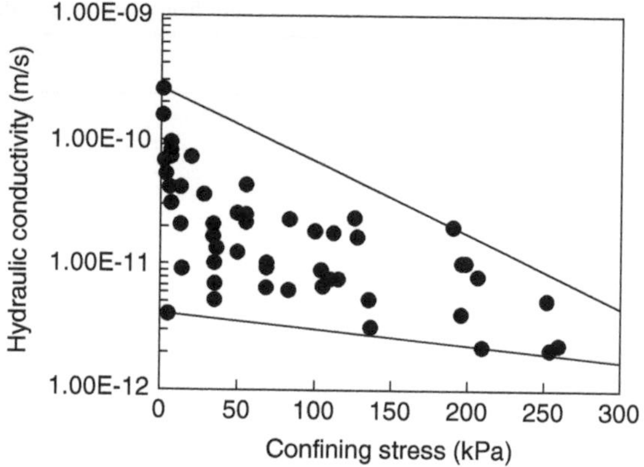


Figure 13. Hydraulic conductivity versus confining stress (Bouazza 2002).

solutions were permeated. However, limited data have been reported to clarify the advantages of calcium bentonites, and further research is needed.

### 4.3 *Testing conditions*

Testing conditions, such as the confining stress and the hydraulic gradient, are important issues which are sometimes dominant in controlling the hydraulic conductivity. If hydraulic conductivity tests are performed using a flexible-wall permeameter, the principal effective stress (the average stress across the specimen), except for an extremely low confining pressure, can be controlled by the cell pressure and the hydraulic gradient. Some types of rigid-wall permeameters, such as fixed-ring permeameters, are able to control the confining stress as well.

For general soils, the hydraulic conductivity of compressible soils or soils containing fractures and macro-pores are very sensitive to changes in effective stress. Higher levels of effective stress reduce the hydraulic conductivity. Thus, the selection of an appropriate stress level is important. The most appropriate way is to conduct the tests under stress conditions that represent the situation in the field.

A decrease in the hydraulic conductivity with an increase in the confining pressure is also observed for GCLs. Bouazza (2002) collected data from various sources and plotted them as shown in Figure 13. As shown in this figure, the hydraulic conductivity generally decreases as the confining stress increases. This decrease in hydraulic conductivity is limited to 1 to 1.5 orders of magnitude, and factors other than the confining stress play more dominant roles in controlling the hydraulic conductivity. These factors may include the type of bentonite used, the type of permeants applied (i.e., chemical composition), and the sequence of permeants followed.

A high hydraulic gradient is often used to reduce the test period. The negative effects of a high hydraulic gradient are thought to be that (1) a large seepage force results in the consolidation of the specimen and a decrease in the hydraulic conductivity and (2) the piping of fine particles due to the large seepage force increases the hydraulic conductivity. ASTM D 5084 recommends a maximum hydraulic gradient of 30 for media with low hydraulic conductivity ($k \leq 10^{-7}$ cm/s).

There have been various results reported on the effects of the hydraulic gradient on the hydraulic conductivity of clay liners and GCLs. For general clay liners, Fox (1996) derived equations to obtain the pore water pressure and the effective stress distributions across the specimen during hydraulic conductivity testing, which indicated that excessive levels of hydraulic gradients can cause reductions in the measured hydraulic conductivity, in particular, for normally consolidated soils with high compressibility, such as soft clays and soil-bentonite slurries. Imamura et al. (1996) conducted a series of long-term rigid-wall permeameter tests on compacted sand/bentonite specimens which are expected to be used as barrier materials for radioactive waste storage facilities. The test results indicated that a two order increase in the hydraulic conductivity ($1 \times 10^{-9}$ cm/s to $1 \times 10^{-7}$ cm/s) occurred after more than 25 pore volumes of flow of a $Ca(OH)_2$ solution (600 ppm as $Ca^{++}$ concentration) under a hydraulic gradient of 800. Furthermore, a decrease in the hydraulic gradient

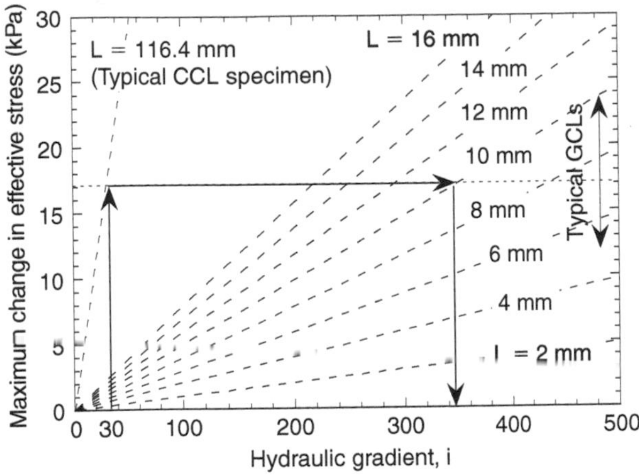


Figure 14. Changes in the effective stress due to the applied hydraulic gradient for different levels of specimen thickness (Shackelford et al. 2000).

from 800 to 30 resulted again in the recovering decrease in hydraulic conductivity ($1 \times 10^{-7}$ cm/s to $1 \times 10^{-9}$ cm/s). Thus, Imamura et al. (1996) concluded that this increase in hydraulic conductivity under a high hydraulic gradient may be attributed to the piping of fine particles. Similar experiments have been conducted on GCLs by Rowe and Orsini (2002). They evaluated the effect of the internal erosion using a fixed-ring hydraulic conductivity apparatus by applying an extremely high hydraulic gradient. An adverse increase in hydraulic conductivity due to an extremely high hydraulic gradient occurred for conventional woven or nonwoven carrier geotextiles, while a scrim-reinforced nonwoven geotextile carrier had the capability of keeping the hydraulic gradient up to 7000, which is equivalent to a water head of approximately 70 m.

Although ASTM D 5084 recommends a hydraulic gradient lower than 30, hydraulic gradients used for measuring the hydraulic conductivity of GCLs typically ranged from 50 to 550. Rad et al. (1994) reported that the hydraulic conductivity of a GCL was not affected by a hydraulic gradient of 2800. The hydraulic gradient has a relatively minor effect on the hydraulic conductivity if the mean effective stress is constant and internal erosion does not occur (Petrov et al. 1997a; Shackelford et al. 2000; Rowe and Orsini 2002). The recommendation that the hydraulic gradient be lower than 30 is probably effective for ordinary compacted clay specimens with a specimen length or thickness of 116 mm (the height of the Proctor mold), and Shackelford et al. (2000) indicated that higher hydraulic gradients can be applied for thinner specimens. Figure 14 indicates increases in the effective stress at the effluent end of the specimen due to the hydraulic gradient for various levels of specimen thickness. For a compacted clay specimen with a length of 116 mm, a hydraulic gradient of 30 causes an increase in effective stress of 17 kPa at the effluent end. For the same increase in effective stress, a hydraulic gradient of 342 can be applied to a GCL with a thickness of 10 mm. However, this high hydraulic gradient can only be applied to GCLs, and it does not necessarily assure the use of a high hydraulic gradient for extremely thinly cut specimens of compacted clay which are usually used as thick layers in practice. Since higher hydraulic gradients across GCLs installed in the field might occur, because they are much thinner than CCLs, the use of such high hydraulic gradients for laboratory tests is considered reasonable to some extent for thin liner materials.

## 5 OVERLAPPING GCLS AND OTHER SPECIAL SITUATIONS

### 5.1 *Overlapping GCLs*

In addition to the hydraulic performance of intact GCLs, the hydraulic performance of unavoidable overlapping is also important. Water flowing through intact GCLs versus overlapping GCLs is

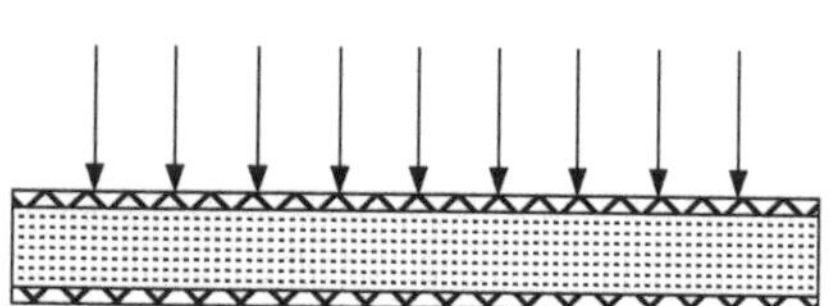
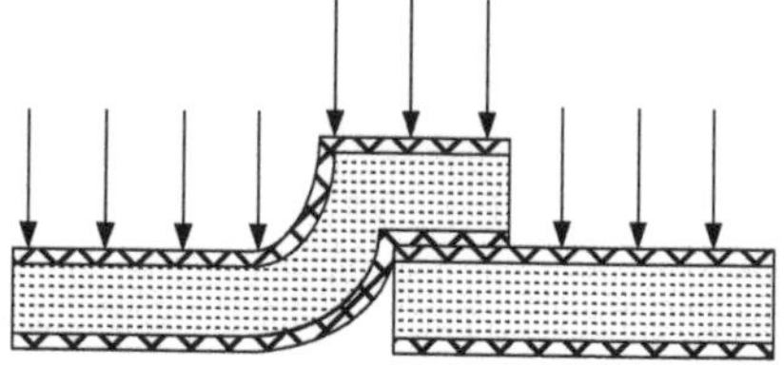

Figure 15. Water flow through intact GCLs (left) and overlapping GCLs (right).

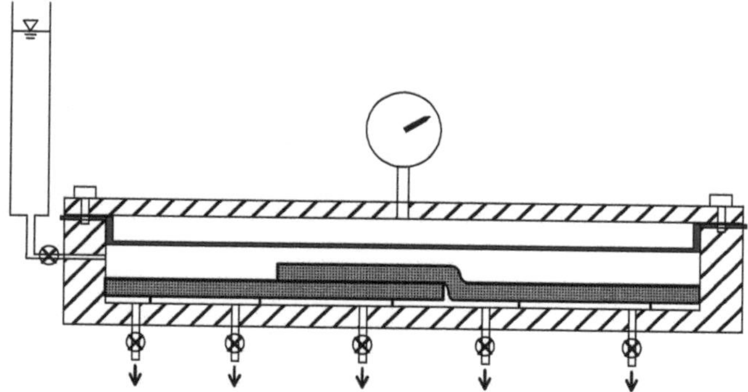

Figure 16. Flow box experimental apparatus (modified from Daniel et al. 1997b).

drawn in Figure 15. Leakage through overlapping GCLs was experimentally evaluated by Estornell and Daniel (1992), Daniel et al. (1997b), and Lagatta et al. (1997). For geotextile-encased GCLs, bentonites are placed on the overlapping area of the geotextile so that the water migration is restricted due to the swelling of the placed bentonites. Daniel et al. (1997b) also presented experimental apparatus to measure the hydraulic performance of overlapping GCLs. Prior to Daniel et al. (1997b), the tank was used to determine the leakage through overlapping GCLs and GCLs overlain by a geomembrane with defects (Estornell and Daniel 1992). Daniel et al. (1997b) proposed a GCL flow box (Figure 16) to determine the leakage. A GCL flow box is simpler than the other testing apparatus for testing overlapping GCLs; it can be applied to a wide range of normal pressure levels and it is capable of providing some environmental conditions such as freezing and thawing.

The differential settlement may have adverse effects on the hydraulic performance of overlapping GCLs. Lagatta et al. (1997) and Viswanadham et al. (1999) discussed this effect. Large overlapping areas may prevent significant increases in hydraulic conductivity.

For geomembrane-glued GCLs, the bentonite portion in an overlapping area plays an important role in the hydraulic performance of the entire area of GCLs. This is because intact GCL panels are practically impervious due to the geomembrane and only the overlapping area provides space for the water flow. Some equations for calculating the rate of leakage or hydrated area through overlapping geomembrane-glued GCLs have been proposed (GSE 2002; Giroud et al. 2002).

### 5.2 *Durability against environmental effects*

The effect of freeze-thaw cycles on the hydraulic performance of clay liners is an anticipated concern when the cycles are applied to cover systems in cold regions. Experimental and field studies have been reported for compacted clay liners (CCLs) (Benson and Othman 1993; Othman and Benson 1993; Benson et el. 1995; Kraus et al. 1997), and it has been concluded that (1) 3 to 5 freeze-thaw cycles resulted in an increase in magnitude of the hydraulic conductivity by almost two orders and (2) freeze-thaw cycles did not have any effect on the hydraulic conductivity of compacted mixtures of bentonite and soil. Any increases in the hydraulic conductivity of CCLs are considered to be due to the formation of ice lenses. It was found that the lower the initial hydraulic conductivity, the faster the frozen front, and the lower the confining stress, the more significant the effects of

Table 4. Hydraulic conductivity of GCLs frozen and thawed in the laboratory (data from Kraus et al. 1997).

| | Hydraulic conductivity (m/s) | |
|---|---|---|
| Sample | Initial | After 20 freeze-thaw cycles |
| Bentofix-1 | $3.0 \times 10^{-11}$ | $3.2 \times 10^{-11}$ |
| Bentofix-2 | $1.6 \times 10^{-11}$ | $2.2 \times 10^{-11}$ |
| Bentofix-3 | $1.7 \times 10^{-11}$ | $2.5 \times 10^{-11}$ |
| Bentomat-1 | $2.9 \times 10^{-11}$ | $1.7 \times 10^{-11}$ |
| Bentomat-2 | $1.7 \times 10^{-11}$ | $1.9 \times 10^{-11}$ |
| Bentomat-3 | $1.8 \times 10^{-11}$ | $1.9 \times 10^{-11}$ |
| Claymax-1 | $2.9 \times 10^{-11}$ | $3.4 \times 10^{-11}$ |
| Claymax-2 | $2.4 \times 10^{-11}$ | $2.1 \times 10^{-11}$ |
| Claymax-3 | $3.5 \times 10^{-11}$ | $2.4 \times 10^{-11}$ |
| Claymax-4 | $4.1 \times 10^{-11}$ | $3.3 \times 10^{-11}$ |

Table 5. Hydraulic conductivity of GCLs subjected to drying-wetting cycles (data from Lin and Benson 2000).

| | Hydraulic conductivity (cm/s) | |
|---|---|---|
| Permeant | Initial | After 5 drying-wetting cycles |
| De-ionized water | $1.2 \times 10^{-9}$ | $2.9 \times 10^{-9}$ |
| 0.0125 mol/L $CaCl_2$ solution | $3.8 \times 10^{-9}$ | $2.8 \times 10^{-7}$ |
| DI water (for 1st wetting) and $CaCl_2$ solution (for subsequent wetting cycles) | $1.6 \times 10^{-9}$ | $1.1 \times 10^{-9}$ ($5.8 \times 10^{-6}$ cm/s after 8 cycles) |
| Tap water (for 1st wetting) and $CaCl_2$ solution (for subsequent wetting cycles) | $1.7 \times 10^{-9}$ | $2.1 \times 10^{-7}$ |

the freeze-thaw cycles. If bentonites are mixed in the clay liner, the ice lenses may close up due to their swelling during the thaw cycle even after the ice lenses have formed in the freezing process. It was concluded that bentonites are resistive against freeze-thaw cycles.

The effects of freezing and thawing on the hydraulic performance of GCLs were reported by Kraus et al. (1997) and Hewitt and Daniel (1997). Kraus et al. (1997) used three types of GCLs for freeze-thaw tests in the laboratory and in the field. As shown in Table 4, which tabulates a summary of the experimental results, the GCLs are so durable against the effects of the freeze-thaw cycles that the hydraulic conductivity values did not change even after 20 freeze-thaw cycles. Similar to compacted clay liners mixed with bentonites, hydrated bentonites are soft and deformable enough that the ice lenses formed in the freezing process close up during the thawing process. Similar results were reported by Hewitt and Daniel (1997). They concluded that GCLs are expected to have an excellent performance against the effects of freezing and thawing.

Unlike the excellent performance of GCLs against the effects of freezing and thawing, deterioration in the hydraulic performance of GCLs subjected to drying and wetting cycles has been addressed (e.g., James et al. 1997; Lin and Benson 2000). Lin and Benson (2000) conducted hydraulic conductivity tests on GCLs subjected to drying and wetting. They conducted drying and wetting cycles very carefully; GCLs were permeated for at least 1 month as a wetting cycle, and they were air-dried under 50% humidity until the weight of the GCLs became constant. They used permeants that were de-ionized water, tap water, and $CaCl_2$ solutions, as listed in Table 5. Tap water had an ionic strength of 0.087 mol/L. Note that the ionic concentrations of not only the tap water, but also the $CaCl_2$ solutions used in their experiments, were low. Precipitated waters are likely to contain these levels of concentrations. Significant increases in the hydraulic conductivity were obtained after 4 drying-wetting cycles for GCLs permeated with $CaCl_2$ solutions and GCLs permeated with tap water first and then with $CaCl_2$solutions, although this concentration (0.0125 mol/L) did not substantially increase the hydraulic conductivity. Even if the GCLs were permeated with de-ionized water for the first wetting cycle, the hydraulic conductivity was increased after 8 drying-wetting cycles. The drying process may have the effect of providing desiccation cracks and of increasing the cation concentrations in the pore water, such that subsequent wetting processes may not be able to provide enough swelling to close up the cracks. The field study reported by James et al. (1997) also revealed that desiccation cracks did not close up during the rewetting cycles when the GCLs were exposed to drying and wetting cycles.

Some researchers have attempted to evaluate the field performance of GCLs including their durability against environmental effects (e.g., Blumel et al. 2002; Henken-Mellies et al. 2002). From their lysimeter experiments conducted in Northern Germany by Henken-Mellies et al. (2202) on cover systems consisting of GCLs containing calcium bentonite, the GCLs had an excellent hydraulic barrier performance and only 0.5% precipitation permeated through GCLs. A significant amount of evaporation (68.5%) in the water balance at this site indicated that the configuration of the cover system and the effect of vegetation, as well as the hydraulic performance of the GCL itself, are key factors to limiting the water infiltration into the waste layer.

## 6 CHEMICAL COMPATIBILITY

### 6.1 *Clay-chemical and bentonite-chemical interactions*

Extremely low levels of the hydraulic conductivity of GCLs are attributed to the swelling of the bentonites contained in the GCLs, as was discussed in Chapter 4. Since swelling is sensitive to chemicals, chemical compatibility is a concern when GCLs are applied to waste containment barriers, in particular, to bottom liners.

Knowledge of clay-chemical interactions is an effective way to evaluate the chemical effects on GCLs. The adverse effect of chemicals on the properties of clays that are used for landfill barriers is an important subject. Many researchers reported that permeation with chemical solutions will result in an increase (sometimes a significant increase) in hydraulic conductivity, even though the hydraulic conductivity is low when the clays are permeated with pure water. These effects can be explained by the changes in soil fabric and are categorized into (1) the dissolution of the clay particles and the chemical compounds, (2) the restriction of the development of a diffuse double layer, and (3) the restriction of osmotic swelling for smectite clay. Mitchell and Madsen (1987), Shackelford (1994), and Shackelford et al. (2000) summarized the clay-chemical interactions and the chemical compatibility of clay.

The dissolutions of soil components result from strong acids and bases. Strong acids promote the dissolving of carbonates, iron oxides, and alumina octahedral layers of clay minerals. Bases promote the dissolving of the silica tetrahedral layers. These effects can cause an increase in hydraulic conductivity, although re-precipitation of the dissolved compounds might clog the pores and decrease the hydraulic conductivity (Mitchell and Madsen 1987).

Mesri and Olson (1971) showed that the hydraulic conductivity is largest for nonpolar fluids, smaller for polar fluids with low dielectric constants (ethyl alcohol and methyl alcohol), and lowest for water, which is polar and has a high dielectric constant (~80). Similar results, which explain the effects of dielectric constants, were reported by other researchers (e.g., Bowders and Daniel 1987; Mitchell and Madsen 1987; Fernandez and Quigley 1988; Acar and Olivieri 1989; Shackelford 1994). However, they showed that no significant increase in hydraulic conductivity occurred when the concentration of organic chemicals was lower than 50%. This is because dilutions with water lead to increases in the dielectric constant.

In practice, direct permeation with organic chemicals is not likely to happen where clay liners are applied. Fernandez and Quigley (1985) conducted two different hydraulic conductivity tests on clayey soils; one was permeated first with water and then benzene and the other was permeated first with water, then with ethanol, and finally with benzene. The former test results showed that the benzene concentration of the effluent reached 50% after only a 0.28 pore volume of benzene flow, and reached 100% at two pore volumes of flow. Only 8% of the water that initially existed in the soil was replaced by polar benzene. Organic liquids flowed dominantly through the large pores, and most of the water in the micro-pores remained in the soil. The increase in hydraulic conductivity was negligible. The other case (permeated with water, ethanol, and benzene) exhibited different results. A large portion of the water was replaced, and there was a significant increase in the hydraulic conductivity. This is attributed to the solubility of ethanol to water. Benzene can easily replace the dilute ethanol in soil pores, although it cannot replace the water.

Osmotic swelling is an important phenomenon for smectite clay. When dry smectite is hydrated, water molecules are strongly attracted to the internal and the external clay surfaces during the hydration phase. After this hydration process, if the exchange cations of the smectite clay are monovalent, the region of the interlayer may retain numerous layers of water molecules during the osmotic phase (Norrish and Quirk 1954). This condition can be observed as a significant amount of swelling, and it is typically observed when Na-bentonites are hydrated with deionized water. When the smectite having monovalent cations at the exchange site is permeated with the solution containing a low concentration of monovalent cations, a large fraction of the water is attracted to the clays, less mobile water is available for the water flow, and the hydraulic conductivity is low. When polyvalent cations exist at exchange sites, the osmotic phase does not occur, and less swelling is observed. Divalent and trivalent cations in permeants result in a significant increase in the hydraulic conductivity of GCLs, much greater than monovalent cations (Shackelford et al. 2000).

Chemically-resistant clays have recently been developed to improve the compatibility of clays, in particular of bentonite (Lo et al. 1994 and 1997; Onikata et al. 1996 and 1999). For example, Onikata et al. (1999) developed a chemically resistant bentonite, called multi-swellable bentonite (MSB), which is made from a natural bentonite mixed with propylene carbonate as the swelling activation agent. Propylene carbonate can enter the interlayer region of the smectite, attract numerous water molecules, and consequently, result in a strong swelling power even if the permeant contains polyvalent cations or a high concentration of monovalent cations. A series of hydraulic conductivity tests showed that MSB yields hydraulic conductivity that is one or two orders of magnitude lower than natural bentonite when the permeant contains calcium cations, as shown in Section 6.6 (Katsumi et al. 2001 and 2008a).

## 6.2 *Consideration of the testing conditions*

Special consideration must be given to the testing conditions in order to conduct the hydraulic conductivity tests with chemical solutions, that is, the "chemical compatibility tests." As discussed in Section 3.2, the selection of the confining stress is important. For the chemical compatibility tests, there are two philosophies for selecting the confining stress. One is that the confining stress is selected in the same way as the field conditions. The other is that a low confining stress is selected so that the effects of the chemicals can be observed more easily. Many experimental data reported in publications were obtained under a low confining stress even though GCLs are considered to be applied to landfill bottom barriers.

The important key variables are (1) the sequence of permeant liquids and (2) the chemical equilibrium. In addition, the confining stress may have an effect, but it is probably minor compared to the other effects. A long duration is usually needed to conduct hydraulic conductivity tests on low-permeable materials, such as GCLs. If a GCL specimen has a thickness of 10 mm, a hydraulic conductivity of $1 \times 10^{-9}$ cm/s, an effective porosity of 0.5, and it is tested under a hydraulic gradient of 200, a test duration of longer than 100 days is needed to obtain only one pore volume of flow through the specimen. In addition, among the soil properties, none varies over such a wide range as the hydraulic conductivity. This makes it difficult to decide when the tests should be terminated. Assuming the same hydraulic gradients and the same specimen lengths for two different specimens with different hydraulic conductivities, the test duration required to obtain the same volume of flow is simply proportional to the hydraulic conductivity. Thus, the termination criterion is a critical issue in conducting the hydraulic conductivity tests. Typical termination criteria are (1) the equality of the inflow and the outflow rates (typically, $\leq 25\%$) and (2) the measurement of a steady hydraulic conductivity (four or more consecutive measurements within 25% to 50% of the mean) (Daniel 1994). For chemical compatibility tests, where chemical solutions and waste leachates are used as permeants, the permeation of a minimum of two pore volumes of flow has to be achieved and the similarity between the chemical compositions of the effluent and the influent has to be ensured so that the clay-chemical interaction is equilibrated (Daniel 1994; Shackelford 1994; Shackelford et al. 1999). Establishing a similarity in the chemical compositions of the effluent and the influent sometimes requires an extremely long testing period (Bowders 1988; Imamura et al. 1996; Shackelford et al. 2000). For example, Imamura et al. (1996) reported that the hydraulic

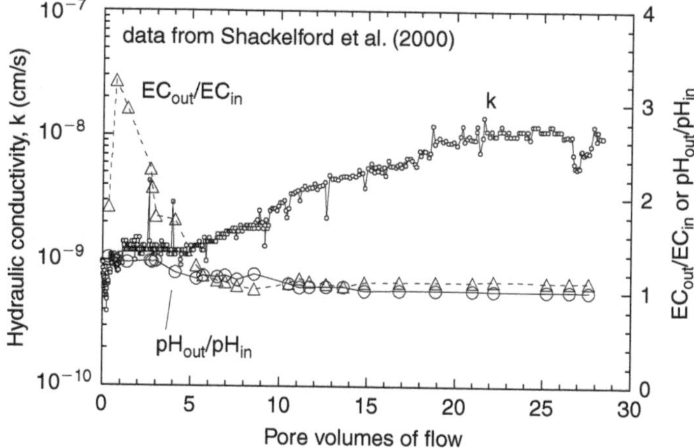


Figure 17. Long-term hydraulic conductivity test results (Shackelford et al. 2000).

conductivity of a compacted sand-bentonite specimen permeated with a $Ca(OH)_2$ solution increases by more than two orders of magnitude after more than 25 pore volumes of flow corresponding to three years of permeation.

For practical indicators of a chemical equilibrium, experimental and theoretical studies by Shackelford et al. (1999) suggest that electric conductivity (EC) can be used as an index of the chemical composition of electrolyte solutions. Measuring the electric conductivity is simple, inexpensive, and rapid. The results of a long-term hydraulic conductivity test on a needle punched GCL with a dilute (0.0125 M) $CaCl_2$ solution is shown in Figure 17 (Shackelford et al. 2000). This data illustrate how a misunderstanding can occur. The hydraulic conductivity was low and steady for the first 5 pore volumes of flow; this satisfies some of the termination criteria except for the chemical equilibrium. After approximately 30 pore volumes of flow, however, the hydraulic conductivity increased to become 25 times larger. The similarity in EC between the effluent and the influent was not yet achieved at 5 pore volumes of flow. The hydraulic conductivity started to increase after the similarity in EC was achieved, which means that measuring the EC of the outflow is effective for determining the appropriate termination criterion.

The effect of the sequence of permeants will be discussed in Section 6.5.

## 6.3 *Effects of inorganic chemicals – concentration and valence of cations*

Several research projects on the hydraulic conductivity of GCLs with chemical solutions/liquids were conducted (e.g., Ruhl and Daniel 1997; Petrov and Rowe 1997; Shackelford et al. 2000; Jo et al. 2001; Shan and Lai 2002). Most of them used electrolyte solutions, because waste leachates may contain several types of cations which might have dominant effects on the hydraulic conductivity of GCLs. Equation (10) indicates that the development of DDLs will be dependent on the ionic strength of the solutions, and the hydraulic conductivity will change along with the ionic strength. Figure 18 presents the hydraulic conductivity values for GCLs (bentonite granules sandwiched between nonwoven and woven geotextiles), permeated chloride solutions for monovalent cation Li, and Mg, Ca, Zn, and Cu as divalent cations, as given in Jo et al. (2001). Hydraulic conductivity tests were performed using a flexible-wall permeameter with a cell pressure of 20 kPa and a hydraulic gradient of approximately 100. Hydraulic conductivity values for the same GCLs permeated with pure water (de-ionized water) were $1.5 \times 10^{-9}$ cm/s and $9.1 \times 10^{-10}$ cm/s. A significant increase in the hydraulic conductivity occurred when the concentration of divalent cations increased from 0.01 mol/L to 0.1 mol/L. The permeation of a 0.005 mol/L solution of divalent cations provided similar hydraulic conductivity values to those of de-ionized water. When the concentration was 0.01 mol/L for the divalent cations, some specimens exhibited an increase in hydraulic conductivity compared to those of the de-ionized water. The permeation of a 0.1 mol/L solution of divalent chloride resulted in a hydraulic conductivity level of approximately $1 \times 10^{-5}$ cm/s. No significant

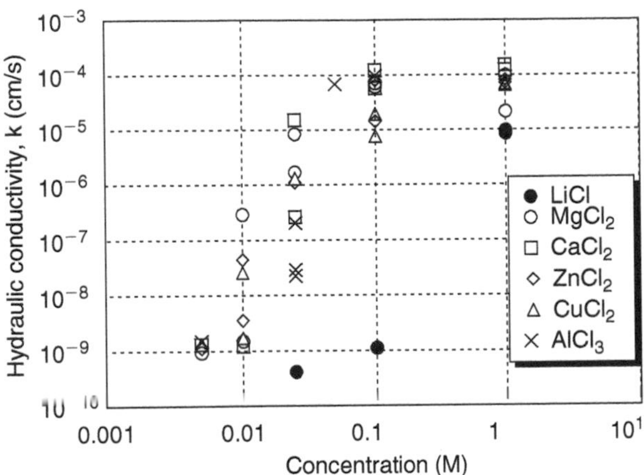


Figure 18. Hydraulic conductivity values of GCLs permeated with inorganic chemical solutions (Jo et al. 2001).

differences were found for solutions with hydraulic conductivity levels between 0.1 mol/L and 1 mol/L. Since monovalent cation Li has less of an effect on the increase in hydraulic conductivity, a hydraulic conductivity of $1 \times 10^{-9}$ cm/s was achieved when permeated with a 0.1 mol/L LiCl solution.

A strong correlation between the swelling power (swell volume according to ASTM D 5890) of bentonite and the hydraulic conductivity of GCLs was observed for different types of permeants, as shown in Figure 19 (Jo et al. 2001, Kateumi et al. 2008). If the chemical solution provides a swelling power that is greater than 20 mL/2 g-solid, the permeation of the solution may result in a level of hydraulic conductivity that is as low as that permeated with pure (de-ionized) water. Even if the values for hydraulic conductivity are not affected by such dilute chemical solutions, however, the bentonite should be affected when the swelling power decreases. For example, hydraulic conductivity tests permeated with a dilute chemical solution (0.0125 mol/L $CaCl_2$), conducted by Shackelford et al. (2000), indicate that the hydraulic conductivity increased after one year of permeation. This corresponds to approximately five pore volumes of flow. Shackelford et al. (2000) emphasized the importance of continuing to run hydraulic conductivity tests long enough to yield a reaction between the clay and the chemicals, and to achieve a chemical equilibrium between the influent and the effluent. They and Shackelford et al. (1999) also proposed a measurement for electrical conductivity to check whether or not a chemical equilibrium has been attained, as was discussed in Section 6.2.

The difference between granular versus powdered was evaluated by Katsumi et al. (2002) and Katsumi and Kamon (2002). Katsumi et al. (2002) conducted hydraulic conductivity tests on a GCL with powdered bentonite, permeated with NaCl and $CaCl_2$ solutions, and compared the results with those from tests conducted by Kolstad (2000) on a GCL with granular bentonite, permeated with LiCl-$CaCl_2$ solutions. According to the results of the swell volume shown in Figure 20 (left), the two types (powdered and granular) of bentonite have almost the same swelling performance against chemicals (the swell volume is measured using ground bentonite). Figure 20 (right) illustrates the ionic strength of the cations of the permeants versus the hydraulic conductivity values. The figure indicates that both GCLs are affected by the chemical solutions, but that the powdered bentonite may be more compatible than the granular bentonite, particularly with chemically strong solutions. This is probably because the pores between the granules are not blocked due to a lower level of swelling of the bentonite, especially for aggressive chemical solutions. This results in a significant increase in hydraulic conductivity. In contrast, the pores of the powdered bentonite are small even when the swelling is limited, and result in a lower hydraulic conductivity. In conclusion, powdered bentonite GCLs are considered to be more compatible against chemical solutions.

Most studies on the chemical compatibility presented above, including Jo et al. (2001) and Katsumi et al. (2002), applied relatively low confining stress levels (20–30 kPa) using flexible-wall

| Reference | Permeant solution | Confining pressure |
|---|---|---|
| Lin et al. (2000) | DI water, Acid, $CaCl_2$ | 20.0 kPa |
| Jo et al. (2001) | DI water, LiCl, NaCl, KCl, $CaCl_2$, $MgCl_2$, $ZnCl_2$, $CuCl_2$ $LaCl_3$ | 20.0 kPa |
| Shan and Lai (2002) | DI water, Tap water, Sea water, Acid, MSW leachate | 34.5 kPa |
| Kolstad et al. (2004a) | LiCl-$CaCl_2$, NaCl-$MgCl_2$, LiCl-NaCl-$CaCl_2$-$MgCl_2$ | 20.0 kPa |
| Katsumi et al. (2007) | DI water, NaCl, KCl, $CaCl_2$, NaCl-KCl, NaCl-$CaCl_2$, MSW leachate | 29.7 kPa |

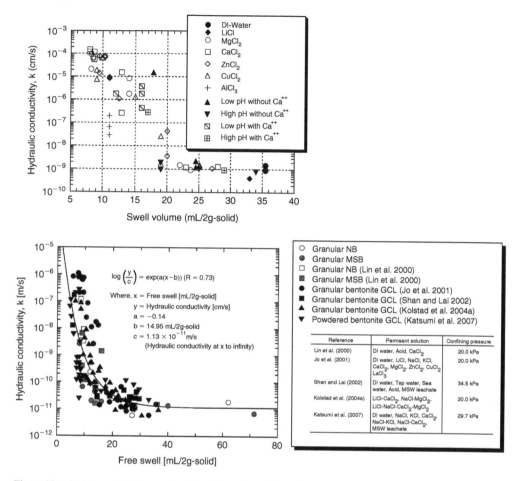

Figure 19. Relationship between the hydraulic conductivity values of GCLs and the swell volume of bentonites (upper: Jo et al. (2001), lower: Katsumi et al. (2008a)).

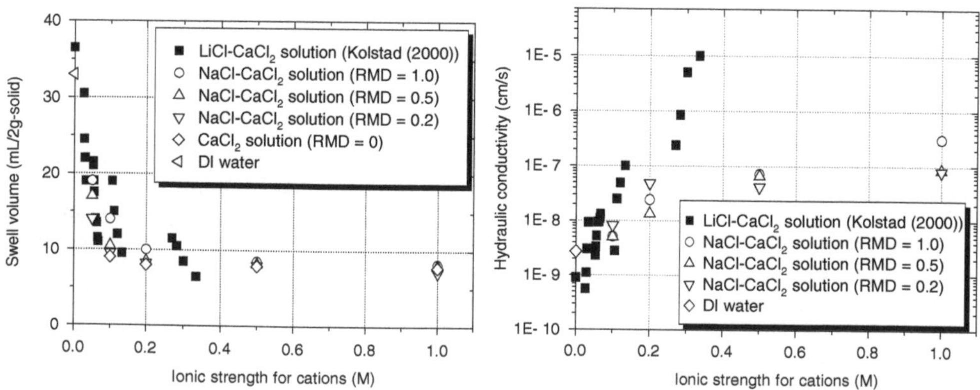


Figure 20. Comparison of granular versus powdered bentonites in GCLs for hydraulic conductivity with chemical solutions: swell volume of the ground bentonite (left) and the hydraulic conductivity (right) (data on the granular bentonite GCLs from Kolstad (2000)); RMD indicates ratio of monovalent to divalent cations, and it is defined as $c_1/(2c_2)^{0.5}$ where $c_1$ and $c_2$ are the concentrations of monovalent and divalent cations, respectively (Katsumi et al. 2002).

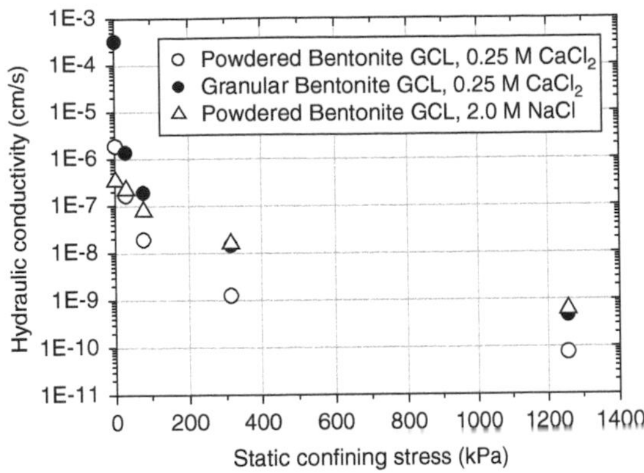


Figure 21. Hydraulic conductivity versus vertical pressure (Katsumi and Fukagawa 2005).

permeameters. However, GCLs used in landfill bottom barriers are subjected to larger overburden pressures due to the load of the waste layer. There are limited published reports on the effects of the confining stress. For example, Petrov and Rowe (1997) used only NaCl solutions under vertical pressures up to 120 kPa. Katsumi and Fukagawa (2005) conducted hydraulic conductivity tests using $CaCl_2$ and NaCl solutions as permeants under vertical pressures ranging from 30 to 1200 kPa using a consolidation cell. After the geotextile-encased needle-punched GCL was hydrated with a given permeant and vertical pressure was applied for 24 hours to achieve the primary consolidation, a hydraulic gradient of 80–90 was applied. Concentrations for the solutions are 0.25 mol/L for $CaCl_2$ and 2.0 mol/L for NaCl. The reason these concentrations were selected was because these concentrations resulted in the substantial increase in hydraulic conductivity of the powdered bentonite GCLs compared to pure water; they provided almost similar hydraulic conductivity values when a low confining stress (30 kPa) was applied in the flexible-wall permeameter. GCLs containing granular bentonites were also used. As indicated in Figure 21, hydraulic conductivity values decreased significantly as the vertical pressure was increased. Compare this result with Figure 13 which illustrates the effect of vertical pressure on the hydraulic conductivity permeated with pure water. The changes in hydraulic conductivity due to the vertical pressure ranged within two orders of magnitude when permeated with water. In contrast, chemical permeation yielded a more significant effect of vertical pressures on hydraulic conductivity values. The powdered bentonite GCL permeated with an NaCl solution and the granular bentonite GCL permeated with a $CaCl_2$ solution exhibited the hydraulic conductivity of approximately $1 \times 10^{-8}$ cm/s under the vertical pressure of 300 kPa. Even the permeation of the $CaCl_2$ solution provided a hydraulic conductivity of $1 \times 10^{-9}$ cm/s when a vertical pressure of 300 kPa was applied. It can be concluded that GCLs are effective in landfill bottom barriers because a relatively large overburden pressure, to which GCLs placed in the bottom of the landfill are also subjected, will result in low hydraulic conductivity even if the GCLs are subjected to waste leachates containing chemicals which might affect the hydraulic conductivity.

Such effects of the overburden pressure on the hydraulic conductivity are considered dependent on the types of permeants and the properties of the bentonites employed. Under a cell pressure of 30 kPa, the powdered bentonite GCL provided almost similar hydraulic conductivity values ($2 \times 10^{-7}$ cm/s) for 2.0 mol/L NaCl and 0.25 mol/L $CaCl_2$ solutions. The effect of vertical stress on the decrease in hydraulic conductivity, however, was more significant for the $CaCl_2$ solution than the NaCl solution. Figure 22 indicates the hydraulic conductivity values versus the bulk void ratio for the same data. The data reported by Petrov and Rowe (1997) are also plotted in Figure 22. Originally, Mesri and Olson (1971) showed the linear relationship between the void ratio ($e$) and the logarithm of the hydraulic conductivity ($k$) for kaolinite, illite, and montmorillonite. Petrov and Rowe (1997) indicated that this relationship can also be observed for GCLs permeated with NaCl solutions with different concentrations, but that the $e$-log $k$ lines are identical for each concentration of NaCl. In conclusion, hydraulic conductivity values are dependent on the void ratio and the concentration of a given chemical (Petrov and Rowe 1997; Shackelford et al. 2000). The void ratio indicates

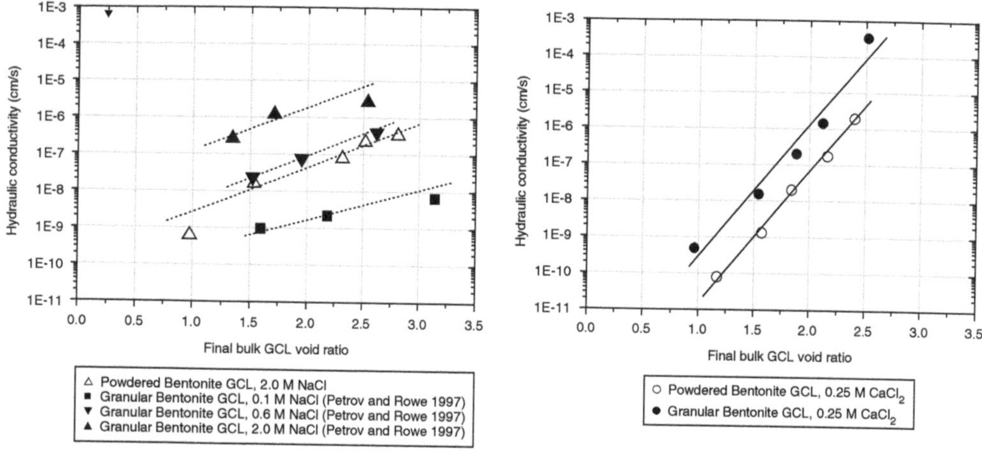


Figure 22. Hydraulic conductivity values versus void ratio (Katsumi and Fukagawa. 2005).

the total void space, which is affected by the effective stress applied, while the concentration of chemicals correlates to the thickness of the DDLs. From Figure 22, the inclination of the *e*-log *k* lines is independent on whether the bentonites are granular or powdered and on the concentrations of NaCl. The inclinations of the $CaCl_2$ solutions are different from those of the NaCl solutions, and they are much larger. The powdered bentonite GCL provided a level of hydraulic conductivity one-order of magnitude lower than the granular bentonite GCL for the same void ratios. Therefore, the type of GCL, the chemical composition, and the expected overburden pressure are key variables to evaluating the hydraulic barrier performance of GCLs.

## 6.4 *Effects of waste leachates*

Real or simulated waste leachates were used for hydraulic conductivity testing in several research works (e.g., Ruhl and Daniel 1997; Gaidi and Alimi-Ichola 2002; Katsumi and Fukagawa 2005, Katsumi et al. 2007). Katsumi et al. (2007) evaluated the possibility of estimating the hydraulic conductivity permeated with waste leachates from electrical conductivity. They used four different waste leachates for hydraulic conductivity on needle-punched GCL encasing powdered bentonites. As previously shown in Figure 20, the RMD (ratio of monovalent cations to divalent cations) does not have a significant effect on the hydraulic conductivity. Although the concentration of all chemicals dissolved in waste leachates is practically impossible, electrical conductivity was proposed as an index to evaluate the aggressiveness of the chemicals because there is a correlation between the electrical conductivity and the ionic strength. From Figure 23, which indicates the swelling and the hydraulic conductivity versus the electrical conductivity for given waste leachates, as well as from the multi-species chemical solutions of NaCl-$CaCl_2$, it is seen that if the electrical conductivity of a waste leachate is low, a large amount of swelling and a low level of hydraulic conductivity will be obtained for the given leachate. The swell versus electrical conductivity data for all four waste leachates fall within the data for the NaCl-$CaCl_2$ solutions. The hydraulic conductivity value for one waste leachate limiting the swelling was larger than that for pure water, and even one order of magnitude larger than that of the GCLs permeated with the NaCl-$CaCl_2$ solutions having the same electrical conductivity. The difference in the hydraulic conductivity values between the waste leachate and the NaCl-$CaCl_2$ solutions can probably be attributed to the effect of the organic chemicals and the biological effect, which cannot be evaluated by electrical conductivity. For practical purposes, however, the electrical conductivity can be sufficiently used to estimate the hydraulic performance of GCLs against waste leachates.

## 6.5 *Effect of the sequence of permeants*

The sequence of permeants has a significant effect on the hydraulic performance, as was briefly indicated in Section 6.1. Values for the hydraulic conductivity, when permeated with sequential

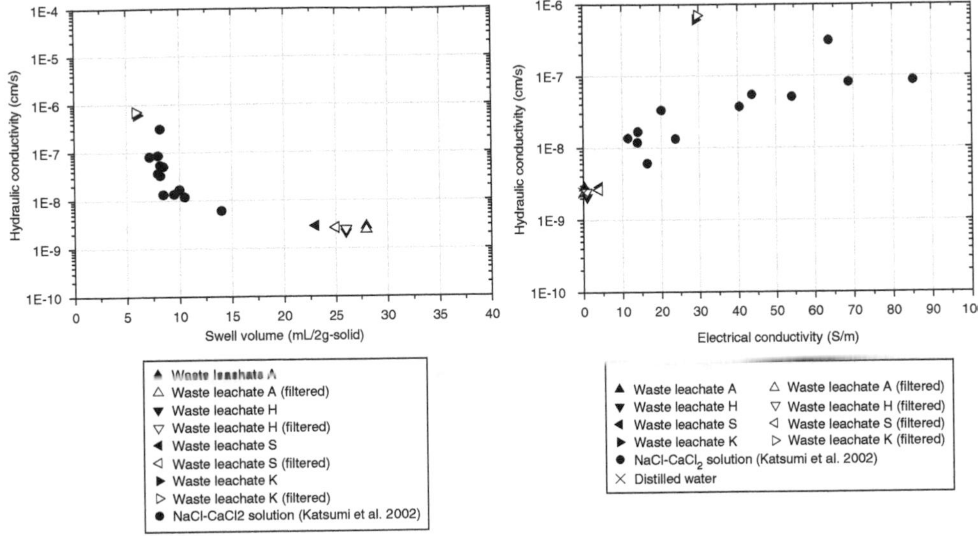


Figure 23. Swelling and hydraulic conductivity for inorganic chemical solutions and waste leachates (Katsumi and Fukagawa 2005).

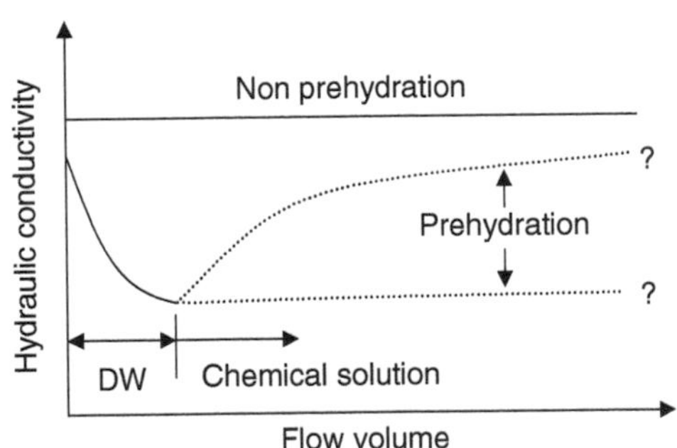


Figure 24. Difference in the hydraulic performance between the non-prehydration and the prehydration conditions.

liquids, are dominantly affected by the first wetting liquid. This is referred to as the "first exposure effect" (Shackleford 1994). Even if a large hydraulic conductivity is obtained for a given chemical solution or waste leachate, the hydraulic conductivity might be low when the GCL is permeated first with pure water and then with a given permeant, as shown in Figure 24.

It is likely that, in practice, GCLs are prehydrated prior to the permeation of a waste leachate. The water content of the underlain base soil affects the increases in the water content of the GCLs. From the experimental work conducted by Bonaparte et al. (1996), GCLs underlain by a base soil with a water content of −4% to +4% of the optimum water content, exhibit a water content ranging from 40–100%. Thus, the chemical compatibility of prehydrated GCLs is also an issue which needs to be addressed. Prehydration (permeation with water prior to the chemical solutions) versus non-prehydration (direct permeation of the chemical solutions without water permeation) conditions were compared in Figure 25. The figure replots the data obtained by Petrov et al. (1997) and Ruhl and Daniel (1997). The permeant liquids were NaCl, HCl, and NaOH solutions and a waste leachate. The levels of ionic strength for these permeants are plotted along the x-axis, while the levels of hydraulic conductivity with the permeant relative to the one permeated with the water are plotted along the y-axis. The non-prehydration condition has a more significant effect on the hydraulic conductivity and is one to three orders of magnitude greater in hydraulic conductivity than the prehydration condition.

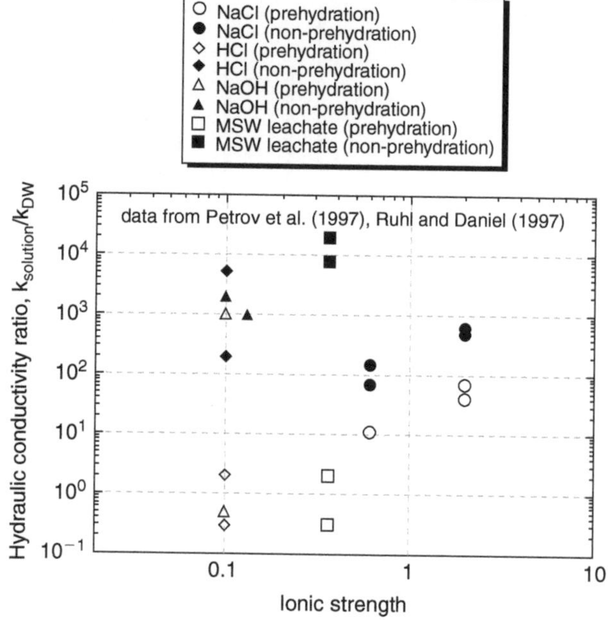


Figure 25. Hydraulic conductivity of GCLs permeated with chemical solutions ($k_{solution}$) relative to the hydraulic conductivity with de-ionized water ($k_{DW}$) and the ionic strength of the chemical solutions.

In order to evaluate the effect of the prehydration water contents, Vasco et al. (2001) conducted hydraulic conductivity tests on GCLs with several different levels of initial (prehydration) water contents permeated with different concentrations of $CaCl_2$ solutions. As shown in Figure 26, if the concentration of $CaCl_2$ is 0.025 mol/L, the hydraulic conductivity yields around $1 \times 10^{-7}$ cm/s regardless of the prehydration water content of the GCL. If the $CaCl_2$ concentration is higher than 0.1 mol/L, the prehydration water content influences the hydraulic conductivity. An increase in the initial water content from 9% to 200% results in a decrease in the hydraulic conductivity of two orders of magnitude. However, even a prehydration water content of 200% results in a level of hydraulic conductivity two orders of magnitude larger than the water permeation, and does not provide hydraulic conductivity lower than $1 \times 10^{-7}$ cm/s. Daniel et al. (1993) conducted similar experiments in which GCLs with different levels of prehydration water content were permeated with five different organic liquids including benzene, gasoline, methanol, methyl tertiary-butyl ether (MTBE), and trichloroethylene (TCE). They showed that the hydraulic conductivity could be lower than $1 \times 10^{-7}$ cm/s (and lower than $1 \times 10^{-8}$ cm/s for benzene and gasoline) if the prehydration water content is greater than 100%. The difference between the results of Daniel et al. (1993) and those of Vasco et al. (2001) might be attributed to the type of bentonite and the nature of the liquids employed. Daniel et al. (1993) used an organic liquid, with which difficulties could be encountered when penetrating the pore water due to its low solubility to water, while the inorganic chemical solutions used by Vasco et al. (2001) are thought to be able to penetrate the pore water with relative ease.

In Vasco et al. (2001), a uniform prehydration over the entire GCL was attempted and suction working in filter paper was effectively used to hydrate the GCLs. Katsumi et al. (2004 and 2008b) conducted hydraulic conductivity tests on GCLs which were prehydrated by being placed on compacted base soil (decomposed granite) by trying to simulate more actual conditions; in contrast, they did not necessarily attempt to attain uniform prehydration. The GCLs used are the needle-punched GCLs encased between nonwoven and woven geotextiles. After a certain period of prehydration, the GCLs were cut into circular shapes, 10 cm in diameter, and were placed on the compacted decomposed granite soil. The GCLs were placed with a woven geotextile as the bottom and with a nonwoven geotextile as the top. This is not consistent with the prehydration tests presented by Vasko et al. (2001), who put GCLs with the nonwoven geotextile as the bottom in an attempt to obtain uniform prehydration over the entire area, while Katsumi et al. (2004 and 2008b)

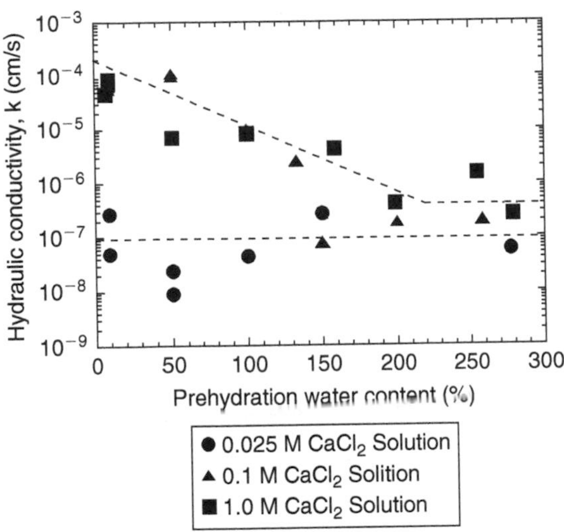


Figure 26. Hydraulic conductivity of GCLs permeated with $CaCl_2$ solutions and a prehydration water content (Vasko et al. 2001).

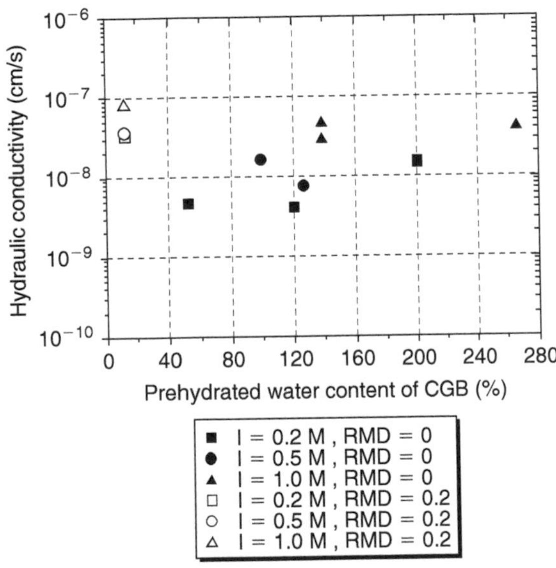


Figure 27. Hydraulic conductivity values versus the prehydration water content (data for non-prehydration from Katsumi et al. (2004)).

tried to simulate the actual case in this study. The hydraulic conductivity values obtained are plotted in Figure 27 with the prehydration water content. The prehydration might provide a hydraulic conductivity that is only 0-1 order of magnitude lower than that under the non-prehydration conditions. The increase in the prehydration water content does not necessarily cause a decrease in the hydraulic conductivity, as is shown in the data (Figure 26) by Vasko et al. (2001). It can be concluded, therefore, that prehydration does not effectively contribute to maintaining a low hydraulic conductivity. This is thought to be because the prehydration water content obtained in this study is not uniform, although Vasko et al. (2001) tried to make their prehydration uniform. From a practical viewpoint, the woven geotextile side of the GCL is put as a bottom and is in contact with the base soil. Therefore, the experimental conditions in this study might simulate an actual field

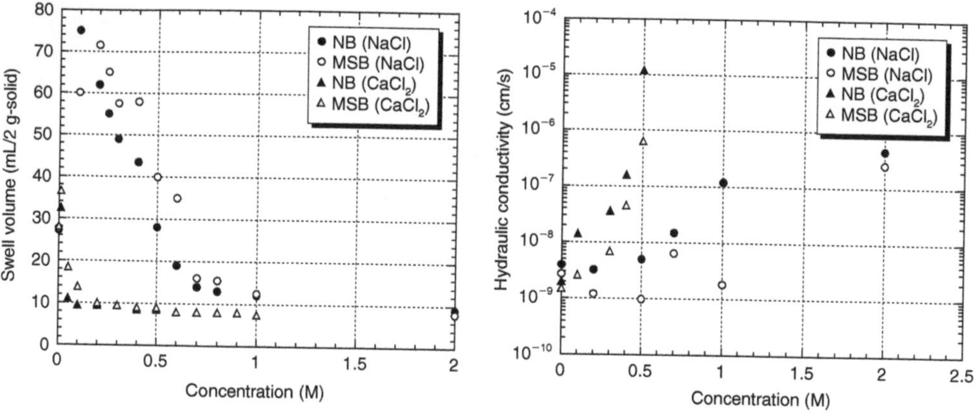


Figure 28. Swell volume and hydraulic conductivity of chemical resistant bentonite (multi-swellable bentonite: MSB) versus natural bentonite (NB) for NaCl and $CaCl_2$ solutions (Katsumi et al. 2001 and 2008a).

rather than the conditions presented by Vasko et al. (2001). Katsumi et al. (2008b) also discussed the heterogeneity of prehydration and its effect on the hydraulic conductivity.

Another important aspect obtained from these results is the magnitude of the hydraulic conductivity values. The hydraulic conductivity values of prehydrated powdered bentonite GCLs ranged from $4 \times 10^{-9}$ to $6 \times 10^{-8}$ cm/s regardless of any prehydration water content or $CaCl_2$ concentrations of permeants. These hydraulic conductivity values were significantly lower than those obtained for the granular bentonite GCLs shown in Figure 26 conducted by Vasko et al. (2001). As discussed in Section 6.3 on non-prehydration conditions, powdered bentonite GCLs are more compatible against chemical solutions for non-prehydration conditions. Similar to the non-prehydration conditions, powdered bentonites are more chemically compatible for prehydration conditions as well.

## 6.6 *Modified bentonites*

The use of a chemically resistant bentonites is considered to be a greatly anticipated counter measure against chemical attack. Several types of chemically resistant bentonites have been developed to enhance their capability and performance (e.g., Onikata et al. 1996 and 1999; Lo et al. 1997).

Multi-swellable bentonite (MSB), developed by Onikata et al. (1999), is bentonite which has been mixed with propylene carbonate (PC) to activate the osmotic swelling capacity. Propylene carbonate is placed in the interlayer of the smectite and attracts numerous water molecules. This results in a strong swelling power even if the permeant contains polyvalent cations or a high concentration of monovalent cations. Figure 28 shows values for the swell volume and the hydraulic conductivity of solutions of natural bentonite (NB) and MSB with NaCl and $CaCl_2$ as pore liquids (Katsumi et al. 2001 and 2008a). For hydraulic conductivity tests using flexible-wall permeameters, NB or MSB granules were placed between the top and the bottom pedestals to achieve a thickness of approximately 1 cm. This condition simulates the cases where these bentonites are used as GCLs, as well as simply to evaluate the chemical compatibility of the bentonites. Each specimen had a dry density of 0.79 g/cm$^3$. Under a cell pressure of 20–30 kPa, the bentonite specimens were exposed to the permeant liquid for longer than 24 hours. Then, a hydraulic gradient of 80–90 was applied to the specimens. MSB exhibits higher levels of swell volume than NB for all concentration levels, and is particularly excellent for concentrations lower than 0.5 mol/L. Most values for the hydraulic conductivity of MSB are one to two orders in magnitude lower than the values for that of NB for the same concentration levels. A hydraulic conductivity level lower than $1 \times 10^{-7}$ cm/s can be achieved for $CaCl_2$ concentrations that are lower than 0.5 mol/L. This means that the applicability of MSB as a landfill barrier material can be encouraged.

## ACKNOWLEDGMENTS

The author of this chapter has been provided with helpful comments and discussions by many individuals, in particular, Professor Craig H. Benson (University of Washington), Professor Charles D. Shackelford (Colorado State University), Professor Masashi Kamon (Kyoto University), Mr. Mitsuji Kondo (Hojun Co., Ltd.), Dr. Masanobu Onikata (Hojun Co., Ltd.), and Dr. Hiroyuki Ishimori (Ritsumeikan University). Their contributions are greatly appreciated.

## REFERENCES

Acar, Y.B. and Olivieri, I. 1989. Pore fluid effects on the fabric and hydraulic conductivity of laboratory-compacted clay, *Transportation Research Record 1219,* TRB, pp. 144–159.

Benson, C.H. and Othman, M.A. 1993. Hydraulic conductivity of compacted clay frozen and thawed in situ, *Journal of Geotechnical Engineering,* ASCE, Vol. 119, No. 2, pp. 276–294.

Benson, C.H., Hardinato, F.S., and Motan, E.S. 1994. Representative specimen size for hydraulic conductivity assessment of compacted soil liners, *Hydraulic Conductivity and Waste Contaminant Transport in Soils, ASTM STP 1142,* D.E. Daniel and S.J. Trautwein (eds.), ASTM, pp. 3–29.

Benson, C.H., Abichou, T.H., Olson, M.A., and Bosscher, P.J. 1995. Winter effects on the hydraulic conductivity of compacted clay, *Journal of Geotechnical Engineering,* ASCE, Vol. 121, No. 1, pp. 69–79.

Blumel, W., Muller-Kirchenbauer, A., Reuter, E., Ehrenberg, H., and von Baubeuge, K. 2002. Performance of geosynthetic clay liners in lysimeters, *Clay Geosynthetic Barriers,* H. Zanzinger, R.M. Koerner, and E. Gartung (eds.), Balkema, pp. 287–294.

Bonapare, R., Othman, M.A., Rad, N.R., Swan, R.H., and Vander Linde, D.L. 1996. Evaluation of various aspects of GCL performance, *Report of 1995 Workshop on Geosynthetic Clay Liners,* EPA/600/R-96/149, F-1-F-34.

Bowders, J.J. 1988. Termination criteria for clay permeability testing (Discussion), *Journal of Geotechnical Engineering,* ASCE, Vol. 114, No. 8. pp. 947–949.

Bowders, J.J. and Daniel, D.E. 1987. Hydraulic conductivity of compacted clay to dilute organic chemicals, *Journal of Geotechnical Engineering,* Vol. 113, No. 12, pp. 1432–1448.

Bouazza, A. 2002. Geosynthetic clay liners, *Geotextiles and Geomembranes,* Vol. 20, pp. 3–17.

Daniel, D.E. 1994. State-of-the-art: Laboratory hydraulic conductivity tests for saturated soils, *Hydraulic Conductivity and Waste Contaminant Transport in Soils, ASTM STP 1142,* D.E. Daniel and S.J. Trautwein (eds.), ASTM, pp. 30–78.

Daniel, D.E., Bowders, J.J., and Gilbert, R.B. 1997a. Laboratory hydraulic conductivity testing of GCLs in flexible-wall permeameters, *Testing and Acceptance Criteria for Geosynthetic Clay Liners, ASTM STP 1308,* L.W. Well (ed.), ASTM, pp. 208–226.

Daniel, D.E., Traetwein, S.J., Goswami, P.K. 1997b. Measurement of hydraulic properties of geosynthetic clay liners using a flow box, *Testing and Acceptance Criteria for Geosynthetic Clay Liners, ASTM STP 1308,* L.W. Well (ed.), ASTM, pp. 196–207.

Darcy, H. 1856. Les Fontaines Publiques de la Ville de Dijon, Dalmont, Paris.

Egloffstein, T.A. 2002. Bentonite as sealing material in geosynthectic clay liners – Influence of the electrolytic concentration, the ion exchange and ion exchange with simultaneous partial desiccation on permeability, *Clay Geosynthetic Barriers,* H. Zanzinger, R.M. Koerner, and E. Gartung (eds.), Swets & Zeitlinger, Lisse, The Netherlands, pp. 141–153.

Fernandez, F. and Quigley, R.M. 1985. Hydraulic conductivity of natural clays permeated with simple liquid hydrocarbons, *Canadian Geotechnical Journal,* Vol. 22, pp. 205–214.

Fernandez, F. and Quigley, R.M. 1988. Viscosity and dielectric constant controls on the hydraulic conductivity of clayey soils permeated with water-soluble organics, *Canadian Geotechnical Journal,* Vol. 25, pp. 582–589.

Fox, P.J. 1996. Analysis of hydraulic gradient effects for laboratory hydraulic conductivity testing, *Geotechnical Testing Journal,* ASTM, Vol. 19, No. 2, pp. 181–190.

Gaidi, L. and Alimi-Ichola, I. 2002. Study of the hydraulic behaviour of geosynthetic clay liners subjected to a leachate infiltration, *Clay Geosynthetic Barriers,* H. Zanzinger, R.M. Koerner, and E. Gartung (eds.), Balkema, pp. 233–246.

Giroud, J.P., Badu-Tweneboah, and Soderman, K.L. 1997. Comparison of leachate flow through compacted clay liners and geosynthetic clay liners in landfill liner systems, *Geosynthetic International,* IFAI, Vol. 4, Nos. 3-4, p. 391–431.

Giroud, J.-P., Thiel, R.S., Kavazanjian, E., and Lauro, F.J. 2002. Hydrated area of a bentonite layer encapsulated between two geomembranes, *Proceedings of the Seventh International Conference on Geosynthetics,* Ph. Delmas, J.P. Gourc, and H. Girard (eds.), Balkema, Lisse, The Netherlands, Vol. 2, pp. 827–832.

Gleason, M.H., Daniel, D.E., and Eykholt, G.R. 1997. Calcium and sodium bentonite for hydraulic containment applications, *Journal of Geotechnical and Geoenvironmental Engineering,* ASCE, Vol. 123, No. 5, pp. 438–445.

GSE. 2002. *The GSE GundSeal GCL Design Manual.*

Henken-Mellies, W.U., Zanzinger, H., and Gartung, E. 2002. Performance of GCL and drainage geocomposite in a landfill cover system, *Proceedings of the Seventh International Conference on Geosynthetics,* Ph. Delmas, J.P. Gourc, and H. Girard (eds.), Balkema, Lisse, The Netherlands, Vol. 2, pp. 833–836.

Hewitt, R. and Daniel, D.E. 1997. Hydraulic conductivity of GCLs after freeze-thaw, *Journal of Geotechnical and Geoenvironmental Engineering,* ASCE, Vol. 123, No. 4, pp. 305–313.

Imamura, S, Sueoka, T., and Kamon, M. 1996. Long term stability of bentonite/sand mixtures at L.L.R.W. storage, *Environmental Geotechnics,* M. Kamon (ed.), Balkema, pp. 545–550.

James, A.N., Fullerton, D., and Drake, R. 1997. Field performance of GCL under ion exchange conditions, *Journal of Geotechnical and Geoenvironmental Engineering,* ASCE, Vol. 123, No. 10, pp. 897–901.

Kamon, M. and Katsumi, T. 2001. Clay liners for waste containment, *Clay Science for Engineering,* K. Adachi and M. Fukue (eds.), Balkema, Rotterdam, pp. 29–45.

Katsumi, T., Onikata, M., Hasegawa, S., Lin, L., Kondo, M., and Kamon, M. 2001. Chemical compatibility of modified bentonite permeated with inorganic solutions, *Geoenvironmental Engineering – Geoenvironmental Impact Management,* R.N. Yong and H.R. Thomas (eds.), Thomas Telford, London, pp. 419–424.

Katsumi, T. and Kamon, M. 2002. Management of contaminated sites (Lecture for Plenary Session No. 4), *Environmental Geotechnics,* L.G. de Mello and M.S.S. Almeida (eds.), Balkema, Lisse, The Netherlands, Vol. 2, pp. 1013–1038.

Katsumi, T., Ogawa, A., Numata, S., Benson, C.H., Kolstad, D.C., Jo, H.Y., Edil, T.B., and Fukagawa, R. 2002. Hydraulic conductivity of GCLs permeated with multi-species inorganic chemical solutions, *Proceedings of the 37th Japan National Conference on Geotechnical Conference,* JGS (in Japanese).

Katsumi, T., Ogawa, A., Fukagawa, R. 2004. Effect of chemical solutions on hydraulic barrier performance of clay geosynthetic barriers, *Proceedings of the Third European Geosynthetics Conference – Geotechnical Engineering with Geosynthetics,* R. Floss, G. Braeu, M. Nussbaumer, and K. Laackmann (eds.), DGGT and TUM-ZG, pp. 701–706.

Katsumi, T. and Fukagawa, R. 2005. Factors affecting the chemical compatibility and the barrier performance of GCLs, *Sixteenth International Conference on Soil Mechanics and Geotechnical Engineering,* Millpress Science Publishers, Rotterdam, Netherlands, Vol.4, pp. 2285–2288.

Katsumi, T., Ishimori, H., Ogawa, A., Yoshikawa, K., Hanamoto, K., and Fukagawa, R. 2007. Hydraulic conductivity of nonprehydrated geosynthetic clay liners permeated with inorganic solutions and waste leachates, *Soils and Foundations,* JGS, Vol.47, No.1, pp. 79–96.

Katsumi, T., Ishimori, H., Onikata, M., and Fukagawa, R. 2008a. Long-term barrier performance of modified bentonite materials against sodium and calcium permeant solutions, *Geotextiles and Geomembranes,* Elsevier, Vol.26, No.1, pp. 14–30.

Katsumi, T., Ishimori, H., Ogawa, A., Maruyama, S., and Fukagawa, R. 2008b. Effects of water content distribution on hydraulic conductivity of prehydrated GCLs against calcium chloride solutions, *Soils and Foundations,* JGS, Vol.48, No.3, pp. 407–417.

Kolstad, D.C. 2000. *Compatibility of Geosynthetic Clay Liners (GCLs) with Multi-Species Inorganic Solutions,* MS Thesis, University of Wisconsin-Madison.

Kraus, J.F, Benson, C.H., Erickson, A., and Chamberlain, E.J. 1997. Freeze-thaw and hydraulic conductivity of bentonite barriers, *Journal of Geotechnical and Geoenvironmental Engineering,* ASCE, Vol. 123, No. 3, pp. 229–238.

Lagatta, M.D., Boardman, B.T., Cooley, B.H., and Daniel, D.E. 1997. Geosynthetic clay liners subjected to differential settlement, *Journal of Geotechnical and Geoenvironmental Engineering,* ASCE, Vol. 123, No. 5, pp. 402–410.

The Landfill System Technologies Research Association (ed.). 2000. *Landfill Sites in Japan 2000,* The Journal of Waste Management.

Lin, L-C. and Benson, C.H. 2000. Effect of wet-dry cycling on swelling and hydraulic conductivity of GCLs, *Journal of Geotechnical and Geoenvironmental Engineering,* ASCE, Vol. 126, No. 1, pp. 40–49.

Mitchell, J.K. 1993. *Fundamentals of Soil Behavior, 2nd Edition,* John Wiley & Sons, Inc., New York, 437 p.

Mitchell, J.K. and Madsen, F.T. 1987. Chemical effects on clay hydraulic conductivity, *Geotechnical Practice for Waste Disposal'87,* R.D. Woods (ed.), ASCE, pp. 87–116.

Norrish, K. and Quirk, J. 1954. Crystalline swelling of montmorillonite, use of electrolytes to control swelling, *Nature,* Vol. 173, pp. 255–257.

Onikata, M., Kondo, M., Hayashi, N., and Yamanaka, S. 1999. Complex formation of cation-exchanged montmorillonites with propylene carbonate: Osmotic swelling in aqueous electrolyte solutions, *Clays and Clay minerals,* Vol. 47, No. 5, pp. 672–677.

Othman, M.A. and Benson, C.H. 1993. Effect of freeze-thaw on the hydraulic conductivity and morphology of compacted clay, *Canadian Geotechnical Journal,* Vol. 30, pp. 236–246.

Petrov, R. and Rowe, R. 1997. Geosynthetic clay liner (GCL) – Chemical compatibility by hydraulic conductivity testing and factors impacting its performance, *Canadian Geotechnical Journal,* Vol. 34, pp. 863–885.

Petrov, R.J., Rowe, R.K., and Quigley, R.M. 1997a. Comparison of laboratory-measured GCL hydraulic conductivity based on three permeameter types, *Geotechnical Testing Journal,* ASTM, Vol. 20, No. 1, pp. 49–62.

Petrov, R.J., Rowe, R.K., and Quigley, R.M. 1997b. Selected factors influencing GCL hydraulic conductivity, *Journal of Geotechnical and Geoenvironmental Engineering,* ASCE, Vol. 123, No. 8, pp. 683–695.

Rowe, R.K. and Orsini, C. 2002. Internal erosion of GCLs placed directly over fine gravel, *Clay Geosynthetic Barriers,* H. Zanzinger, R.M. Koerner, and E. Gartung (eds.), Swets & Zeitlinger, Lisse, The Netherlands, pp. 187–197.

Rowe, R.K., Mukunoki, T., and Bathurst, R.J. 2004. Compatibility of a GCL with arctic diesel before and after freeze and thaw, *Proceedings of the Third European Geosynthetics Conference – Geotechnical Engineering with Geosynthetics,* R. Floss, G. Braeu, M. Nussbaumer, and K. Laackmann (eds.), DGGT and TUM-ZG, pp. 193–196.

Ruhl, J.L. and Daniel, D.E. 1997. Geosynthetic clay liners permeated with chemical solutions and leachates, *Journal of Geotechnical and Geoenvironmental Engineering,* ASCE, Vol. 123, No. 4, pp. 369–381.

Shackelford, C.D. 1994. Waste-soil interactions that alter hydraulic conductivity, *Hydraulic Conductivity and Waste Contaminant Transport in Soils, ASTM STP 1142,* D.E. Daniel and S.J. Trautwein (eds.), ASTM, pp. 111–168.

Shackelford, C.D., Malusis, M.A., Majeski, M.J., and Stern, R.T. 1999. Electrical conductivity breakthrough curves, *Journal of Geotechnical and Geoenvironmental Engineering,* ASCE, Vol. 125, No. 4, pp. 260–270.

Shackelford, C.D., Benson, C.H., Katsumi, T., Edil, T.B., and Lin, L. 2000. Evaluating the hydraulic conductivity of GCLs permeated with non-standard liquids, *Geotextiles and Geomembranes,* Elsevier, Vol. 18, Nos. 2–3, pp. 133–161.

Shan, H.-Y. and Lai, Y.-J. 2002. Effect of hydrating liquid on the hydraulic properties of geosynthetic clay liners, *Geotextiles and Geomembranes,* Elsevier, Vol. 20, No. 1, pp. 19–38.

Shang, J., Lo, K., and Quigley, R. 1994. Quantitative determination of potential distribution in Stern-Guoy double-layer model. *Canadian Geotechnical Jounral.* Vol. 31, pp. 624–636.

Viswanadham, B.V.S., Jessberger, H.L., and Rao, G.V. 1999. Discussion to 'Geosynthetic clay liners subjected to differential settlement,' *Journal of Geotechnical and Geoenvironmental Engineering,* ASCE, Vol. 125, No. 2, pp. 159–160.

# CHAPTER 5

# Contaminant transport through GCL-based liner systems

C.B. Lake
*Dalhousie University, Halifax, Canada*

R.K. Rowe
*Queen's University, Kingston, Canada*

## 1 INTRODUCTION

The migration of dissolved contaminants/solutes from landfills into the underlying geological environment can be mitigated with proper design of the landfill barrier system (Rowe et al. 2004). Traditionally, these low hydraulic conductivity barrier systems were composed of compacted clay liners (CCLs), in-situ clays or geomembranes (GMs). In many government jurisdictions, a GM and CCL are considered "standard" liner components of municipal solid waste (MSW) landfill environmental regulations (MOE, 1998; USEPA, 1998). Figure 1 shows a (MSW) landfill liner system that is required by environmental guidelines in the province of Nova Scotia, Canada (NSDEL, 1997).

With increasing amounts of technical literature, experience and awareness, many landfill designers now consider Geosynthetic Clay Liners (GCLs) to be complementary additions to these traditional bottom liner systems. Many international regulatory bodies allow for consideration of "alternative" liners (i.e. GCLs) to be incorporated in the liner system as long as it can be demonstrated that the alternative liner that is being proposed is equivalent to the component it is replacing. The question then becomes: "How does the engineer or hydrogeologist assess whether or not the proposed GCL based system is "equivalent" to a regulated case"? In essence, as discussed by Rowe et al. (1997b) and Rowe (1998), a proper comparison of equivalency should incorporate a contaminant transport modeling assessment of the regulated case versus the proposed GCL alternative.

Fundamental to this equivalency assessment is knowledge of the liner system contaminant transport parameters. The intention of this chapter is to: i) review general principles related to modeling contaminant transport of dissolved solutes through GCLs in landfill bottom liners, with an emphasis on diffusion and sorption (hydraulic principals have been covered in Chapter 4), ii) describe techniques for measuring GCL diffusion coefficients for inorganic and organic contaminants, iii) discuss factors that can influence GCL diffusion coefficients, iv) summarize various GCL diffusion and sorption coefficients that can be used for design and, v) provide results from an example case in which contaminant transport principles are utilized to infer equivalency between a CCL and GCL based liner system.

## 2 MODELING CONTAMINANT TRANPORT THROUGH GCLs

Modeling of contaminant transport through porous media can be generally described by the advection-dispersion equation (Bear, 1972; Freeze & Cherry, 1979). For simplified boundary conditions (e.g. an infinitely thick porous media, constant concentration in the landfill throughout its life, etc.), analytical solutions (Ogata & Banks, 1961) are available and are relatively easy to implement into computer programs or spreadsheets. However, for most landfill applications, proper modeling must incorporated multiple layers in the liner system and underlying hydrogeological units (each with different contaminant transport parameters). In addition modeling should consider the effect of (a) leachate collection systems that remove a portion of the contaminant during its service life, and (b) changing landfill leachate concentrations throughout its lifespan. These factors, inter alia, necessitate the use of numerical solutions to the general advection-dispersion

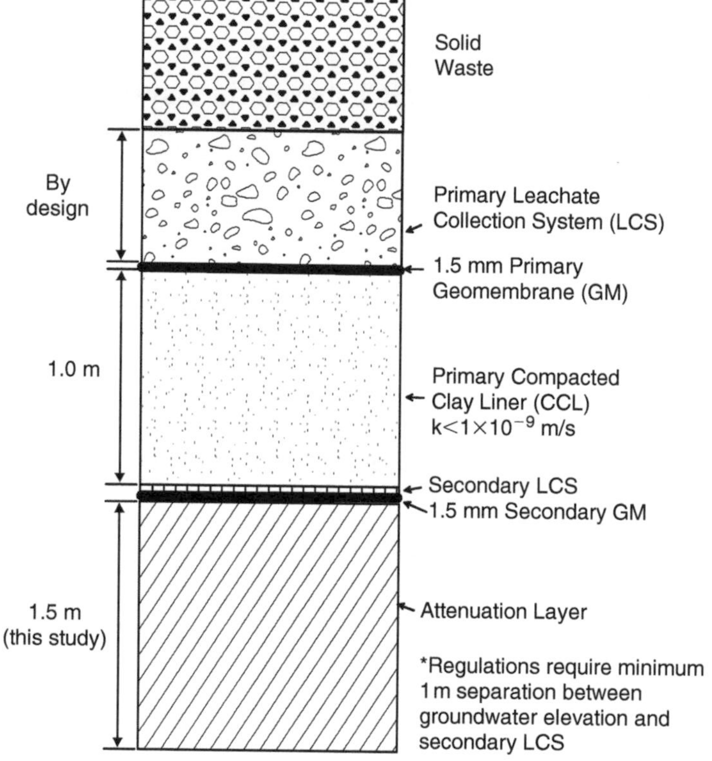


Figure 1. Schematic of regulated municipal solid waste landfill liner system for Nova Scotia, Canada (modified from NSDEL, 1997). *Some geotextiles, etc. have been left out of schematic for clarity.

equation. Although there are many techniques to solve the advection-dispersion equation numerically (finite element, finite difference), a one-dimensional model developed by Rowe & Booker (1984) uses a finite layer technique in which the Laplace transform is inverted using a numerical technique proposed by Talbot (1979). This finite layer technique can be utilized in combination with a variety of boundary conditions which are realistic to municipal solid waste landfills. The technique has been implemented into the computer model POLLUTE (Rowe & Booker, 2004), which is used throughout this chapter to solve the advection-dispersion equation.

Rowe et al. (2004) have described the procedures necessary for modeling contaminant transport through an entire landfill liner system. If a GCL is part of that liner system, the contaminant transport through the saturated GCL can be represented by a slightly modified form of the advection-dispersion equation (Rowe et al. (2000) for a single reactive solute (no degradation):

$$n_t \frac{\partial c}{\partial t} = \left( n_t D_t \frac{\partial^2 c}{\partial z^2} - n_t \bar{v} \frac{\partial c}{\partial z} \right) - \rho K_d \frac{\partial c}{\partial t} \tag{1}$$

where:
$c$ = concentration in the GCL at depth z and time t [$ML^{-3}$]; $n_t$ = total porosity of the GCL [−]; $D_t$ = diffusion coefficient deduced from total porosity [$L^2T^{-1}$]; $\bar{v}$ = average linearized groundwater velocity [$LT^{-1}$]; $\rho$ = dry density [$ML^{-3}$]; $K_d$ = partitioning coefficient [$M^{-1}L^3$].

Solving the partial differential equation given above allows an estimate of the concentration, c, at any time, t, and depth z, in the GCL. POLLUTE (Rowe & Booker, 2004) will readily solve equation (1) and calculate the contaminant transport through a GCL when it is incorporated as part of a landfill liner system. An example of this is provided in section 6 of this chapter. A discussion of the main processes involved in contaminant transport through GCLs (and other barrier systems) is provided below.

## 2.1 *Advection*

The average linearized groundwater velocity, $\bar{v}$, sometimes referred to as the seepage velocity, represents the advective flow through a GCL. The product of $n_t\bar{v}$ is the Darcy velocity, $v_a$, which depends on the hydraulic conductivity of the GCL. As discussed in Chapter 4, many factors can influence the hydraulic conductivity of a GCL and hence it is important to select a hydraulic conductivity that is representative of anticipated field conditions (hydration, permeating fluid, stress level, etc.). If a GCL is being utilized in combination with a geomembrane to form a composite liner, the Darcy velocity or leakage through the GM/GCL composite system can be calculated using methods outlined by Rowe et al. (2004). Based on limited experimental results (Wilson-Fahmy & Koerner, 1995), a GCL underlying a GM has the potential to reduce leakage through a composite liner (relative to a CCL) due to the improved contact conditions with the GM. The Darcy velocity through the landfill liner system is required for solution of equation 1.

## 2.2 *Diffusion*

If $v_a$ is relatively low through a soil (as is often the case for well designed municipal solid waste (MSW) landfills liners) the diffusive migration of contaminants may be the dominant transport mechanism through the liner system. Diffusion, the migration of a contaminant from areas of high concentration to areas of low concentration (Dutt & Low, 1962; Kemper & van Schiak, 1966), has been observed as a significant transport process through low hydraulic conductivity natural clay deposits (Goodall & Quigley, 1979) as well as low hydraulic conductivity compacted clay liners (King et al., 1993). As discussed in Chapter 4, the hydraulic conductivity of a GCL can be similar or lower than that of a compacted clay liner (provided similar testing conditions are employed) and thus the diffusive migration through GCLs will also need to be considered in any contaminant transport assessment involving GCLs. This assessment requires estimates of the GCL diffusion coefficient ($D_t$). A discussion of the various methods for obtaining GCL diffusion coefficients as well as the various factors affecting GCL diffusion coefficients are provided in sections 3 and 4 of this chapter.

## 2.3 *Sorption*

Sorption is a general term used to describe a process by which chemical species initially present in the soil pore fluid interacts with the soil-water system, resulting in a loss of mass or concentration of that chemical species from the soil pore fluid (Rowe et al., 2004). Yong et al. (1992) suggest that often it is difficult to distinguish between mechanisms such as physical adsorption, chemical adsorption and/or precipitation and hence the term sorption often "lumps" these many different processes together. Sorption parameters required for predicting contaminant migration are usually evaluated by batch, column, or diffusion testing (Yong et al., 1992; Rowe et al., 2004). Sorption of different contaminants to GCLs may be dependant on many different factors, a discussion of which is beyond the scope of this chapter. For practical applications, sorption is often represented mathematically as a linear process in which $K_d$, the partitioning coefficient, can be used in modeling of contaminant migration through the GCL. A discussion of sorption of contaminants to GCLs is provided in sections 3 and 4 of this chapter.

# 3 ESTABLISHING DIFFUSION AND SORPTION PARAMETERS FOR GCLS

## 3.1 *Experimental procedures and data interpretation*

For dissolved contaminants often found in MSW leachate, a significant amount of research has been performed related to establishing diffusion and sorption coefficients for GCLs (Rowe et al. 2000; Lake & Rowe, 2000a & b, Lake & Rowe, 2004; Rowe et al. 2005; Lake et al. 2007; Lange et al. 2008). Two different apparatus to perform such tests are shown in Figure 2 (Specified Volume Diffusion Test) and Figure 3 (Constant Stress Diffusion Test). The differences between these two apparatus are the method of GCL hydration prior to diffusion testing and the method of effective stress application during diffusion testing. The Specified Volume Diffusion (SVD) test and Constant Stress Diffusion (CSD) test are described in sections 3.2 and 3.3, respectively.

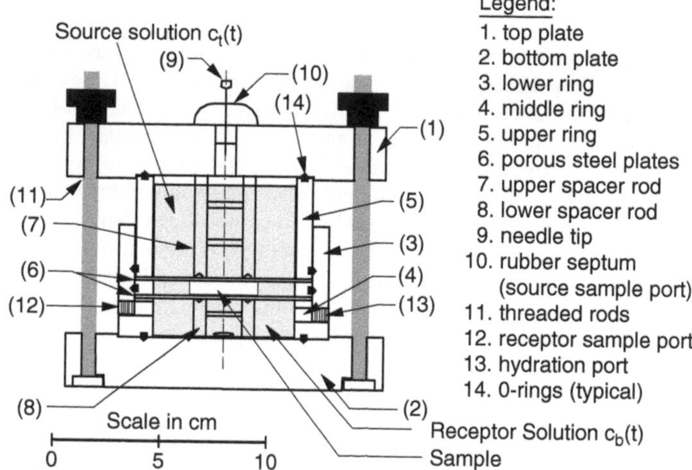


Figure 2. Schematic of cell used for specified volume diffusion testing (SVD) (modified from Rowe et al., 2000).

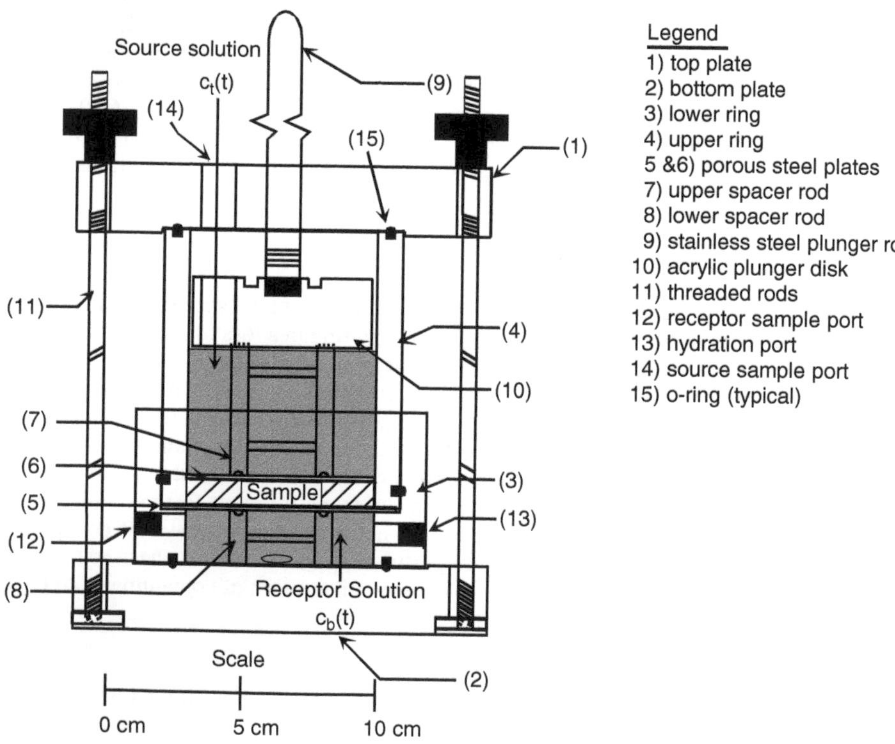


Figure 3. Schematic of cell used for constant stress diffusion testing (CSD) (modified from Rowe et al., 2000).

Essentially, GCL diffusion testing involves placing a contaminant solution above the previously hydrated GCL (source) and "clean" water below the GCL (receptor), as described by Rowe et al. (2000). As the diffusion test proceeds, the source and receptor solutions are sampled regularly and concentrations for the contaminant of interest are then experimentally tested using the appropriate chemical analytical technique. Figure 4 shows an example of source and receptor concentration profiles plotted in terms of the normalized concentration (i.e. source and receptor values are divided

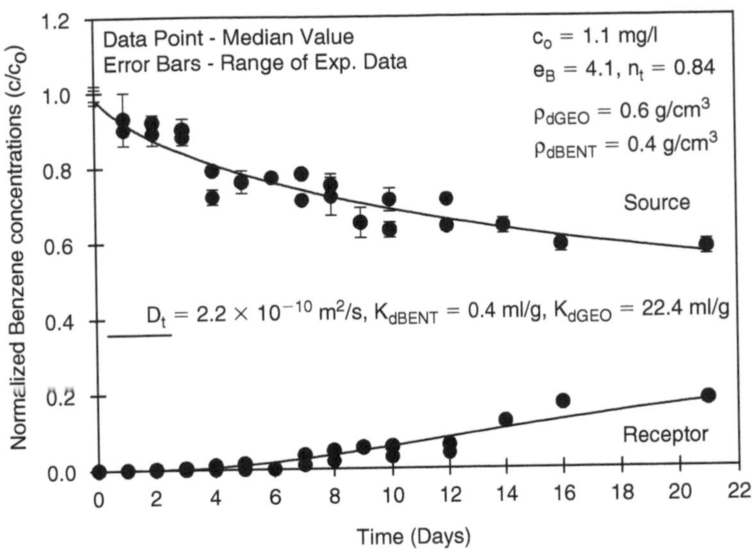


Figure 4. Theoretical best-fit POLLUTE curves (solid lines) fit to benzene experimental diffusion test data (solid and outlined circles). Modified from Lake and Rowe (2004).

by the initial source concentration, $c_o$). The solid circles in the figure represent the experimentally measured concentrations for benzene. The top data shows the observed decreases in concentration in the source solution due to diffusive flux into the GCL and sampling. The bottom data shows the observed concentration increase in the receptor. This net increase in mass of contaminant in the receptor is the difference between the increase in mass due to diffusive flux from the sample minus the decrease in mass due to sampling and sorption to the GCL.

To obtain diffusion and sorption coefficients from this experimental data, equation 1 is solved numerically by invoking boundary conditions utilized for the test. The boundary conditions for this type of test are of the "finite mass" type (Rowe et al., 2000) and for the source compartment, the concentration at any time, $c_t(t)$, is given by:

$$c_t(t) = c_o - \frac{1}{H_r}\int_o^t f_t(t)dt - \frac{q_c}{H_r}\int_o^t c_t(t)dt \tag{2}$$

where: $c_o$ = initial concentration in the source solution [$ML^{-3}$]; $H_r$ = height of source fluid (volume of source fluid per unit area) [L]; $f_t(t)$ = mass flux of contaminant into the soil at any time t [$ML^{-2}T^{-1}$]; $q_c$ = source fluid collected for sampling (replaced with distilled water) per unit area, per unit time [$LT^{-1}$]. The receptor concentration at any time, $c_b(t)$, can be expressed similarly as:

$$c_b(t) = c_{bo} + \frac{1}{h_b}\int_0^t f_b(t)dt - \frac{q_c}{h_b}\int_0^t c_b(t)dt \tag{3}$$

where: $c_{bo}$ = initial concentration in the receptor solution [$ML^{-3}$]; $h_b$ = height of receptor (volume of receptor reservoir per unit area) [L]; $f_b(t)$ = mass flux of contaminant into the receptor reservoir at any time t [$ML^{-2}T^{-1}$]; $q_c$ = receptor fluid collected for sampling (replaced with distilled water) per unit area, per unit time [$LT^{-1}$].

In this type of GCL diffusion testing, the parameters, $D_t$, and often $K_d$, are unknown for the GCL. Therefore, an iterative technique matches theoretical source and receptor reservoir curves for a best fit value of $D_t$ and $K_d$ for the GCL diffusion test with observed experimental data. These theoretical curves for the source and receptor are shown as the two solid lines in Figure 4. Values of $K_d$ obtained from diffusion testing are often used in combination with $K_d$ values obtained from batch testing with the GCL and the contaminant solution of interest, as described by Rowe et al.

(2004). It should be noted that for Figure 4, sorption of benzene was observed with the bentonite ($K_{dBENT}$) and geotextile ($K_{dGEO}$) of the GCL. As discussed by Rowe et al. (2005), this sorption of the geotextile and bentonite components of the GCL can be expressed as an equivalent sorption parameter, $K_{deq}$, which is more amenable to practical GCL modeling situations.

With no advective transport though the sample (which is often the case for diffusion testing), the flux, f, at any point in the GCL can be expressed in terms of the porous media diffusion coefficient $D_p = n_e D_e$,

$$f = -n_e D_e \frac{dc}{dz} = -D_p \frac{dc}{dz} = -n_t D_t \frac{dc}{dz} \tag{4}$$

It was found by Rowe et al. (2000) that the actual values of the effective porosity, $n_e$, and effective diffusion coefficient, $D_e$, are not required to predict transport through a single GCL provided the range of values of $n_e$ and $D_e$ correspond to the same product $D_p = n_e D_e$. This approach of using $D_p$ is convenient since generally only the total porosity, $n_t$, is known for a GCL. As shown in equation [4] above, $D_t$, is a diffusion coefficient deduced from the experimental data for the total porosity, $n_t$(i.e. $D_p = n_t D_t$) In actuality, $D_t$, is more representative of a mass transfer coefficient, as opposed to an effective diffusion coefficient, $D_e$. It should be noted that while this $D_t$ approach is suitable for one GCL approximately 1 cm thick, it may not be suitable for multiple layers of GCL.

### 3.2 *Specified volume diffusion (SVD) test: Inorganic contaminants*

Figure 2 shows the specified volume diffusion (SVD) test cell that can be used to obtain inorganic diffusion coefficients for a GCL where the height (volume) to which the GCL can swell is specified. The SVD testing apparatus shown in Figure 2 consists of a lower ring (receptor reservoir), a middle ring (GCL sample holder) and an upper ring (source reservoir), all made of acrylic. Rigid porous steel plates, the same diameter as the outside edge of the middle ring, are placed above and below the middle ring to prevent the GCL from swelling beyond the thickness of the middle ring. After the diffusion cell is assembled with the top and bottom plates, threaded rods and spacer rods (see Figure 2), the GCL is hydrated. GCL swelling begins with the spacer rods, placed in the upper and lower reservoirs, assisting in constraining the movement of the porous steel plates. This allows control of the total porosity and the final bulk GCL void ratio, $e_B$ (the ratio of the volume voids to the volume of solids in the GCL (geotextiles included) at the end of hydration, as defined by Petrov et al. (1997)). By manufacturing different ring heights (5 mm, 7 mm, 9 mm, and 11 mm), one can investigate the influence of $e_B$ on the diffusion coefficient. After hydration of the GCL to a specified $e_B$, a diffusion test is then performed under a zero hydraulic gradient condition (pure diffusion). Experimental source and receptor concentrations are then used to obtain diffusion and sorption coefficients using the methodology described above.

Details of the SVD test apparatus and procedures can be found in Rowe et al. (2000).

### 3.3 *Constant stress diffusion (CSD) apparatus: Inorganic contaminants*

Similar to the SVD apparatus, the constant stress diffusion (CSD) test apparatus shown in Figure 3 consists of an acrylic lower ring which forms the receptor reservoir and an acrylic upper ring which forms the source reservoir. The GCL sample is supported by a porous steel plate, which rests on a ridge in the lower ring and the lower spacer rod. The top porous steel plate (same diameter as the upper reservoir) sits on top of the GCL sample and is free to move when the sample hydrates (or contracts due to changes in pore fluid composition). The purpose of the upper spacer rod, which sits on the upper porous steel plate, is to support a plunger that extends through a hole in the top of the apparatus. A load frame rests on the plunger and allows application of a stress to the sample during hydration and diffusion testing of the GCL. Details of the constant stress diffusion test procedures are given by Rowe et al. (2000). The application of an effective stress on the GCL during diffusion testing is most similar to conditions expected in the field and is an appropriate test method for evaluating diffusion parameters for design applications. Experimental source and receptor concentrations are used to obtain diffusion and sorption coefficients using the methodology described above.

### 3.4 *Volatile organic compound (VOC) diffusion testing*

Past VOC diffusion testing with natural clayey soil (Barone et al., 1992; Donahue et al., 1999) and geomembranes (Rowe 1998; Sangam & Rowe 2001) have shown that the type of material chosen for a diffusion testing apparatus will depend, inter alia, on the chemical and physical properties of the VOC being tested. Lake & Rowe (2004) used a glass variation of the SVD apparatus discussed in section 3.2 to estimate GCL VOC diffusion coefficients for a mixture of aromatic and chlorinated compounds. Tests by Lake & Rowe (2004) were performed at relatively high final bulk GCL void ratios in order to keep stresses generated by the GCL during hydration (and subsequently onto the glass apparatus) to a minimum. Thus, the results can be considered an upper bound to the diffusion coefficient at lower void ratios (relative to those expected in the field).

Rowe et al. (2005) describe a stainless steel variation of the CSD test apparatus discussed in section 3.3, utilizing springs to apply the stress to the GCL during diffusion testing with a BTEX solution. Depending on the desired stress level, the appropriate spring stiffness can be chosen to simulate the applied stress in the field. This stainless steel testing cell was also used to examine the effect of low temperature on GCL solute transport through GCLs (see section 4.2.3). Test methods utilized for such organic compound tests on GCLs are similar to that described in section 3.1 and as discussed by Lake & Rowe (2004).

## 4 FACTORS INFLUENCING DIFFUSION AND SORPTION COEFFICIENTS

There are several factors that can potentially influence the diffusion and/or sorption coefficient for a particular GCL and contaminant: a) the final bulk GCL void ratio, $e_B$, b) the chemical composition of the fluid in contact with the GCL, and, c) the type of test performed to establish the GCL diffusion coefficient. Therefore, to ensure representative parameters for design, it is important that diffusion and sorption coefficients are obtained using contaminant solutions, stress levels and $e_B$ values similar to that expected in the field.

### 4.1 *Effect of $e_B$ on diffusion coefficient: Inorganic contaminants*

The final bulk GCL void ratio, $e_B$, represents the void ratio of the entire GCL (ratio of volume of voids to volume of solids for both the geotextile and bentonite in the GCL). Depending on the effective stress applied to the GCL and the type of GCL manufacture (Lake & Rowe, 2000b)), $e_B$ will vary for a hydrated GCL. Petrov & Rowe (1997) showed that for a given $e_B$ for a needlepunched GCL, the nature of the permeating fluid strongly influenced hydraulic conductivity values. This was also observed for inorganic diffusion coefficients deduced for a GCL as shown by Lake & Rowe (2000b & 2002). To demonstrate the influence of $e_B$ on chloride diffusion coefficients, Figure 5 shows the results of various diffusion tests performed at typical field values of $e_B$ for a thermally treated, needlepunched, GCL (samples hydrated with distilled water). For 0.05 M and 0.08 M NaCl solutions, chloride diffusion coefficients varied linearly with the final bulk GCL void ratio, with approximately an order of magnitude difference in diffusion coefficients observed for the range of $e_B$ values considered. Test results included three different types of GCL and two different test methods (CSD and SVD). Also included in this data is a test that was performed to simulate a GCL hydrating unconfined (3 kPa applied stress) and then being subsequently consolidated due to a higher stress (145 kPa). Even though the sample was initially hydrated to a high bulk GCL void ratio under the 3 kPa stress, it was the lower $e_B$ after consolidation ($e_B = 2.6$) which appeared to "control" the diffusion process (i.e. the data point from this test fell within the linear relationship shown in Figure 5).

Based on the test data presented by Lake & Rowe (2000b), an empirical equation (for source solutions of NaCl of less than 0.08 M, only) can be written for the diffusion coefficient, $D_t$, in terms of final bulk GCL void ratio, $e_B$, viz:

$$D_t(Cl^-) = (1.18\, e_B - 0.78) \times 10^{-10} m^2/s \text{ [for } 1 < e_B < 3.5] \quad (5)$$

$$D_t(Na^+) = (1.82\, e_B - 1.06) \times 10^{-10} m^2/s \text{ [for } 1 < e_B < 3.5] \quad (6)$$

These equations can be used in preliminary design and for designs that are relatively insensitive to $D_t$. However if the choice of diffusion coefficient is critical to the suitability of a given design

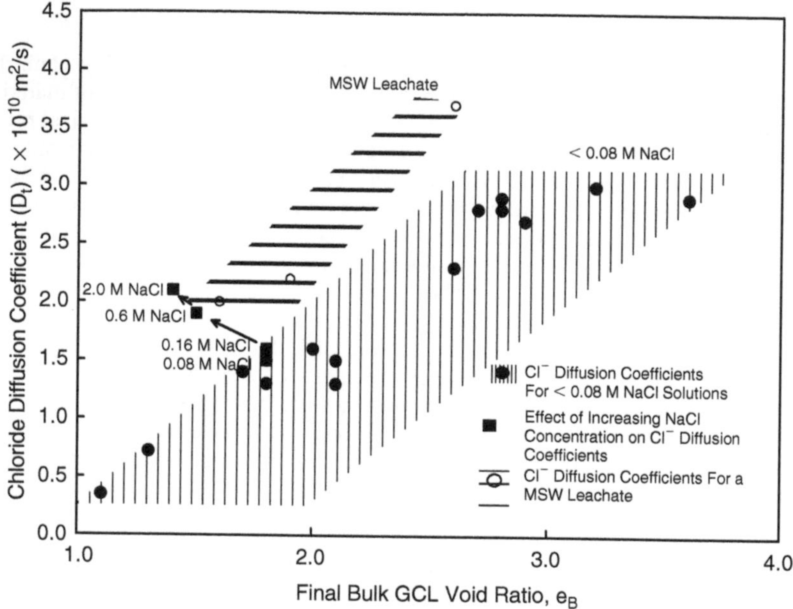


Figure 5. $Cl^-$ diffusion coefficients ($D_t$) obtained with <0.08 M NaCl concentrations compared to $Cl^-$ diffusion coefficients obtained with increasing NaCl concentrations and MSW leachate (modified from Rowe & Lake, 2000).

then test should be performed to simulate as close as possible the actual conditions anticipated in the field.

### 4.1.1 *Effect of type of GCL manufacture on inorganic diffusion coefficients*

Three different types of Bentofix needlepunched, thermally treated GCLs were tested by Lake & Rowe (2000b). The Bentofix NW GCL (referred to as NWNW in this chapter) employs a nonwoven, scrim reinforced carrier geotextile and a nonwoven cover geotextile which encapsulates granular sodium bentonite. The Benotfix NS GCL (referred to as WNW in this chapter) employs a woven carrier geotextile and a nonwoven cover geotextile which also ecapsulates granular sodium bentonite. The Bentofix BFG5000 (referred to as WNWB in this chapter) employs a woven carrier geotextile and a nonwoven cover geotextile which encapsulates powdered sodium bentonite. Bentonite is also impregnated in the top geotextile. The results of diffusion tests performed on these three GCLs are shown in Figure 5. The observation that there is generally a linear relationship for all the samples tested with less than 0.08 M NaCl solutions suggests that for the three GCLs and range of bulk GCL void ratios examined, the type of GCL manufacture has little effect on deduced GCL diffusion coefficients, particularly when compared to the effect of fluid type. However, each of these GCLs tested were needlepunched GCLs with thermally treated fibres. A test performed on a non-thermally treated needpunched GCL exhibited a slightly higher diffusion coefficient at low confining stresses as shown by Lake & Rowe (2000b). As discussed by Lake & Rowe (2000a), the thermal treatment of the needlepunching has the ability to restrict the swelling of the GCL at low confining stresses and therefore limit $e_B$ during hydration.

## 4.2 *Different types of dissolved contaminants in contact with the GCL*

### 4.2.1 *Inorganic contaminants*

The relationships given by equations [5] and [6] were for 0.08 M, or lower, NaCl solutions. To investigate the influence of increasing NaCl concentration on GCL diffusion coefficient, Lake (2000) performed a five phase GCL diffusion test on the NWNW GCL at a constant applied stress of 145 kPa in which the source solution was replaced with different concentrations of NaCl (phase I,

0.08 M NaCl; phase V, 2.0 M NaCl) after depletion from the previous phase of the diffusion test. As one can see from Figure 5 (solid squares), the Cl diffusion coefficient increased by a factor of 1.4 from phase I ($1.5 \times 10^{-10}$ m$^2$/s) to phase V ($2.2 \times 10^{-10}$ m$^2$/s). This increase in NaCl concentration also caused $e_B$ to decrease from 1.8 to 1.4, due to osmotic consolidation (see Figure 5). As was demonstrated by Lake & Rowe (2000b) this increase in diffusion coefficient can be significantly greater if a confining stress is not applied during testing. However, for most municipal solid waste landfill applications, it is anticipated that significant confining stress will be present on the GCL prior to contact with high salt solutions.

Diffusion testing with a relatively complex synthetic leachate solution compared to the NaCl solutions above results in many different levels of ion exchange, double layer contraction, or c-axis contraction to occur with the bentonite in the GCL. Two diffusion tests were performed by Lake & Rowe (2002) to examine the influence of leachate on diffusion coefficients for the same NWNW GCL. Figure 5 shows chloride diffusion results for the synthetic leachate solution at confining stresses of 22 kPa and 145 kPa. As can be seen from Figure 5, results plot above the trend line area suggested by equation 5. As discussed by Lake & Rowe (2002), adsorption of $K^+$ and $Mg^{++}$ and possibly $NH_4^+$ from the leachate to the bentonite in the GCL at the expense of $Na^+$ and $Ca^{++}$ was most likely causing double layer contraction of the clay. The increased salt concentration of the leachate also likely caused similar double layer contraction in the bentonite. Even though the diffusion coefficient with the synthetic leachate resulted in a chloride diffusion coefficient of approximately 1.5 to 2 times higher than that of the 0.08 NaCl solutions, these leachate values are still lower than that of a typical compacted clay liner ($\sim 6 \times 10^{-10}$ m$^2$/s). Lange et al.(2008) similarly noted faster diffusive migration rates for anions that included $Cl^-$ and $SO_4^{2-}$ from acidic mine drainage water compared to other mining solutions; the high metal loading (charges +2, +3) contributed to the active release of Na and contraction of the clay's double layers.

The important point to be made here is that the deduced diffusion coefficient depends both on $e_B$ and the level of potential interaction of the leachate with the bentonite under the anticipated field conditions. Hence tests should be performed using leachate solutions and applied stress conditions as close as practical to that anticipated in the field application.

#### 4.2.2 *Organic contaminants*

Obtaining diffusion and sorption coefficients of volatile organic compounds for GCLs and other landfill barrier systems are critical for proper design of such systems. Several papers have been published regarding GCL VOC diffusion coefficients. Lo (1992) performed hydraulic conductivity testing with GCLs and monitored 1,4 dichlorobenzene (DCB) effluent concentrations during flexible wall hydraulic conductivity testing. The effective diffusion coefficient for dichlorobenzene for a Claymax GCL was reported as $9.8 \times 10^{-11}$ m$^2$/s. Other DCB results are reported by Lo (1992) for GCLs containing organically modified bentonites (1.2 to $1.4 \times 10^{-10}$ m$^2$/s).

Specified volume GCL diffusion testing (needlepunched, thermally treated GCL) by Lake & Rowe (2004) utilized a VOC multi-component solution of dichloromethane (DCM), 1,2 dichloroethane (DCA), trichloroethene (trichloroethylene, TCE), benzene and toluene at concentrations below 3.2 mg/L for chlorinated VOCs and 1.1 mg/L for aromatic hydrocarbons. Diffusion coefficients deduced from testing ranged from approximately $2 \times 10^{-10}$ m$^2$/s to $3 \times 10^{-10}$ m$^2$/s from and showed that the rate of contaminant migration proceeded through the hydrated GCL in the order of DCM>DCA>benzene>TCE and toluene. This was attributed to varying degrees of sorption of DCA, benzene, TCE and toluene to the geotextile component of the GCL as well as to the bentonite present in the GCL. Sorption tests performed with only the geotextiles assisted in examining the sorption of VOCs during diffusion testing. The results reported by Lake & Rowe (2004) were performed at relatively high final bulk GCL void ratios and therefore represent conditions where the GCL fully hydrates under less than 0.5 m of cover material. If the diffusion coefficient of the VOCs decrease with lower GCL bulk void ratios (as was shown for chloride and sodium in Lake & Rowe, 2000b), then it would be reasonable to expect that the diffusion coefficients obtained in this study provide an upper bound for GCL VOC diffusion coefficients for a range of stress conditions expected in the field (for the five VOCs examined).

Diffusion test modeling, as well as geotextile sorption test results presented by Lake & Rowe (2004) suggests that there is varying amounts of VOC sorption to the polypropylene geotextiles present in the NWNWT GCL (Lake & Rowe, 2004). The amount of VOC sorption is likely to vary from one GCL product to another due to differences in polymers used for geotextiles as well as mass of geotextiles. As previously mentioned in section 3.1, the VOC sorption of a GCL can be represented by $K_{deq}$.

Due to the hydration energy of exchangeable cations such as $Na^+$ and $Ca^{++}$, the surface of the bentonite clay in the GCL is quite hydrophilic (i.e. water loving) and this is not conducive for sorption of non-polar organic compounds to mineral surfaces. Usually the sodium bentonite used in GCLs has a very low fraction of organic carbon and hence partitioning of non-polar organic compounds would be expected to be small. Smith & Jaffe (1994) reported a bentonite $K_d$ value of 0.2 mL/g for benzene for a bentonite while Lo et al. (1997) reported "insignificant" sorption of BTEX compounds with bentonite. Organoclays or organophilic clays may be used to provide greater sorption of VOCs than can obtained from normal sodium bentonite (Boyd et al., 1988; Smith & Jaffe, 1994; Jaynes & Vance, 1996). As indicated by Lake & Rowe (2005), activated carbon or organoclays could potentially be used as a GCL bentonite additive to improve sorption of some VOCs. When 2% of powdered activated carbon was mixed with powdered bentonite, the sorption of the mixture was much higher than several organoclays for TCE, benzene, and toluene. DCM sorption to the same mixture showed a slight decrease relative to that of some of the organoclays. Lorenzetti et al. (2005) discussed the potential increase in hydraulic conductivity to inorganic solutions when organoclays were present at more than 20% of the amount of bentonite in the GCL.

#### 4.2.3 *Effect of temperature on volatile organic compound diffusion and sorption coefficients*

To examine the effect of low temperatures on VOC diffusion and sorption parameters, Rowe et al. (2005) subjected a Bentofix NW GCL to diffusion and sorption testing with a mix of benzene, ethyl benzene, and xylenes (BTEX), each at concentrations of 2500 μg/L. The BTEX partitioning coefficients for the geotextiles were less at 5°C than at 22°C implying less retardation at lower temperatures (see Table 1). However the diffusion coefficient also decreased in value from 22°C to 5°C (see Table 1). Despite reduced retardation at 5°C, the mass transport of BTEX through the GCL at 5°C was less than that at 22°C indicating that the effect of a lower diffusion coefficient dominated over the effect of a lower sorption coefficient. These tests were performed to assess the contaminant transport performance of the GCL for a cold region application in Canada's north (Li et al. 2002)

### 4.3 *Constant stress vs. specified volume diffusion testing*

The effect of the type of diffusion test performed (CSD vs. SVD) can be indirectly seen by examining Figure 5. For the 0.05 M to 0.08 M NaCl solutions, both SVD and CSD tests were performed to obtain this data. Both SVD and CSD test results fall within the general location of the regression line, suggesting that the type of test performed has little influence on the diffusion coefficient. However as stated in section 4.2.1, if higher concentration salt solutions or complex leachates are utilized for testing, the CSD test condition is more representative of anticipated field conditions, and is recommended for these types of solutions.

## 5 SUMMARY OF DIFFUSION AND SORPTION COEFFICIENTS TO USE FOR PRELIMINARY DESIGNS

For most preliminary GCL landfill liner designs involving a contaminant migration assessment through the proposed GCL barrier system, an estimate of GCL diffusion and sorption coefficients will often suffice. Table 1 summarizes a variety of GCL diffusion and sorption coefficients for different inorganic and organic species that have been published by the authors. Also included with each diffusion coefficient result is the type of GCL tested, the $e_B$ and $n_t$ value at which the GCL was tested, the type of source solution used and the reference of the work. The following should be noted regarding the results in Table 1: a) the results are for tests performed with Bentofix thermally treated, needlepunched GCLs, and b) GCL samples were hydrated with de-aired, de-ionized water.

As discussed in section 4, it is often $e_B$ which "controls" the diffusion coefficient for a given contaminant source solution or leachate. However, for design purposes, often only the proposed stress levels are known for the proposed GCL barrier system, and not $e_B$. To properly obtain the GCL diffusion coefficient, a constant stress diffusion test should be performed with the applied stress level and leachate anticipated to be in contact with the GCL. If this type of detail is not needed

Table 1. Summary of GCL diffusion coefficients reported in the literature.

| Species | GCL | $e_B$ (−) | $n_t$ (−) | $D_t$ ($m^2$/s) | $K_d$ (mL/g) | Source Solution | Ref. |
|---|---|---|---|---|---|---|---|
| $Cl^-$ | NWNW | 1.1 | 0.56 | $3.5 \times 10^{-11}$ | 0 | 0.05 M NaCl | 1 |
| | NWNW | 1.7 | 0.67 | $1.4 \times 10^{-10}$ | 0 | 0.08 M NaCl | 1 |
| | NWNW | 1.8 | 0.68 | $1.5 \times 10^{-10}$ | 0 | 0.08 M NaCl | 1 |
| | NWNW | 1.8 | 0.68 | $1.5 \times 10^{-10}$ | 0 | 0.08 M NaCl | 1 |
| | NWNW | 1.8 | 0.69 | $1.3 \times 10^{-10}$ | 0 | 0.08 M NaCl | 1 |
| | NWNW | 2.1 | 0.73 | $1.5 \times 10^{-10}$ | 0 | 0.05 M NaCl | 1 |
| | NWNW | 2.6 | 0.77 | $2.3 \times 10^{-10}$ | 0 | 0.08 M NaCl | 1 |
| | NWNW | 2.7 | 0.78 | $2.8 \times 10^{-10}$ | 0 | 0.08 M NaCl | 1 |
| | NWNW | 2.8 | 0.78 | $2.9 \times 10^{-10}$ | 0 | 0.08 M NaCl | 1 |
| | NWNW | 3.2 | 0.80 | $3.0 \times 10^{-10}$ | 0 | 0.08 M NaCl | 1 |
| | NWNW | 1.8 | 0.68 | $1.6 \times 10^{-10}$ | 0 | 0.16 M NaCl | 1 |
| | NWNW | 1.5 | 0.65 | $1.9 \times 10^{-10}$ | 0 | 0.6 M NaCl | 1 |
| | NWNW | 1.4 | 0.63 | $2.1 \times 10^{-10}$ | 0 | 2.0 M NaCl | 1 |
| | NWNW | 1.6 | 0.67 | $2.0 \times 10^{-10}$ | 0 | S.L. | 1 |
| | NWNW | 1.9 | 0.70 | $2.2 \times 10^{-10}$ | 0 | S.L. | 1 |
| | NWNW | 2.6 | 0.77 | $3.7 \times 10^{-10}$ | 0 | S.L. | 1 |
| | WNW | 1.3 | 0.59 | $7.2 \times 10^{-11}$ | 0 | 0.08 M NaCl | 1 |
| | WNW | 2.0 | 0.69 | $1.6 \times 10^{-10}$ | 0 | 0.05 M NaCl | 1 |
| | WNW | 2.9 | 0.77 | $2.7 \times 10^{-10}$ | 0 | 0.08 M NaCl | 1 |
| | WNWB | 2.1 | 0.74 | $1.3 \times 10^{-10}$ | 0 | 0.05 M NaCl | 1 |
| | WNWB | 2.8 | 0.79 | $2.8 \times 10^{-10}$ | 0 | 0.08 M NaCl | 1 |
| | WNWB | 3.6 | 0.83 | $2.9 \times 10^{-10}$ | 0 | 0.08 M NaCl | 1 |
| $Na^+$ | NWNW | 1.1 | 0.56 | $6.0 \times 10^{-11}$ | 0 | 0.05 M NaCl | 1 |
| | NWNW | 1.7 | 0.67 | $2.2 \times 10^{-10}$ | 0 | 0.08 M NaCl | 1 |
| | NWNW | 1.8 | 0.68 | $1.5 \times 10^{-10}$ | 0 | 0.08 M NaCl | 1 |
| | NWNW | 1.8 | 0.68 | $1.6 \times 10^{-10}$ | 0 | 0.08 M NaCl | 1 |
| | NWNW | 1.8 | 0.69 | $2.1 \times 10^{-10}$ | 0 | 0.08 M NaCl | 1 |
| | NWNW | 2.1 | 0.73 | $2.5 \times 10^{-10}$ | 0 | 0.05 M NaCl | 1 |
| | NWNW | 2.6 | 0.77 | $4.0 \times 10^{-10}$ | 0 | 0.08 M NaCl | 1 |
| | NWNW | 2.7 | 0.78 | $4.4 \times 10^{-10}$ | 0 | 0.08 M NaCl | 1 |
| | NWNW | 2.8 | 0.78 | $4.8 \times 10^{-10}$ | 0 | 0.08 M NaCl | 1 |
| | NWNW | 3.2 | 0.80 | $5.0 \times 10^{-10}$ | 0 | 0.08 M NaCl | 1 |
| | NWNW | 1.8 | 0.68 | $1.7 \times 10^{-10}$ | 0 | 0.16 M NaCl | 1 |
| | NWNW | 1.5 | 0.65 | $1.9 \times 10^{-10}$ | 0 | 0.6 M NaCl | 1 |
| | NWNW | 1.4 | 0.63 | $2.1 \times 10^{-10}$ | 0 | 2.0 M NaCl | 1 |
| | WNW | 1.3 | 0.59 | $1.1 \times 10^{-10}$ | 0 | 0.08 M NaCl | 1 |
| | WNW | 2.0 | 0.69 | $2.7 \times 10^{-10}$ | 0 | 0.05 M NaCl | 1 |
| | WNW | 2.9 | 0.77 | $4.0 \times 10^{-10}$ | 0 | 0.08 M NaCl | 1 |
| | WNWB | 2.1 | 0.74 | $2.5 \times 10^{-10}$ | 0 | 0.05 M NaCl | 1 |
| | WNWB | 2.8 | 0.79 | $4.4 \times 10^{-10}$ | 0 | 0.08 M NaCl | 1 |
| | WNWB | 3.6 | 0.83 | $4.3 \times 10^{-10}$ | 0 | 0.08 M NaCl | 1 |
| $K^+$ | NWNW | 1.6 | 0.67 | $2.5 \times 10^{-10}$ | 2.5 | S.L. | 2 |
| | NWNW | 2.6 | 0.77 | $4.6 \times 10^{-10}$ | 3.1 | S.L. | 2 |
| $Mg^{+2}$ | NWNW | 2.6 | 0.77 | $2.2 \times 10^{-10}$ | 3.3 | S.L. | 2 |
| DCM | NWNW | 4.1 | 0.84 | $3.2 \times 10^{-10}$ | 0** | VOC Mix. | 3 |
| | NWNW | 4.6 | 0.86 | $3.3 \times 10^{-10}$ | 0** | VOC Mix. | 3 |
| DCA | NWNW | 4.1 | 0.84 | $3.2 \times 10^{-10}$ | 0.6** | VOC Mix. | 3 |
| | NWNW | 4.6 | 0.86 | $3.3 \times 10^{-10}$ | 0.5** | VOC Mix. | 3 |
| TCE | NWNW | 4.1 | 0.84 | $2.1 \times 10^{-10}$ | 7.9** | VOC Mix. | 3 |
| | NWNW | 4.6 | 0.86 | $2.1 \times 10^{-10}$ | 6.1** | VOC Mix. | 3 |

*(Continued)*

Table 1. Summary of GCL diffusion coefficients reported in the literature (*continued*).

| Species | GCL | $e_B$ (−) | $n_t$ (−) | $D_t$ ($m^2/s$) | $K_d$ (mL/g) | Source Solution | Ref. |
|---|---|---|---|---|---|---|---|
| Benzene | NWNW | 4.1 | 0.84 | $2.5 \times 10^{-10}$ | 2.3** | VOC Mix. | 3 |
| | NWNW | 4.6 | 0.86 | $2.2 \times 10^{-10}$ | 2.6** | VOC Mix. | 3 |
| | NWNW | 4 | 0.80 | $3.7 \times 10^{-10}$ | 4.4 | BTEX Mix. | 4 |
| | NWNW | 4 | 0.78 | $2.2 \times 10^{-10}$ | 2.6 | BTEX Mix.* | 4 |
| Toluene | NWNW | 4.1 | 0.84 | $2.1 \times 10^{-10}$ | 7.7** | VOC Mix. | 3 |
| | NWNW | 4.6 | 0.86 | $2.2 \times 10^{-10}$ | 8.2** | VOC Mix. | 3 |
| | NWNW | 4 | 0.80 | $3.1 \times 10^{-10}$ | 15.0 | BTEX Mix. | 4 |
| | NWNW | 4 | 0.78 | $1.8 \times 10^{-10}$ | 8.7 | BTEX Mix* | 4 |
| Ethylbenzene | NWNW | 4 | 0.80 | $2.9 \times 10^{-10}$ | 36. | BTEX Mix. | 4 |
| | NWNW | 4 | 0.78 | $1.7 \times 10^{-10}$ | 22 | BTEX Mix* | 4 |
| m&p xylenes | NWNW | 4 | 0.80 | $2.5 \times 10^{-10}$ | 42 | BTEX Mix. | 4 |
| | NWNW | 4 | 0.78 | $1.5 \times 10^{-10}$ | 25 | BTEX Mix* | 4 |
| oxylene | NWNW | 4 | 0.80 | $2.6 \times 10^{-10}$ | 27 | BTEX Mix. | 4 |
| | NWNW | 4 | 0.78 | $1.5 \times 10^{-10}$ | 14 | BTEX Mix* | 4 |

Notes:
- One should consult the following references to understand the limitations of results presented in Table 1: [1] Lake & Rowe, 2000b; [2] Lake & Rowe, 2002; [3] Lake & Rowe, 2004; [4] Rowe et al. 2005
- Test temperature ~ 20°C unless otherwise noted; * refers to tests being performed at 7°C
- ** Estimated based on $K_{deq}$ from Rowe et al. (2005)
- DCM: Dichloromethane; DCA: 1,2 Dichloroethane; TCE: Trichloroethylene; S.L.: Synthetic Leachate; VOC Mix.: DCM, DCA, TCE, benzene and toluene water-based solution each at concentrations each below 1 mg/L; BTEX mix: benzene, toluene, ethylbenzene, m,p,&o xylene water-based solution each at concentrations below 2.5 mg/L

for preliminary design, the following relationship between applied stress and $e_B$ for the three GCLs in Table 1 (reported by Rowe & Lake, 1999) may be used as an initial estimate for these GCLs:

$$e_B \cong \frac{3.25 - Log_{10}\sigma'v}{0.57} \quad \textit{For the NWNW GCL} \tag{7}$$

$$e_B \cong \frac{3.5 - Log_{10}\sigma'v}{0.69} \quad \textit{For the WNW GCL} \tag{8}$$

$$e_B \cong \frac{2.9 - Log_{10}\sigma'v}{0.32} \quad \textit{For the WNWB GCL} \tag{9}$$

## 6 CONTAMINANT MIGRATION ASSESSMENTS OF GCL LINER SYSTEMS

### 6.1 *Equivalency of GCL and CCL liner systems*

GCLs are often utilized as a replacement for a CCL and the questions becomes, "when is a GCL equivalent to a CCL?" Equivalency issues often arise in municipal solid waste landfill design applications in which the cost to transport suitable clay to construct a CCL based liner system is more than that of a GCL based liner system. It may be that the clay present at the proposed landfill site has a marginal hydraulic conductivity and the cost of improving the hydraulic conductivity by importing a lower hydraulic conductivity clay from another site or utilizing a sand-bentonite liner system makes the liner system uneconomic relative to a GCL based liner system. Many regulatory agencies that specify a CCL system with a hydraulic conductivity of $1 \times 10^{-9}$ m/s to be used for a municipal solid waste landfill will allow substitution of an alternative liner system such as a GCL or amended soil-bentonite liner, provided the alternate system (i.e. GCL) can be proven to be "equivalent".

For liquid containment applications in which contamination by the liquid is not an issue, comparisons of equivalency by comparing advective fluxes or travel times is often sufficient for evaluating

Table 2. Cases considered for comparing NSDEL liner system with alternative GCL based liner systems.

| Case No. | Description | Primary Liner Composition | Total Primary Liner Thickness (m) |
|---|---|---|---|
| 1 (Figure 1) | NSDEL Regulations | GM/CCL1 | 1.0015 |
| 2 (Figure 6) | Alternative to NSDEL Regulations | GM/GCL/CCL2 | 1.0015 |
| 3 (Figure 7) | Alternative to NSDEL Regulations | GM/GCL/CCL3 | 0.6015 |

Notes
GM: See Table 3 for material parameters
CCL1: CCL specified by NSDEL (see Table 3)
CCL2: Marginal clay ($k = 5 \times 10^{-9}$ m/s) used for CCL (see Table 3), thickness, 0.99 m
CCL3: Marginal clay ($k = 5 \times 10^{-9}$ m/s) used for CCL (see Table 3), thickness, 0.59 m

equivalency between CCL and GCL liner systems (Richardson, 1997 and Giroud et al., 1997). However, if GCLs (or CCLs and GMs) are to be utilized for containment of contaminants in applications such as the liner systems of municipal solid waste landfills, a detailed contaminant transport analysis should be performed. As discussed by Rowe et al (1997b), Rowe (1998), Lake & Rowe (1999), Rowe and Brachman (2004), and Rowe et al (2004), this contaminant transport assessment should includes factors such as diffusion, advection (including leakage between the geomembrane and liner contact), sorption (if present), biodegradation (if present), and finite service lives of engineered components. When each of these factors are considered in conjunction with the landfill characteristics (i.e. size and leachate characteristics) and the hydrogeological setting, a proper comparison of equivalency between a CCL and GCL can be made.

### 6.2 *Assessing GCL/CCL equivalency for municipal solid waste landfills: Example*

To demonstrate incorporation of GCLs into contaminant transport assessments, as well as to assess GCL/CCL equivalency, three slightly different example cases are presented in this section with respect to a municipal solid waste landfill barrier application. A summary of the differences between the three cases used in the equivalency assessment are shown in Table 2.

Case 1, as was shown in Figure 1, is essentially the regulated barrier system prescribed by NSDEL (1997). Case 2, as shown in Figure 6, is a variation of Case 1, where the compacted clay does not meet the hydraulic conductivity specification for the primary CCL ($5 \times 10^{-9}$ m/s versus the regulation of $1 \times 10^{-9}$ m/s; see Table 3), but the hypothetical material is readily available on-site for construction. A GCL has been added beneath the primary GM, to improve its hydraulic characteristics and make the harmonic hydraulic conductivity ($\bar{k}$) of the GCL/CCL "composite clay liner" meet regulatory requirements for primary barrier's hydraulic conductivity ($\bar{k} = 8 \times 10^{-10}$ m/s). Case 3 represents a variation of Case 2 in which the CCL portion of the primary liner is proposed to be "thinned" to 0.6 m, from 1.0 m. Both Cases 2 and 3 represent possible design scenarios which may be present during initial assessment of a landfill barrier system. From a practical perspective, the marginal clay used in Cases 2 and 3 may represent a very feasible economic alternative to importing $1 \times 10^{-9}$ m/s hydraulic conductivity clay.

A hypothetical landfill and geological environment was chosen for the contaminant transport assessment. The characteristics of the landfill are shown in Table 4. Two different contaminants are examined in the contaminant transport assessment, chloride and dichloromethane (DCM). Chloride is a relatively conservative contaminant which is often considered in contaminant transport assessment while DCM is an organic contaminant often found in low concentrations in MSW leachate (Rowe, 1995). As shown in Table 4, the DCM half-life in the leachate was assumed to be 10 years. A DCM half-life of 50 years was assumed for the soil based on the results of Rowe et al. (1997a). No sorption of DCM to the soil layers was considered in the analysis (a conservative assumption for DCM). Other pertinent parameters for the GMs, CCLs, GCL and attenuation layer are also provided in Table 4.

The service lives (see Rowe 2005 for a discussion of service lives) of the various landfill components were also considered in the analyses. It was assumed that the primary leachate collection

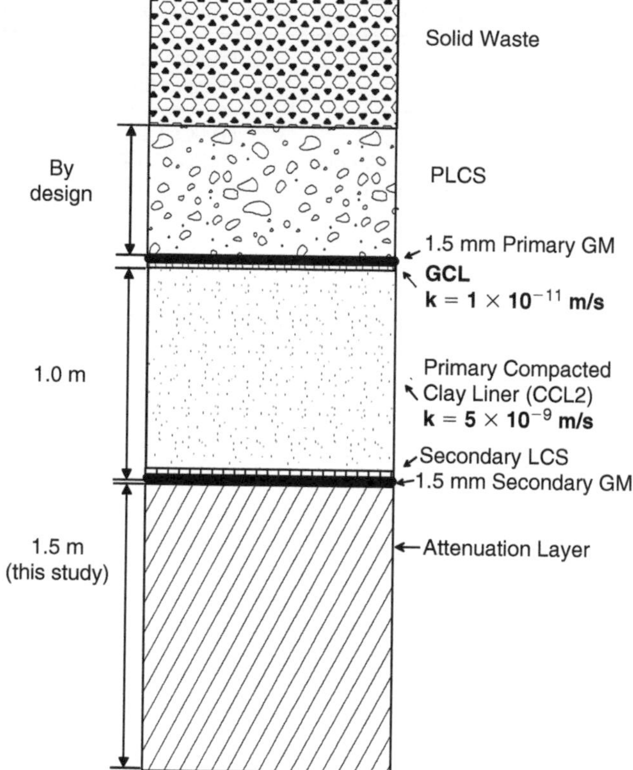


Figure 6. Alternative design to NSDEL (1997) regulations (CCL2 $k = 5 \times 10^{-9}$ m/s). *Some geotextiles, etc. have been left out of schematic for clarity.

system was functioning as designed (design leachate level of 0.3 m) and removing leachate for a period of 60 years. At this time, the primary leachate collection system underwent a gradual "failure"; a leachate mound instantaneously developing above the primary liner system to 50% of its maximum height (5 m). At 70 years, it reached its maximum height of 10 m. At 150 years, the primary geomembrane was assumed to instantaneously fail, causing the leachate mound height to decrease to the point where all infiltration coming into the landfill was migrating through the primary liner system. At 350 years, the secondary geomembrane and secondary leachate collection system was assumed to undergo instantaneous "failure" and all infiltration coming into the landfill (0.15 m/a) was being transferred into the underlying hydrogeological system. Leakage rates for the three cases considered were calculated using the methods outlined by Rowe, 1998.

Figure 8 shows modeling results for chloride. Chloride is representative of conservative inorganic contaminants since it undergoes no sorption/degradation in the landfill or during its migration through the landfill barrier system. For each of the three cases shown in Figure 8, aquifer chloride concentrations are well below the typical chloride drinking water objective of 250 mg/L for the size of the landfill considered. This is not surprising for a double geomembrane lined barrier system. As noted in Table 3, the diffusion coefficient of chloride is very low and for the low leakage rates that occur with only a few small holes in the geomembrane (even for a 10 m high mound) the impact in the aquifer prior to 350 years (secondary geomembrane failure) was approximately zero for each of the three cases. After secondary geomembrane failure, flushing of leachate from the landfill occurs causing a maximum chloride concentration of approximately 17 mg/L for each of the three cases to occur at approximately 360 years. It is interesting to note that the results of each of the three cases considered for chloride are essentially indistinguishable from each other (i.e. the graphs of cases 1 to 3 plot on top of each other). This implies that for the conditions examined herein, the use of a primary "composite clay" liner of GCL and CCL of marginal hydrau lic conductivity of

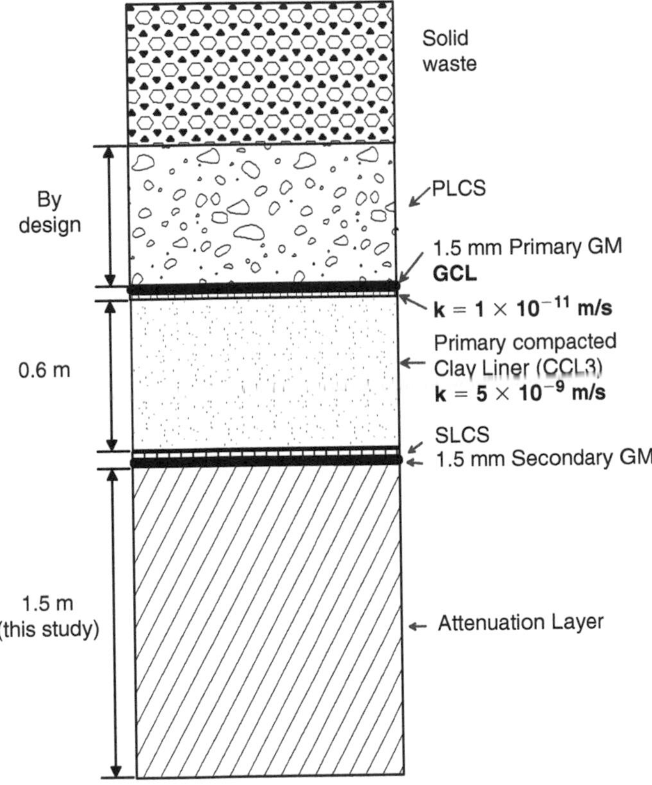


Figure 7. Alternative design to NSDEL (1997) regulations (CCL3 is 0.6 m thick and $k = 5 \times 10^{-9}$ m/s). *Some geotextiles, etc. have been left out of schematic for clarity.

Table 3. Barrier parameters used in contaminant migration assessment.

| | Prim. & sec. GM | CCL1 | CCL2 and CCL3 | GCL | Attenuation layer |
|---|---|---|---|---|---|
| Thickness (m) | 0.0015 | 1.0 | See Table 2 | 0.01 | 1.5 |
| Diffusion Coefficient Chloride ($m^2/s$) | $1 \times 10^{-13}$ | $6 \times 10^{-10}$ | $7 \times 10^{-10}$ | $2 \times 10^{-10}$ | $7 \times 10^{-10}$ |
| Diffusion Coefficient DCM ($m^2/s$) | $1 \times 10^{-12}$ | $6 \times 10^{-10}$ | $7 \times 10^{-10}$ | $2 \times 10^{-10}$ | $7 \times 10^{-10}$ |
| Henry's Coefficient Chloride, $S_{gf}$, (–) | $8 \times 10^{-4}$ | – | – | – | – |
| Henry's Coefficient DCM $S_{gf}$ (–) | 2.3 | – | – | – | – |
| No of holes/ha | 2.5 | – | – | – | – |
| Hole radius (mm) | 0.005 | – | – | – | – |
| Service life (a) | | $\infty$ | $\infty$ | $\infty$ | $\infty$ |
| Primary GM | 150 | – | – | – | – |
| Secondary GM | 350 | – | – | – | – |
| Hydraulic Conductivity (m/s) | – | $1 \times 10^{-9}$ | $5 \times 10^{-9}$ | $1 \times 10^{-11}$ | $1 \times 10^{-8}$ |
| Geomembrane-Clay Transmissivity ($m^2/s$) | – | $1.6 \times 10^{-8}$ | – | $2 \times 10^{-10}$ | $8.6 \times 10^{-8}$ |
| Sorption, $\rho_d K_d$ (–) | – | 0 | 0 | 0 | 0 |
| Porosity | – | 0.35 | 0.35 | 0.70 | 0.40 |

Table 4. Hypothetical landfill characteristics.

| Landfill Properties | |
|---|---|
| Length (m) | 1000 |
| Mass of Waste/unit area (t/m$^2$) | 15 |
| Proportion of chloride in waste (mg/kg) | 1800 |
| Proportion of DCM in waste (mg/kg) | 2.3 |
| Initial concentration in leachate | |
| Chloride, $c_o$ (mg/L) | 2500 |
| DCM, $c_o$, (mg/L) | 3.3 |
| Percolation through waste (m/a) | 0.15 |
| Chloride, $t_{1/2}$(a) | $\infty$ |
| DCM, half-life in landfill, $t_{1/2}$ (a) | 10 |
| DCM, half-life below GM, $t_{1/2}$ (a) | 50 |
| Aquifer Properties | |
| Thickness Modeled, (m) | 3 |
| Porosity (-) | 0.1 |
| Base Darcy Flux (horizontal), $v_b$ (m/a) | 1 |

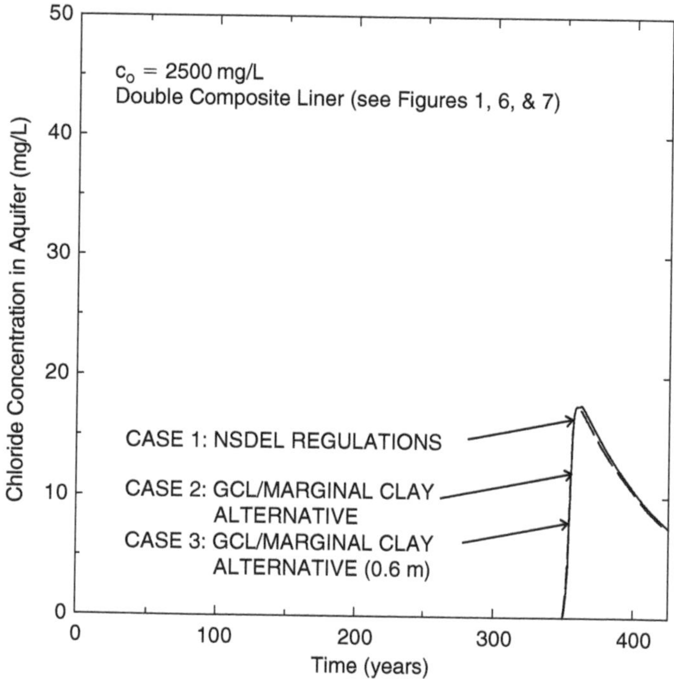


Figure 8. Comparison of aquifer chloride concentrations for NSDEL regulatory barrier and two alternatives (case 1 versus cases 2 and 3).

$5 \times 10^{-9}$ m/s is "equivalent" with respect to chloride contaminant transport relative to the regulatory specified Case 1 (where the compacted clay liner has a hydraulic conductivity of $1 \times 10^{-9}$ m/s).

What is also interesting is that for the same GCL/CCL primary composite liner used in case 3 (a smaller combined thickness of 0.6 m), an equivalent impact of chloride in the aquifer is produced. Essentially, the contaminant migration of chloride into the aquifer for these three cases is controlled by the relative low leakage through the secondary geomembrane. With this small leakage rate, the decrease in chloride mass within the landfill is controlled by the rate of leachate collection and hence by the time the secondary geomembrane fails at 350 years, the mass of chloride in the landfill

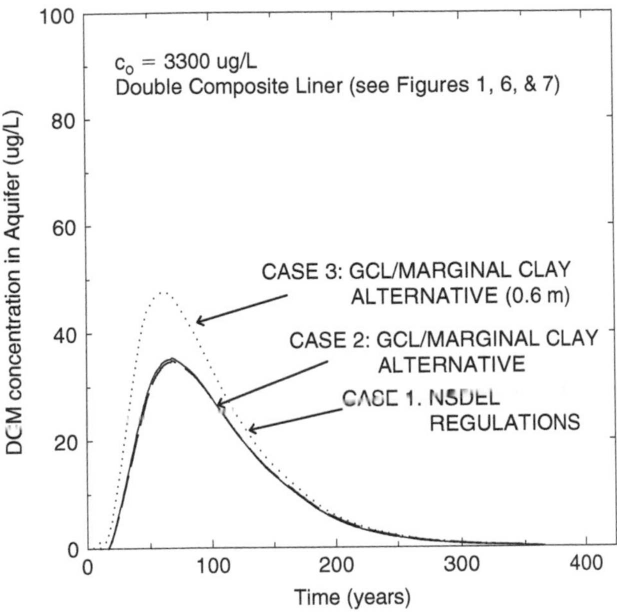


Figure 9. Comparison of aquifer DCM concentrations for NSDEL regulatory barrier and two alternatives (case 1 versus cases 2 and 3).

is similar for all three cases. Therefore at 350 years, the flushing of chloride out of the landfill (same rate for all three cases) results in similar chloride impacts. Since the diffusive flux of chloride is relatively low through the geomembrane, the very small amount of chloride that does migrate through the secondary geomembrane by leakage is diluted by the aquifer. If chloride was the only concern for contamination in this particular landfill, it could be argued Cases 2 and 3 are equivalent to the regulated Case 1. Moreover, the thickness of the marginal clay could be reduced at least to 0.6 m (probably even more) when utilized with the GCL and still be equivalent to the regulated Case 1. Case 2 and 3 could result in considerable cost savings for the landfill construction for the owners since both of these cases allow an on-site material to be used (cheaper compared to importing $1 \times 10^{-9}$ m/s CCL) while still being equivalent to the regulated barrier system prescribed by Case 1.

Volatile organic compounds such as DCM are often present at low levels in municipal solid waste leachate (Rowe, 1995). Small VOC molecules such as DCM will migrate more readily through HDPE GMs (Sangam & Rowe, 2001) than chloride. Contaminant migration of DCM through the double lined systems of Cases 1 to 3 will not be controlled by leakage through GMs as for chloride but by diffusion and degradation of the compound as it migrates through the barrier system. Figure 9 shows results for the three cases examined for DCM. Relative to the chloride results in Figure 8, it can be seen that peak DCM impact in the aquifer occurs relatively quickly (approximately 60 years) for all three cases examined. The regulated barrier system of Case 1 limits DCM peak concentrations to slightly less than the Maximum Acceptable Concentration (MAC) of 50 μg/L specified by many drinking water guidelines. Figure 9 also shows that Case 2 of the GCL/CCL primary liner system essentially plots on top of the regulated case of Case 1, indicating that for all practical purposes, the alternative GCL lined barrier system is equivalent to that of the regulated case for DCM (and chloride). It is interesting to note that the Case 3 barrier system results in peak DCM impacts in the aquifer approximately 40% greater than Cases 1 and 2. Because the thickness of the marginal CCL in case 3 is only 0.6 m, the diffusion of the DCM from the landfill to the aquifer occurs faster than Cases 1 and 2 which is noted by the quicker breakthrough time in the aquifer of DCM in case 3. This quicker rate of contaminant transport does not allow biodegradation to occur to the same extent of cases 1 and 2 and hence peak impacts of DCM in case 3 are higher. Therefore, unlike chloride, the three cases examined are not equivalent with respect to DCM. However, for the conditions modeled, it could be argued that although case 3 is not equivalent to cases 1 and 2 for DCM, the "thinner" GCL based liner system still ensures DCM aquifer impacts are less than 50 μg/L. It should be noted that even a 0.6 m thick layer of CCL with $k = 1 \times 10^{-9}$ m/s would also not be "equivalent" to Case 1 for DCM using similar modeling techniques. Regardless of the case 3 modeling results, it has been shown for these examples that if a GCL is used in combination with a compacted clay with marginal hydraulic conductivity (that does not meet regulations), the system can be considered equivalent with respect to DCM provided the total thickness of the primary composite liner does not change relative to that specified in the regulations.

Three cases have been presented which have demonstrated how one can assess equivalency of CCL and GCL based liner systems. Each equivalency assessment will depend on the parameters used for the analysis and hence the results presented for Cases 1 to 3. Similar methods can be utilized to examine other barrier systems (e.g., see Rowe and Brachman 2004). The important thing with any type of equivalency comparison is to consider the factors discussed above. Often the equivalency of a CCL and GCL based liner system will be more complicated than only considering that of advective travel times and obtaining GCL diffusion and sorption coefficients as described in this chapter will be required for any assessment.

## 7 CONCLUSIONS

A number of factors influencing solute contaminant transport through GCL based liner systems have been presented in this chapter. It was shown that contaminant transport through GCL based liner systems can be modeled using similar approaches to other porous media materials, with diffusion through a GCL being a necessary parameter to include in any contaminant transport assessment. Two different test apparatus, the SVD test apparatus and the CSD test apparatus can be utilized to estimate GCL diffusion coefficients, if required for design. A summary of diffusion test results involving various solutions and leachates have been provided as data for initial design purposes. Factors such as the bulk GCL void ratio, $e_B$, can have an effect on the GCL diffusion coefficient for a particular fluid in contact with the GCL. Generally speaking, higher values of $e_B$ for a GCL result in higher diffusion coefficients, all other factors being equal. Higher values of $e_B$ can result when the effective confining stress on the GCL is decreased during hydration. It was also discussed that although low temperatures tend to decrease the amount of BTEX sorption for a GCL, the lower diffusion coefficient at this same temperature results in a lower flux of BTEX contaminant across the GCL.

An example problem of a hypothetical landfill and hydrogeological environment was presented in section 6 in order to demonstrate an equivalency comparison between a CCL and GCL liner system. A regulatory specified double liner system for the province of Nova Scotia, Canada was examined as the base case in which a $1 \times 10^{-9}$ m/s compacted clay liner is specified for the primary GM composite liner system. It was shown that provided the total barrier system thickness remains the same for this particular system (with a marginal hydraulic conductivity clay being used in place of the $1 \times 10^{-9}$ m/s clay), a GM/GCL/marginal CCL composite liner system will be equivalent to that of the regulated case for both chloride and DCM contaminants. However, if this same primary barrier is reduced to 0.6 m thick (removing 0.4 m of marginal clay), 40% higher impacts in the underlying aquifer result for DCM (chloride impacts remain the same). This example highlighted the fact that equivalency comparisons between CCL and GCL base liner systems should incorporate a contaminant transport impact assessment using the method outlined by Rowe (1998). These results suggest that GCLs have the potential to provide equivalent contaminant impacts relative to a low hydraulic conductivity CCLs.

## REFERENCES

Barone, F.S., Rowe, R.K., & Quigley, R.M. 1992. A Laboratory estimation of diffusion and adsorption coefficients for several volatile organics in a natural clayey soil. *Journal of Contaminant Hydrology* 10: 225–250.

Bear, J. 1972. *Dynamics of Fluids In Porous Media*, American Elsevier.

Boyd, S.A., Mortland, M.M., & Chiou, C.T. 1988. Sorption characteristics of organic compounds on hexadecyltrimethylammonium-smectite. *Soil Science Society of America Journal* 52:652–657.

Donahue, R.B., Barbour, S.L., & Headley, J.V. 1999. Diffusion and adsorption of benzene in Regina clay. *Canadian Geotechnical Journal* 36: 430–442.

Dutt, G.R. & Low, P.F. 1962. Diffusion of alkali chlorides in clay-water systems, Soil Science, 93: 233–240.

Freeze, R.A. & Cherry, J.A. 1979. *Groundwater*, Prentice Hall.

Giroud, J.P., Badu-Tweneboah, K. & Soderman, K.L. 1997. Comparison of leachate flow through compacted clay and geosynthetic clay liners in landfill liner systems. *Geosynthetic International*, 4(3–4):391–431.

Goodall, D.C. & Quigley, R.M., 1977. Pollutant migration from two sanitary landfill sites near Sarnia, Ontario, *Canadian Geotechnical Journal* 14: 223–236.

Jaynes, W.F. & Vance, G.F. 1996. BTEX sorption by organo-clays: cosorptive enhancement and equivalence of interlayer complexes. *Soil Science Society of America Journal* 60:1742–1749.

Kemper, W.D. & van Schaik, J.C., 1966. Diffusion of salts in clay-water systems. Soil Science Soc. Amer. Proc., 30, 534–540.

Kim, J.Y., Edil, T.B., & Park, J.K. 2001. Volatile organic transport through compacted clay, *Journal of Geotechnical and Geoenvironmental Engineering*, ASCE, 127(2): 126–134.

King, K.S., Quigley, R.M., Fernandez, F., Reades, D.W., & Bacopoulos, A. 1993. Hydraulic conductivity and diffusion monitoring of the Keele Valley Landfill liner, Maple, Ontario, *Canadian Geotechnical Journal* 30: 124–134.

Lake, C.B. 2000. *Contaminant transport through geosynthetic clay liners and a composite liner system*. Ph.D. Dissertation, The University of Western Ontario, London, Ontario, Canada.

Lake, C.B. & Rowe, R.K. 1999. The role of contaminant transport in two different geomembrane/geosynthetic clay liner composite liner designs. *Geosynthetics '99 Proceedings*, Boston, USA: 661–670.

Lake, C.B. & Rowe, R.K. 2000a. Swelling of needle-punched, thermal locked GCLs. *Geotextiles and Geomembranes*, 18:77–101.

Lake, C.B. & Rowe, R.K. 2000b. Diffusion of sodium and chloride through geosynthetic clay liners. *Geotextiles and Geomembranes*, 18:103–131.

Lake, C.B. & Rowe, R.K. 2002. Migration of leachate constituents through geosynthetic clay liners by diffusion. *Proceedings of International Symposium on Clay Geosynthetic Barriers*, Nuremberg, Germany: 177–186.

Lake, C.B. & Rowe, R.K. 2004. Volatile organic compound diffusion and sorption coefficients for a needlepunched GCL. *Geosynthetics International (Special Issue on GCLs)*, 11(4): 257–272.

Lake, C.B. & Rowe, R.K. 2005. A comparative assessment of volatile organic compound (VOC) sorption to GCL bentonites, Geotextiles and Geomembranes, Vol. 23(4), 323–347.

Lake, C.B., Cardenas, C., Goreham, V., and Gagnon, G.G. 2007. Aluminium migration through a geosynthetic clay liner, *Geosynthetic International*, 14(4): 201–210.

Lange, K., Rowe, R.K., Jamieson, H. 2008. Diffusion of metals in geosynthetic clay liners, *Geosynthetics International*, (in press)

Li, H.M., Bathurst, R.J., & Rowe, R.K. 2002. Use of GCLs to control migration of hydrocarbons in severe environmental conditions. *International Symposium on Geosynthetic Clay Barriers*, Nuremberg, Germany, April: 187–198.

Lo, I.M.C 1992. *Development and evaluation of clay-liner materials For Hazardous Waste Sites*. Ph.D. Dissertation, The University of Texas at Austin.

Lo, I.M.C, Mak, R.K.M., & Lee, S.C.H. 1997. Modified clays for waste containment and pollutant attenuation. *Journal of Environmental Engineering*, ASCE 123:25–32.

Lorenzetti, R.J., Bartelt-Hunt, S.L., Burns, S.E., Smith, J.A. 2005. Hydraulic conductivities and effective diffusion coefficients of geosynthetic clay liners with organobentonite amendments. Geotextiles and Geomembranes, 23(5):385–400.

MOE, Ministry of the Environment, Ontario, Canada, 1998. Landfill Standards – A Guideline on the regulatory and approval requirements for new or expanding landfilling sites. Queens Printer for Ontario.

Myrand, D., Gilliam, R.W., Sudicky, E.A., O'Hannesin, S.F., & Johnson, J.L. 1992. Diffusion of volatile organic compounds in natural clay deposits: laboratory tests. *Journal of Contaminant Hydrology* 10:.159–177.

NSDEL, Nova Scotia Department of Environment and Labour 1997. Guidelines for disposal of contaminated solids in landfills. Government of Nova Scotia, Canada.

Ogata, A. & Banks, R.B. 1961. A solution of the differential equation of longitudinal dispersion in porous media, *US Geological Survey, Professions Paper*: 411–A.

Petrov, R.J. & Rowe, R.K. 1997. GCL – chemical compatibility by hydraulic conductivity testing and factors impacting its performance. *Canadian Geotechnical Journal* 34(6):863–885.

Petrov, R.J., Rowe, R.K., & Quigley, R.M. 1997. Selected factors influencing GCL hydraulic conductivity. *Journal of Geotechnical and Geoenvironmental Engineering, ASCE* 123:683–695.

Richardson, G.N. 1997. GCLs: Alternative subtitle D liner systems, *Geotechnical Fabrics Report*, May: 36–42.

Rowe, R.K., 1995. Leachate characterization for MSW landfills. *Proceedings Sardinia 95, Fifth International Landfill Symposium*, S. Margherita di Pula, Cagliari, Italy 2: 327–344.

Rowe, R.K. 1998. Geosynthetics and the minimization of contaminant migration through barrier systems beneath solid waste – keynote lecture. *Proceedings, 6th International Conference on Geosynthetics, Atlanta*: 1: 27–102.

Rowe, R.K. 2005. Long-term performance of contaminant barrier systems. 45th Rankine Lecture, Geotechnique, **55** (9): 631–678.

Rowe, R.K. & Barone, F.S. 1991. Diffusion tests for chloride and dichloromethane In Halton till. *Report of the Geotechnical Research Centre*, The University of Western Ontario, London, Ontario.

Rowe, R.K. & Booker, J.R., 1984. 1-D Pollutant migration in soils of finite depth, *Journal of Geotechnical* Engineering, ASCE: 111(4): 479–499.

Rowe, R.K. & Booker, J.R. 2004. POLLUTE v.7 – 1-D pollutant migration through a non-homogeneous soil©. Distributed by GAEA GAEA Technologies Ltd., 87 Garden St., Whitby, Ontario, Canada L1N 9E7, mfraser@gaea.ca, www.gaea.ca

Rowe, R.K. and Brachman, R.W.I. 2004. Assessment of equivalency of composite liners. *Geosynthetics International*, 11(4): 273–286

Rowe, R.K. & Lake, C.B. 1999. Geosynthetic clay liner research and design applications. *Proceedings of 7th International Landfill Symposium*. S. Margherita di Pula, Cagliari, Sardinia, October, 3:181–188.

Rowe, R.K., Hrapovic, L., Kosaric, N. & Cullimore, D.R. 1997a. Anaerobic degradation of dichloromethane diffusing through clay. *Journal of Geotechnical and Geoenvironmental Engineering*, 123(12):1085–1095.

Rowe, R.K., Lake, C., von Maubeuge, K. & Stewart, D. 1997b. Implications of diffusion of chloride through geosynthetic clay liners, *Geoenvironment '97*, Melbourne, Australia, November: pp. 295–300.

Rowe, R.K., Lake, C.B. & Petrov, R.J. 2000. Apparatus and procedures for assessing inorganic diffusion coefficients through geosynthetic clay liners, *ASTM Geotechnical Testing Journal*: 23(2):206–214.

Rowe, R.K., Mukunoki, T., & Sangam, P.H 2005. Effect of temperature on BTEX diffusion and sorption for a geosynthetic clay liner, *ASCE Journal of Geotechnical and Geo-environmental Engineering,* 131(10): 1211–1221.

Rowe, R.K., Quigley, R.M., Brachman, R.W.I. & Booker, J.R. 2004. *Barrier Systems for Waste Disposal Facilities*, E & FN Spon, London, 579p

Sangam, H.P. & Rowe, R.K, 2001. Migration Of Dilute Aqueous Organic Pollutants Through HDPE Geomembranes. *Geotextiles and Geomembranes* 19(6):329–357.

Smith, J.A. & Jaffe, P.R., 1994. Benzene transport through landfill liners containing organophilic bentonite. *Journal of Environmental Engineering*, ASCE 120(6):1559–1577.

Talbot, A. 1979. The accurate numerical integrations of Laplace transforms, *Journal of Institute Mathematics Applications*, 23: 97–120.

USEPA (United States Environmental Protection Agency) 1998. Solid waste disposal facility criteria, Technical Manual, EPA530-R-93-017.

Voice, T.C. 1988. *Activated carbon adsorption, in Standard Handbook of Hazardous Waste Treatment and Disposal, H.H. Freeman (ed)*: 6.3–6.21. McGraw-Hill.

Wilson-Fahmy, R.F. & Koerner, R.M. 1995. Leakage rates through holes in geomembranes overlying geosynthetic clay liners, *Proceedings of Geosynthetics '95, Industrial Fabrics Association International:* 655–668.

Yong, R.N., Mohamed, A.M.O., & Warkentin, B.P. 1992. *Principals of Contaminant Transport in Soils, Developments in Geotechnical Engineering, 73*. Elsevier.

# CHAPTER 6

# Chemico-osmosis and solute transport through Geosynthetic clay liners

A. Dominijanni & M. Manassero
*Politecnico di Torino, Torino, Italy*

## 1 INTRODUCTION

Geosynthetic clay liners (GCLs) are hydraulic barriers consisting of a thin layer of bentonite sandwiched between two geotextiles or bonded to a geomembrane. GCLs are used in waste containment facilities because of their low hydraulic conductivity to water ($<10^{-10}$ m/s) and ease of installation. For GCLs without a geomembrane, the bentonite is responsible for the low hydraulic conductivity. Most bentonites used for GCLs contain at least 70% montmorillonite, a clay mineral characterized by high surface area and high negative electric charge, due to isomorphous substitution of lower-valence cations for higher-valence cations within the crystalline structure. The negative charge of clay particles is compensated by the cations (typically sodium) that are contained in the pore solution. The cation exchange capacity (CEC), usually expressed as milliequivalents per 100 grams (meq/100 g), represents the excess of surface charge per unit weight of the minerals. The CEC for montmorillonite typically varies between 80–150 meq/100 g (Mitchell & Soga 2005).

The cations in the pore solution are "attracted" to the clay particles, whereas the anions are repelled, in order to maintain electro-neutrality. The cations are partially located in the Stern layer directly at the surface of the minerals and partially in the diffuse double layer (DDL) or Gouy-Chapman theory. The cations that are located in the Stern layer are not mobile and may be considered as part of the solid skeleton.

When a GCL is placed in contact with a bulk solution, the ion concentrations in the pore solution are discontinuous with respect to the ion concentrations in the bulk solution due to the presence of the charge of the solid skeleton. The concentration of the cations (i.e. counter-ions) is higher than that of the bulk solution, whereas the concentration of the anions (i.e. co-ions) is lower. In Figure 1, a GCL is placed in contact with two bulk solutions containing a 1:1 salt consisting of mono-valent ions (e.g. NaCl or KCl). Due to the electro-neutrality requirement, the cation concentration is equal to the anion concentration in the bulk solutions. At the boundaries with the GCL, the ion concentrations are discontinuous. Within the GCL, electro-neutrality must take into account the charge of the solid skeleton or fixed charge (FC) that is negative. As a result, the cation concentration is higher than the salt concentration in the bulk solution, whereas the anion concentration is lower.

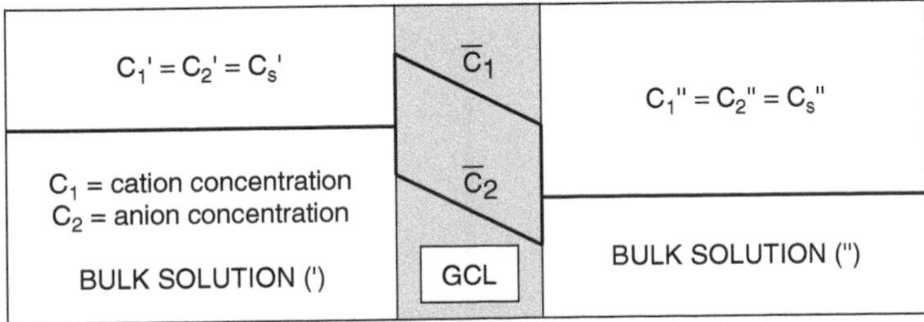


Figure 1. Ion concentration profiles within a GCL in contact with two bulk solutions. $C_s$ is the concentration of a (1:1) electrolyte in the external bulk solutions; $\overline{C}_1, \overline{C}_2$ are the concentration of the cation and the anion, respectively, within the GCL.

The ion partition mechanisms associated with the momentum transfer between the components of the solution and the condition of absence of electric current are responsible for the restricted movement of salt solutions through semi-permeable membranes and chemico-osmosis, i.e. the volumetric flux of the solution in response to a salt concentration gradient.

## 2 THERMODYNAMICS OF IRREVERSIBLE PROCESSES

The coupled transport theory represents the generalization of the classic advective-diffusive theory adopted for modelling the migration of solutes through uncharged porous media. This theory is generally based on the formalism of the Thermodynamics of Irreversible Processes (Staverman 1952, De Groot & Mazur 1962, Katchalsky & Curran 1965, Mitchell & Soga 2005).

The approach developed by Staverman (1952) is based on the strongly simplifying assumption that the semipermeable membrane can be idealized as a "discontinuity" between two bulk solutions. In such an approach, all the variables are referred to the two compartments in contact with the membrane, so it is not necessary to evaluate the phenomena occurring within the membrane. The main advantage of such an approach is that the problem of modelling transport mechanisms through the porous medium is avoided.

In the discontinuous version of the Thermodynamics of Irreversible Processes under isothermal conditions, the membrane is considered as a transition region between two homogeneous compartments having the same temperature, and the differences in the thermodynamic potentials across the membrane that represent the driving forces responsible for the corresponding flows are assumed to be small (see Fig. 2). Under these conditions, and in the absence of chemical reactions, the dissipation function, $\Phi$, defined as the rate of entropy production multiplied by absolute temperature, may be expressed as follows (De Groot & Mazur 1962, Katchalsky & Curran 1965):

$$\Phi = \sum_{i}^{N} J_i \Delta\mu_i^{ec} \tag{1}$$

where N = number of the components of the solution, $J_i$ = molar mass flow of the i-th component and $\Delta\mu_i^{ec}$ = electrochemical potential difference of the i-th component. The electrochemical potential differences are intended as the thermodynamic forces driving the mass fluxes. The finite differences are defined as the values in the "left" bulk solution minus those in the "right" bulk solution, as indicated in Figure 2.

The conventional assumption of the Thermodynamics of Irreversible Processes is that the processes under consideration are sufficiently slow as to make all the fluxes linear functions of all the forces operating in the system. The resulting system of N equations is given as follows:

$$J_i = \sum_{j} L_{ij} \cdot \Delta\mu_i^{ec} \quad i = 1,2,\ldots, N \tag{2}$$

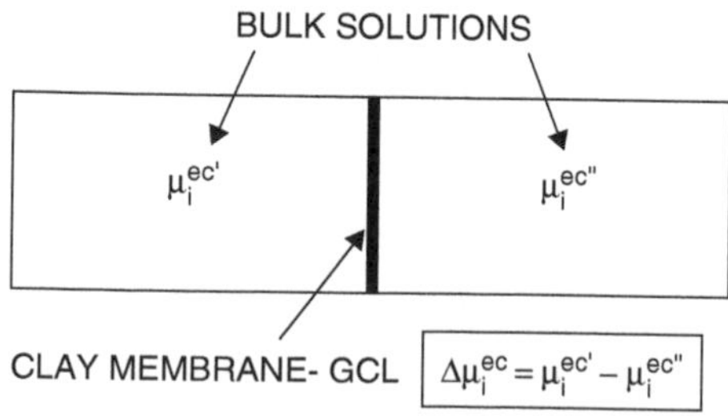


Figure 2. Reference scheme for the application of the discontinuous version of the Thermodynamics of Irreversible Processes.

where $L_{ij}$ = phenomenological coefficients. Based on Onsager's law of reciprocity (Onsager 1931a,b), the matrix of the coefficients $L_{ij}$ is symmetrical, such that

$$L_{ij} = L_{ji}. \tag{3}$$

The typical analysis of osmosis and solute transport through a membrane is restricted to the case of a solution containing a binary electrolyte consisting of a counter-ion (charge polarity opposite that of the membrane charge) and a co-ion (same charge polarity as the membrane) (Katchalsky & Curran 1965, Groenevelt & Bolt 1969), or:

$$(\text{Counter} - \text{ion})^{\nu_1}_{z_1}(\text{Co} - \text{ion})^{\nu_2}_{z_2} \tag{4}$$

Where ($\nu_1$, $z_1$) and ($\nu_2$, $z_2$) are the stoichiometric coefficient and the electrochemical valence of the counter-ion (index 1) and co-ion (index 2), respectively.

The salt in solution is assumed to be completely dissociated such that

$$(\text{Counter} - \text{ion})^{\nu_1}_{z_1}(\text{Co} - \text{ion})^{\nu_2}_{z_2} \rightarrow \nu_1(\text{Counter} - \text{ion})^{z_1} + \nu_2(\text{Co} - \text{ion})^{z_2}.$$

Clay particles generally have a negative charge such that the counter-ions are cations (positive charged ions) and the co-ions are anions (negative charged ions).

The ionic concentrations, $C_i$, in a bulk solution are related to the salt concentration, $C_s$, as follows

$$C_i = \nu_i C_s. \tag{5}$$

Electro-neutrality in the external or bulk solutions implies that:

$$z_1 C_1 + z_2 C_2 = 0. \tag{6}$$

Using Equation 5, the electro-neutrality condition (Eq. 6) reduces to:

$$z_1 \nu_1 + z_2 \nu_2 = 0. \tag{7}$$

For a solution containing a binary electrolyte, the dissipation function (Eq. 1) is given by

$$\Phi = J_w \Delta\mu_w + J_1 \Delta\mu_1^{ec} + J_2 \Delta\mu_2^{ec} \tag{8}$$

where ($J_w$, $\Delta\mu_w$), ($J_1$, $\Delta\mu_1^{ec}$), ($J_2$, $\Delta\mu_2^{ec}$) are the molar mass flux and the electrochemical potential difference of the solvent (water), the counter-ion and the co-ion, respectively. In Equation 8, the electrochemical potential difference of the solvent has been assumed to be coincident with the chemical potential difference because the solvent molecules do not have an electrical charge.

The chemical potential difference of the solvent may be related to the hydraulic pressure difference, $\Delta P$, and the osmotic pressure difference, $\Delta\Pi$, by means of the following relation (Katchalsky & Curran 1965):

$$\Delta\mu_w = \overline{V}_w(\Delta P - \Delta\Pi) \tag{9}$$

where $\overline{V}_w$ = partial molar volume of the solvent.

The electrochemical potential differences of the ions are given by (Katchalsky & Curran 1965):

$$\Delta\mu_i^{ec} = \overline{V}_i \Delta P + \Delta\mu_i^c + z_i F \Delta\varphi \tag{10}$$

where $\overline{V}_i$ = partial molar volume of the i-th ion, $\Delta\mu_i^c$ = chemical part of the chemical potential difference of the i-th ion, F = Faraday constant ($9.648\,7010^4$ C mol$^{-1}$) and $\Delta\varphi$ = electrical potential difference.

The choice of the generalised forces and fluxes is to some extent arbitrary and based on convenience, i.e., based on condition that the product of each conjugate pair has the correct dimensions and the dissipation function is invariant. Substituting Equations 9 and 10 into Equation 8, the dissipation function becomes:

$$\Phi = q\Delta P - J_w \overline{V}_w \Delta\Pi + J_1 \Delta\mu_1^c + J_2 \Delta\mu_2^c + I_e \Delta\varphi \tag{11}$$

where $q = J_w\overline{V}_w + J_1\overline{V}_1 + J_2\overline{V}_2$ is the total volumetric flux of the solution (or Darcy's velocity of the solution) and $I_e = F(z_1J_1 + z_2J_2)$ is the electric current density. The expression of the dissipation function given by Equation 11 is particularly convenient when the bulk solutions are not closed electrically, so that the electric current through the membrane has to equal zero ($I_e = 0$). If the electric current is zero, the last term in Equation 11 drops out and the ionic fluxes are related as follows:

$$\frac{J_1}{\nu_1} = \frac{J_2}{\nu_2} = J_s \tag{12}$$

where $J_s$ = molar mass flux of the salt. In the absence of an applied electric current, the dissipation function may be written in the following form:

$$\Phi = q\Delta P - J_w\overline{V}_w\Delta\Pi + J_s\Delta\mu_s^c \tag{13}$$

where $\Delta\mu_s^c = \nu_1\Delta\mu_1^c + \nu_2\Delta\mu_2^c$ is the chemical part of the chemical potential difference of the salt. The osmotic pressure, $\Delta\Pi$, is related to $\Delta\mu_s^c$ as follows:

$$\Delta\Pi = C_{1,ave}\Delta\mu_1^c + C_{2,ave}\Delta\mu_2^c = C_{s,ave}\Delta\mu_s^c \tag{14}$$

where $C_{1,ave}$, $C_{2,ave}$ and $C_{s,ave}$ are the average concentrations between the two external solutions of the counter-ion, the co-ion and the salt, respectively.

The average concentration $C_{1,ave}$, $C_{2,ave}$ and $C_{s,ave}$ are not uniquely defined since various types of averages may be used to evaluate them (e.g. arithmetic, geometric, logarithmic, harmonic, etc.). However, since the concentration differences between the two external solutions are assumed to be small, the difference between the various types of average also should be small. Arithmetic averages will always be used herein to determine the average concentrations. The average salt concentration is given by:

$$C_{s,ave} = \frac{C_s' + C_s''}{2} \tag{15}$$

where $C_s'$ and $C_s''$ are the salt concentrations in the left and the right cell, respectively.

For dilute solutions, $J_w\overline{V}_w \approx q$ and Equation 11 can be rewritten in the following form:

$$\Phi = q\Delta P + J^d\Delta\mu_s^c \tag{16}$$

where $J_s^d = J_s - qC_{s,ave}$ represents the diffusive part of the salt mass flux. In Equation 16, the number of independent forces and fluxes is reduced from 3 (as in Eq. 8) to 2 under the condition of no-electric current. Transforming Equation 8 into Equation 16, we have changed the generalised flows and forces so as to obtain a more suitable form of the dissipation function for practical applications. In Equation 16, the flows are the volumetric flux, q, and the salt diffusive molar mass flux, $J_s^d$, and the forces are the hydraulic pressure difference, $\Delta P$, and the chemical potential difference of the salt, $\Delta\mu_s^c$. For dilute solutions, $\Delta\mu_s^c$ may be related to the salt concentration difference, $\Delta C_s$, as follows (Katchalsky &d Curran 1965):

$$\Delta\mu_s^c = (\nu_1 + \nu_2)RT\frac{\Delta C_s}{C_{s,ave}} \tag{17}$$

where R = universal gas constant (8.3145 J/mol K) and T = absolute temperature.

Using the generalised flows and forces comparing in Equation 16, the phenomenological equations are:

$$q = \alpha_{11}\Delta P + \alpha_{12}\Delta\mu_s^c \tag{18a}$$

$$J_s^d = \alpha_{21}\Delta P + \alpha_{22}\Delta\mu_s^c. \tag{18b}$$

Based on Onsager's law of reciprocity, the matrix of the coefficients $\alpha_{ij}$ is symmetrical, such that:

$$\alpha_{12} = \alpha_{21}. \tag{19}$$

The phenomenological equations given by Equations 18a,b describe the transport of a solution containing a binary electrolyte through a semipermeable membrane. The phenomenological coefficients $\alpha_{ij}$ have to be measured by means of appropriate experiments. In general, these coefficients may depend on the hydraulic pressure and the salt concentration, in addition to the physical properties of the membrane, such as the porosity, the pore size and the membrane charge density. From a purely phenomenological point of view, only a systematic experimental investigation may characterize a membrane in a complete way, relating the phenomenological parameters to all varying quantities of the system.

The phenomenological coefficients $\alpha_{ij}$ have some special values corresponding to the limiting behaviour of the membrane. If the membrane has no selective ability with respect to the solute, the coefficient $\alpha_{12}$ is equal to zero. In this case, the membrane is not semi-permeable and the flows of solvent and solute are not coupled, such that the volumetric flux is given by:

$$q = \alpha_{11}\Delta P \quad @\, \alpha_{12} = 0 \tag{20}$$

and the salt flux by:

$$J_s = qC_{s,ave} + J_s^d = qC_{s,ave} + \alpha_{22}\Delta\mu_s^c \quad @\, \alpha_{12} = 0. \tag{21}$$

For dilute solutions, $\Delta\mu_s^c$ is given by Equation 17 and the salt flux (Eq. 21) may be written as follows:

$$J_s = qC_{s,ave} + \frac{\alpha_{22}(\nu_1 + \nu_2)RT}{C_{s,ave}}\Delta C_s \quad @\, \alpha_{12} = 0. \tag{22}$$

Comparing Equations 20 and 22 with the flux equations of the advective-diffusive transport theory, the phenomenological coefficients $\alpha_{11}$ and $\alpha_{22}$ may be related to the parameters usually used in environmental geotechnical engineering for the special case of non-membrane behaviour:

$$\alpha_{11} = \frac{k_h}{\gamma_w L_h} \tag{23a}$$

$$\alpha_{22} = nD_s^* \frac{C_{s,ave}}{L_h(\nu_1 + \nu_2)RT} \quad @\, \alpha_{12} = 0 \tag{23b}$$

where $n$ = connected porosity, $k_h$ = hydraulic conductivity, $\gamma_w$ = unit weight of the solution, $L_h$ = thickness of the membrane, $D_s^* = \tau_m D_{s,0}$ = effective salt diffusion coefficient, $\tau_m$ = matrix tortuosity factor and $D_{s,0}$ = free-solution salt diffusion coefficient. $\tau_m$ represents the tortuous nature of the actual diffusive pathways through the porous medium due to the geometry of the interconnected pores. The free-solution salt diffusion coefficient is given by:

$$D_{s,0} = \frac{(\nu_1 + \nu_2)D_{1,0}D_{2,0}}{\nu_1 D_{2,0} + \nu_2 D_{1,0}} = \frac{(|z_1| + |z_2|)D_{1,0}D_{2,0}}{|z_1|D_{1,0} + |z_2|D_{2,0}} \tag{24}$$

where $D_{1,0}$ and $D_{2,0}$ are the free-solution diffusion coefficients of the counter-ion and of the co-ion, respectively.

The relation between $\alpha_{11}$ and $k_h$ (Eq. 23a) is general and is valid also in the case of membrane behaviour, because the hydraulic conductivity of a semipermeable membrane is defined as ratio between the volumetric flux and the hydraulic pressure (head) difference under no-electric current and no-concentration difference conditions. On the contrary, the relation between $\alpha_{22}$ and $D_s^*$ (Eq. 23b) is valid only in the case of non-membrane behaviour, i.e., in the absence of osmotic effects.

When the membrane is an "ideal" or "perfect" semipermeable membrane, the salt flux is completely hindered:

$$J_s = qC_{s,ave} + J_s^d = (\alpha_{11}C_{s,ave} + \alpha_{12})\Delta P + (\alpha_{12}C_{s,ave} + \alpha_{22})\Delta\mu_s^c = 0. \tag{25}$$

Equation 25 implies the following conditions:

$$\alpha_{12} = -\alpha_{11}C_{s,ave} \quad @\ \text{non-membrane behaviour} \tag{26a}$$

$$\alpha_{22} = -\alpha_{12}C_{s,ave} \quad @\ \text{non-membrane behaviour} \tag{26b}$$

Using Equation 26a, the volumetric flux, q, is given by:

$$q = \alpha_{11}(\Delta P - \Delta\Pi) \quad @\ \text{non-membrane behaviour} \tag{27}$$

The actual clay membranes are able to restrict the passage of the solute only partially. In order to characterise completely the transport of the solution through a clay membrane, the three coefficients $\alpha_{11}$, $\alpha_{12}$ and $\alpha_{22}$ should be measured by means of appropriate experiments. However, some other phenomenological coefficients, related to the $\alpha_{ij}$ coefficients by simple relations, are more convenient for the experimental measurements.

The volumetric flux (Eq. 18a) may be rewritten as follows:

$$q = \frac{k_h}{\gamma_w L_h}(\Delta P - \omega \Delta \Pi) \tag{28}$$

where $k_h$ is related to $\alpha_{11}$ by Equation 23a and

$$\omega = -\frac{\alpha_{12}}{\alpha_{11} C_{s,ave}} \tag{29}$$

is the chemico-osmotic efficiency coefficient. The chemico-osmotic efficiency coefficient is also called reflection coefficient and is indicated using the Greek letter $\sigma$. The use of the symbol $\omega$ is preferred in the engineering literature because the symbol $\sigma$ typically represents stress or electric conductance (Malusis et al. 2003).

Using Equation 18a, the hydraulic pressure difference may be expressed as a function of the volumetric flux:

$$\Delta P = \frac{1}{\alpha_{11}}\left(q - \alpha_{12}\frac{\Delta \Pi}{C_{s,ave}}\right). \tag{30}$$

Substituting Equation 30 into Equation 18b, the salt flux, $J_s$, is given by:

$$J_s = (1-\omega) q C_{s,ave} + \frac{n D_\omega^*}{L_h}\Delta C_s \tag{31}$$

where

$$\begin{aligned} D_\omega^* &= \left(\alpha_{22} - \frac{\alpha_{12}^2}{\alpha_{11}}\right)\frac{RT(\nu_1+\nu_2)L_h}{n C_{s,ave}} \\ &= \alpha_{22}\frac{RT(\nu_1+\nu_2)L_h}{n C_{s,ave}} - \omega^2 \frac{k_h}{\gamma_w n} C_{s,ave} RT(\nu_1+\nu_2) \end{aligned} \tag{32}$$

is the osmotic effective salt diffusion coefficient.

The chemico-osmotic efficiency coefficient, $\omega$, may be determined from Equation 28 using one of the following experimental condition:

$$\omega = \left(\frac{\Delta P}{\Delta \Pi}\right)_{q=0, I_e=0} \tag{33a}$$

$$\omega = \frac{L_h \gamma_w}{k_h}\left(\frac{q}{\Delta \Pi}\right)_{\Delta P=0, I_e=0} \tag{33b}$$

The osmotic effective salt diffusion coefficient is defined by the following experimental condition:

$$D_\omega^* = \frac{L_h}{n}\left(\frac{J_s}{\Delta C_s}\right)_{q=0, I_e=0}. \tag{34}$$

With respect to the more conventional case of non-membrane behaviour, the definition of the diffusive coefficient is more ambiguous, because the two boundary conditions $q=0$ and $\Delta P = 0$ cannot be satisfied at the same time in a semipermeable membrane. In Equation 31, it is implicitly stated that we have defined the diffusive coefficient as the coefficient that we can measure when the volumetric flux is zero ($q=0$).

Using the condition $\alpha_{12}=0$ and Equation 23b, a non-semipermeable membrane is characterised by the following special values of the parameters $\omega$ and $D_\omega^*$:

$$\omega = 0 \quad D_\omega^* = D_s^* \tag{35}$$

An ideal membrane, instead, is characterised by the following special values of the parameters ω and $D^*_\omega$:

$$\omega = 1 \quad D^*_\omega = 0. \tag{36}$$

The chemico-osmotic efficiency coefficient, ω, of a clay membrane is a measure of its effectiveness in causing hydraulic flow under an osmotic gradient (see Eq. 28) and of its ability to prevent the passage of ions (see Eq. 31). Such a coefficient is equal to zero ($\omega = 0$) when the membrane is not semipermeable and assumes a maximum value of unity ($\omega = 1$) when the membrane is ideal or perfect. The values of ω are frequently assumed to vary from zero to unity, although there are not thermodynamic restrictions to such a variation such that low negative values also have been measured (Kemper & Quirk 1972).

Based on Equations 35 and 36, the osmotic effective salt diffusive coefficient varies with ω. We will discuss the dependency of $D^*_\omega$ on ω within GCLs on the basis of experimental evidence and theoretical considerations.

## 3 EXPERIMENTAL DATA

The ability of bentonites to behave as semipermeable membranes has been evaluated by means of the experimental measurement of the chemico-osmotic efficiency coefficient, ω (Kemper 1960, Kemper & Rollins 1966, Groenevelt & Bolt 1969, Kemper & Quirk 1972, Hanshaw & Coplen 1973, Kharaka & Berry 1973). Malusis et al. (2001) developed a novel testing apparatus able to impose a condition of no-volumetric flux ($q = 0$) through a GCL specimen in contact with two external solutions maintained at constant salt concentration. An evaluation of ω through Equation 33a can be made by measuring the hydraulic pressure difference that has to be imposed at the boundaries of the specimen in order to prevent the volumetric flux of the solution. Also, the osmotic effective salt diffusion coefficient, $D^*_\omega$, can be determined from salt mass flux measurements using Equation 34.

The experimental results of Figure 3, obtained by Malusis & Shackelford, (2002a) on GCL specimens, indicate clearly that ω decreases as the average salt concentration, $C_{s,ave}$, increases for a given porosity, n. Such results confirm the experimental observations of Kemper & Rollins (1966) reported in Figure 4 and relative to bentonite specimens. The data in Figure 4 indicate also that ω decreases with an increase in cation charge ($Ca^{2+}$ versus $Na^+$) for a given porosity and average salt concentration.

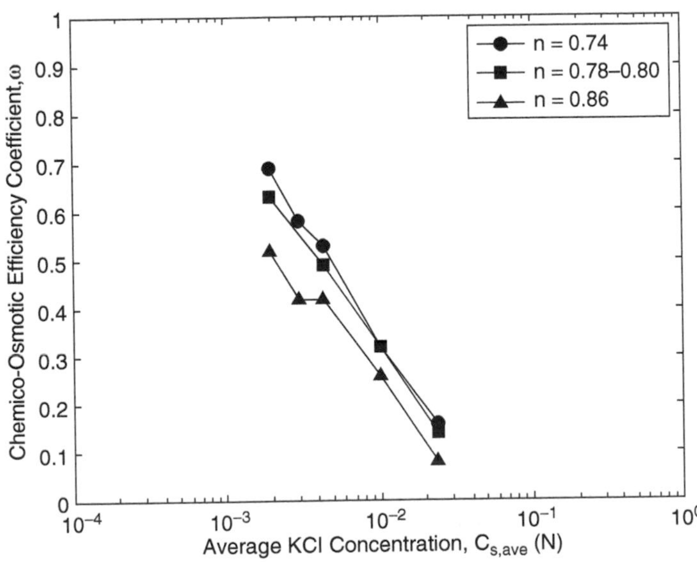


Figure 3. Chemico-osmotic efficiency coefficients as a function of average salt concentration across the specimen and the specimen porosity (n) for a geosynthetic clay liner (data from Malusis & Shackelford 2002a).

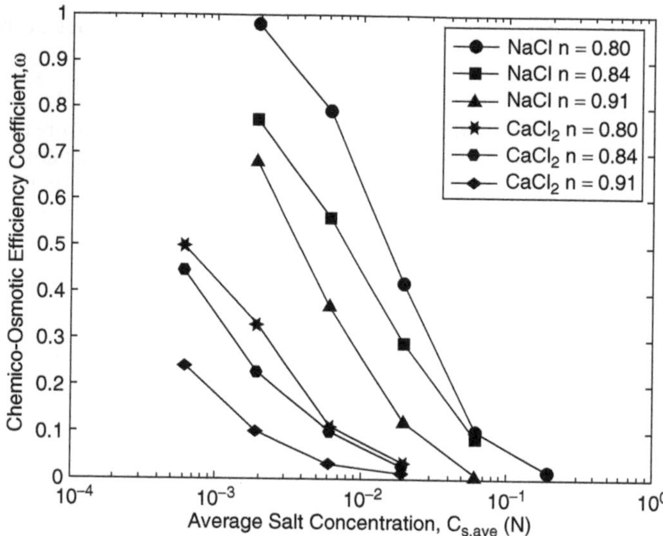


Figure 4. Chemico-osmotic efficiency coefficients as a function of average salt concentration across the specimen and the specimen porosity (n) for bentonite specimens (data from Kemper & Rollins 1966).

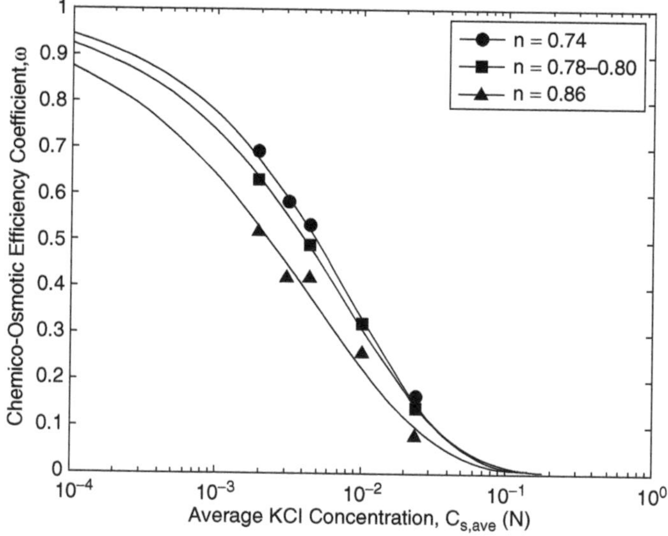


Figure 5. Exponential regression of measured chemico-osmotic efficiency coefficients versus average salt concentration for a geosynthetic clay liner (data from Malusis & Shackelford 2002a).

A possible function for fitting experimental data is:

$$\omega(C_{s,ave}) = \exp(-a \cdot C_{s,ave}^{b}) a, b \geq 0 \tag{37}$$

The advantages of this functional equation are that the function is continuous over the entire domain of definition of $C_{s,ave}$ and explicitly satisfies the following limits:

$$\lim_{C_{s,ave} \to 0} \omega(C_{s,ave}) = 1; \quad \lim_{C_{s,ave} \to \infty} \omega(C_{s,ave}) = 0. \tag{38}$$

Exponential fits to the measured data of Figures 3 and 4 are shown in Figure 5 and 6, and the regressed parameters are summarised in Table 1.

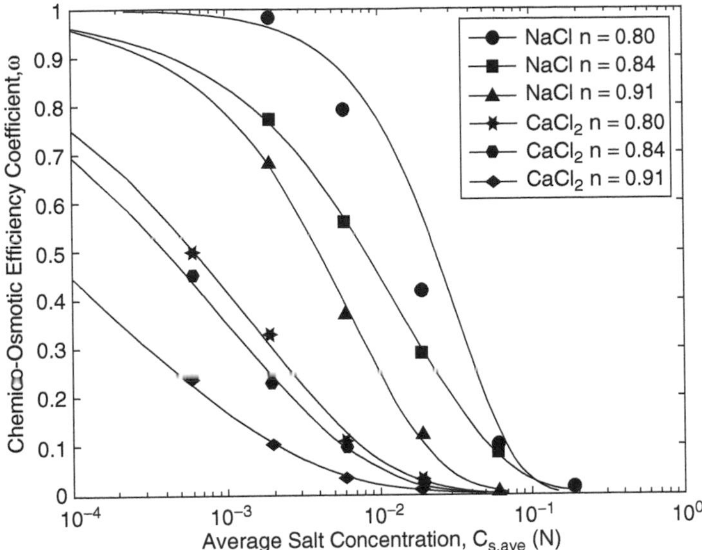


Figure 6. Exponential regression of measured chemico-osmotic efficiency coefficients versus average NaCl and $CaCl_2$ concentration for bentonite specimens (data from Kemper & Rollins 1966).

Table 1. Results of regression analyses.

| Salt | Porosity n | Regression Coefficients for: $\omega(C_{s,ave}) = \exp(-a \cdot C_{s,ave}^{b})^{(1)}$ a | b | $r^2$ |
|---|---|---|---|---|
| NaCl | 0.80 | 49.269 | 1.1410 | 0.9376 |
| | 0.84 | 15.824 | 0.6508 | 0.9989 |
| | 0.91 | 43.118 | 0.7487 | 0.9983 |
| KCl | 0.74 | 20.263 | 0.6331 | 0.9944 |
| | 0.79(2) | 16.943 | 0.5802 | 0.9985 |
| | 0.86 | 16.455 | 0.5219 | 0.9671 |
| $CaCl_2$ | 0.80 | 24.274 | 0.4821 | 0.9935 |
| | 0.84 | 23.745 | 0.4530 | 0.9967 |
| | 0.91 | 18.870 | 0.3417 | 0.9860 |

(1) ω = chemico-osmotic efficiency coefficient; $C_{s,ave}$ is in normality (N); $r^2$ = coefficient of determination.
(2) Average of range: $0.78 \leq n \leq 0.80$

Because membrane behaviour is complex, depending on such factors as the relative mobility of the ions, the valences of the ions, the charge density of the membrane, and the porosity of the membrane, the most appropriate form for the functional relationship between ω and $C_{s,ave}$ is not so straightforward. For example, Equation 37 is based on the assumption that ω cannot be negative; however, low negative values for ω have been reported in the literature (e.g., Kemper and Quirk 1972).

The results obtained by Malusis & Shackelford (2002a, b) also provide the following conclusions: (i) the hydraulic conductivity, $k_h$, corresponding to the low values of the salt concentration for which an osmotic effect was observed, was approximately constant; (*ii*) the osmotic effective salt diffusion coefficient, $D^*_\omega$, is dependent, through the average salt concentration, on the chemico-osmotic efficiency coefficient, ω.

The definition of an ideal membrane implies that the solute flux is zero and $D^*_\omega = 0$ when $\omega = 1$. On this basis, Malusis & Shackelford (2002b) proposed the following decomposition of the osmotic

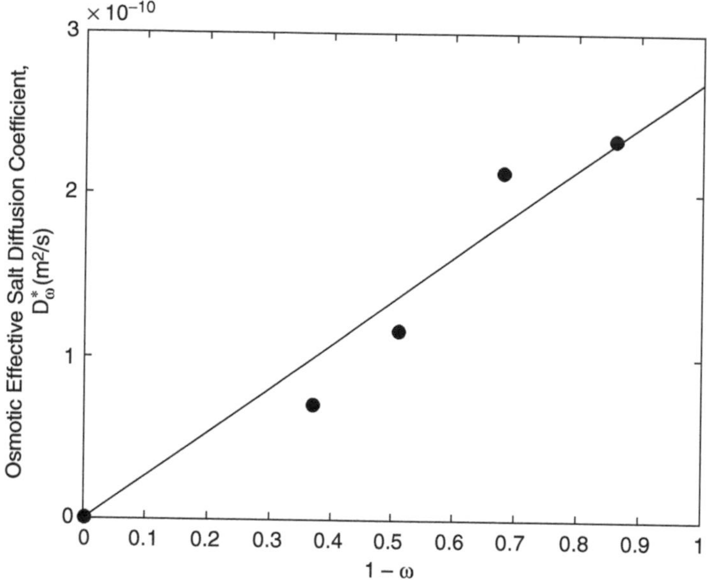


Figure 7. Linear regression of measured effective salt-diffusion coefficients versus $1-\omega$ for a geosynthetic clay liner (data from Malusis & Shackelford 2002b).

effective salt diffusion coefficient:

$$D_{\omega}^{*} = \tau_a D_{s,0} \tag{39}$$

where $\tau_a$ = apparent tortuosity factor and $D_{s,0}$ = free-solution salt diffusion coefficient (see Eq. 24).

The apparent tortuosity coefficient may be defined further as the product of a matrix tortuosity factor, $\tau_m$, representing the tortuous nature of the actual diffusive pathways through the porous medium due only to the geometry of the interconnected pores, and a generalized restrictive tortuosity factor, $\tau_r$, as follows:

$$\tau_a = \tau_m \tau_r \tag{40}$$

where $\tau_r$ is assumed to be a function of the chemico-osmotic efficiency coefficient, so that when $\omega = 0$, $\tau_r$ should be equal to 1 (non restrictive behaviour) and $\tau_a$ reduces to $\tau_m$.

As shown in Figure 7, on the basis of the experimental data of Malusis & Shackelford (2002b) the ratio $D_{\omega}^{*}/(1-\omega)$ is approximately constant. Therefore, as a first approximation, that the following assumption can be proposed:

$$\tau_r \cong 1 - \omega \tag{41}$$

so that $D_{\omega}^{*}$ can be defined as follows (Manassero & Dominijanni, 2003):

$$D_{\omega}^{*} \cong (1-\omega)\tau_m D_{s,0}. \tag{42}$$

The value of $\tau_m$ can be calculated from $D_{\omega}^{*}$ when $\omega = 0$ as follows:

$$\tau_m = \left. \frac{D_{\omega}^{*}}{D_{s,0}} \right|_{\omega=0}. \tag{43}$$

From the linear regression of the osmotic effective salt diffusion coefficients measured by Malusis & Shackelford (2002b) for a GCL having porosity varying between 0.78 and 0.80, the effective salt diffusion coefficient, $D_s^{*}$, is equal to $2.7 \cdot 10^{-10}\,m^2/s$ (i.e. $\tau_m = 0.14$, for $D_{s,0}^{*} = 19.9 \cdot 10^{-10}\,m^2/s$).

As a result, the following macroscopic model for the contaminant transport through clay membranes can be proposed:

$$q = \frac{k_h}{\gamma_w L_h}\left[\Delta P - \omega(C_{s,ave})\Delta\Pi\right] \tag{44a}$$

$$J_s = \left[1 - \omega(C_{s,ave})\right]qC_{s,ave} + n\left[1 - \omega(C_{s,ave})\right]\frac{\tau_m D_{s,0}}{L_h}\Delta C_s \tag{44b}$$

where $\omega$ is the only parameter that is varying with the salt concentration.

## 4 PHYSICAL INTERPRETATION OF TRANSPORT PARAMETERS

One of the most important advantages of the thermodynamic approach is that it doesn't require any specification of the physical phenomena involved in the transport of ions through a semi-permeable membrane. This advantage is even more appreciable if we consider that our knowledge of the physico-chemical mechanisms controlling ion distribution within the pores at the microscopic scale is still partial and controversial. However, the experimental observations suggest the existence of some general relations between the phenomenological coefficients. In particular, the relation between the osmotic effective salt diffusion coefficient, $D^*_\omega$, and the chemico-osmotic efficiency coefficient, $\omega$, may be investigated on the basis of a sufficiently general physical model.

It's well known that clays are able to generate a partition of ions due to the negative electric charge of the surfaces of their particles. At the microscopic scale, such a partition may be modelled by the diffuse double layer (DDL) theory. The DDL theory states that the electric potential distribution within the pores is governed by the Poisson-Boltzmann equation. Moreover, other mechanisms, normally not contemplated in the DDL theory, may play a significant role in partitioning ions within clay pores, such as: (*i*) steric hindrance due to the finite size of ion molecules; (*ii*) dielectric exclusion due to the polarization of matrix/solvent interface; and (*iii*) the increase in the solute solvation energy due to possible changes in the properties of solvent in nanopores.

However, apart from the partition mechanism, the actual ion concentrations in the membrane, $\overline{C}_i$, may be related to the virtual or equivalent salt concentration, $C_s$, by means of a simple relation such as:

$$\overline{C}_i = \nu_i C_s \Gamma_i \quad i = 1, 2 \tag{45}$$

where $\Gamma_i$ = partition coefficient of the i-th ion. $C_s$ represents the salt concentration of a virtual or equivalent bulk solution that would be in thermodynamic equilibrium with an infinitesimal element of the membrane at the generic position x within the membrane (Dormieux et al. 1995, Yaroshchuk 1995). At the boundaries (i.e. $x = 0$ and $x = L_h$), the virtual solution coincides with the bulk solutions that are in contact with the clay membrane. The ion partition coefficients represent the interaction of the ion species with the charged solid skeleton of the clay membrane. If the fixed charge (FC) is negative, the anion species are repelled from solid particles and the coefficient $\Gamma_2$ varies from zero to unity (i.e. $0 \leq \Gamma_2 \leq 1$). Also, the cation is attracted by the solid particles and the coefficient $\Gamma_1$ is larger than unity (i.e. $\Gamma_1 \geq 1$).

If the only partition mechanism considered is that associated to electrostatic repulsion of ions from charged pore walls, the partition coefficients may be related to the macroscopic electric potential of the membrane, $\psi_m$, as follows:

$$\Gamma_i = \exp\left(-\frac{z_i F}{RT}\psi_m\right) \tag{46}$$

where $\psi_m$ is also called Donnan potential.

In order to relate the macroscopic electric potential, $\psi_m$, to the physical and chemical properties of the semipermeable porous medium, Teorell (1935, 1937) and Meyer & Sievers (1936a,b) proposed to adopt an electro-neutrality condition, taking into account the concentration of the FC, $C_X$, uniformly distributed within the porous medium, as follows:

$$z_1\overline{C}_1 + z_2\overline{C}_2 + \varpi C_X = 0 \tag{47}$$

where $\varpi$ is the sign of the fixed charge (for clays with a negative fixed charge, $\varpi = -1$).

The fixed charge concentration, $C_X$, may be assumed proportional to the CEC and inversely proportional to the void ratio, $e = (1 - n)/n$, or:

$$C_X = \phi_X \mathrm{CEC} \rho_s \frac{1}{e} \tag{48}$$

where $\phi_X$ is the fixed charge coefficient and $\rho_s$ is the density of the solid phase.

The fixed charge coefficient, $\phi_X = \phi_X(S, \varepsilon_r, r_1, \ldots)$, varies from zero to unity and may account implicitly for the dependency of the FC concentration on the characteristic dimension of the pores through the specific surface, S, and on the relative dielectric constant of the solvent, $\varepsilon_r$. When $S \to \infty$ or $\varepsilon_r \to \infty$, the coefficient of the FC tends to 1. However, for real clays with a finite and not negligible microscopic characteristic dimension of pores, the reduction of the ion partition ability is taken into account adopting a value of $\phi_X$ smaller than 1. Moreover, the parameter $\phi_X$ accounts for the formation of a Stern layer of immobile cations, specifically adsorbed on the solid surface. The effect of the formation of the Stern layer may be introduced in the macroscopic model by means of an empirical dependency of $\phi_X$ on the type of cations present in solution through the radius of the cation species, $r_1$. The formation of a Stern layer reduces the theoretical FC concentration, so that $\phi_X < 1$. Finally, $\phi_X$ also may be used for taking into account the effect of "imperfections", i.e. preferential transport paths having a microscopic dimension higher relative to the intact matrix and producing a reduction in effectiveness of the membrane in partitioning the ion species.

Neglecting the effect of inertia on the movement of each component of the solution, the momentum equations for water and ions are (Spiegler 1958, Gu et al. 1999, Dominijanni 2005):

$$\overline{C}_w \nabla(-\mu_w) = f_{w1}(v_w - v_1) + f_{w2}(v_w - v_2) + f_{wm}(v_w - v_m) \tag{49a}$$

$$\overline{C}_1 \nabla(-\mu_1^{ec}) = f_{1w}(v_1 - v_w) + f_{12}(v_1 - v_2) + f_{1m}(v_1 - v_m) \tag{49b}$$

$$\overline{C}_2 \nabla(-\mu_2^{ec}) = f_{2w}(v_2 - v_w) + f_{21}(v_2 - v_1) + f_{2m}(v_2 - v_m) \tag{49c}$$

where $\overline{C}_w = C_w$ = water concentration, $v_w$ = water filtration velocity, $v_1$ = cation velocity, $v_2$ = anion velocity, $v_m$ = velocity of the solid skeleton, $f_{ij}$ = friction coefficient between the components i-th and j-th. The terms on the left-hand side of Equations 49 represent the forces for unit volume that drive the movement of the components of the solution. The terms on the right-hand side represent the momentum exchanged between the components of the solution. Under the assumption of binary interactions between the components of the system, the matrix of the friction coefficients is symmetric, such that

$$f_{ij} = f_{ji} \quad i, j = 1, 2, w, m \tag{50}$$

The assumption of a rigid solid skeleton implies that $v_m$ is equal to zero.

For dilute solutions, the water filtration velocity may be assumed equal to the solution filtration velocity (i.e. $v_w \cong q/n$) and the ion species may be assumed to be sufficiently rarefied, such that the prevalent interaction force that they exchange with the other components of the mixture is that with water. As a result, the interaction forces among ions and between ions and the solid skelton may be neglected in comparison with the interaction force between ions and water (i.e. $f_{ij}, f_{im} \ll f_{iw}$ $i, j = 1,2$).

The gradients of the electro-chemical potential in Equations 49 may be referenced to the virtual solution due to the condition of thermodynamic equilibrium. The momentum equations for dilute solutions may be rewritten as follows:

$$\nabla(-P) - \nabla(\Pi) = f_{1w}(v_w - v_1) + f_{2w}(v_w - v_2) + f_{wm} v_w \tag{51a}$$

$$\overline{C}_1 \left( RT \frac{\nabla(-C_1)}{C_1} + z_1 F \nabla(-\varphi) \right) = f_{1w}(v_1 - v_w) \tag{51b}$$

$$\overline{C}_2 \left( RT \frac{\nabla(-C_1)}{C_2} + z_2 F \nabla(-\varphi) \right) = f_{2w}(v_2 - v_w) \tag{51c}$$

where the hydraulic pressure, P, the osmotic pressure, $\Pi = (\nu_1 + \nu_2) RTC_s$, the ion concentrations, $C_i = \nu_i C_s$, and the electric potential, $\varphi$, are referenced to the virtual solution. From Equations 51, the volumetric flux, $q = nv_w$, and the ion mass fluxes, $J_i = n\overline{C}_i v_i$, may be derived as follows:

$$q = nd_h \left[ \nabla(-P) - \nabla(-\Pi) + RT \sum_{i=1}^{2} \Gamma_i \nabla(-C_i) + F\nabla(-\varphi) \sum_{i=1}^{2} z_i \Gamma_i C_i \right] \tag{52a}$$

$$J_i = q\Gamma_i C_i + nD_i^* \Gamma_i \nabla(-C_i) + n\Gamma_i C_i D_i^* z_i \frac{F}{RT} \nabla(-\varphi) \quad i = 1, 2 \tag{52b}$$

where $d_h = 1/f_{wm}$ is the mechanical permeability and $D_i^* = RT\overline{C}_i/f_{iw}$ is the macroscopic effective diffusion coefficient. Neglecting all phenomena due to the coupling between microscopic variations in concentration, velocity and electric potential, the macroscopic effective diffusion coefficient may be assumed constant and related to the free-solution diffusion coefficient, $D_{i,0}$, as follows:

$$D_i^* = \tau_m D_{i,0} \tag{53}$$

where $\tau_m$ is the tortuosity factor.

In the absence of electric current (i.e. $I_e = 0$), the electric potential gradient may be eliminated from Equation 52. For a mono-dimensional problem, the fluxes are given by:

$$q = -\frac{k_{h,\lambda}}{\gamma_w}\left[\frac{dP}{dx} - \omega_\lambda \frac{d\Pi}{dx}\right] \tag{54a}$$

$$J_s = (1 - \omega_\lambda) q C_s - n D_{\omega,\lambda}^* \frac{dC_s}{dx} \tag{54b}$$

where $k_{h,\lambda}$ = local hydraulic conductivity, $\omega_\lambda$ = local chemico-osmotic efficiency coefficient and $D_{\omega,\lambda}^*$ = local osmotic effective diffusion coefficient (Dominijanni & Manassero, 2005). The local phenomenological parameters are given by the following expressions:

$$\frac{k_{h,\lambda}}{\gamma_w} = \frac{n d_h}{1 + d_h RT \dfrac{(\Gamma_2 - \Gamma_1)^2 \nu_1 \nu_2 C_s}{\nu_1 \Gamma_2 D_2^* + \nu_2 \Gamma_1 D_1^*}} \tag{55a}$$

$$\omega_\lambda = 1 - \frac{\nu_1 D_2^* + \nu_2 D_1^*}{\nu_1 \Gamma_2 D_2^* + \nu_2 \Gamma_1 D_1^*}\Gamma_1\Gamma_2 = 1 - \frac{\nu_1 D_{2,0} + \nu_2 D_{1,0}}{\nu_1 \Gamma_2 D_{2,0} + \nu_2 \Gamma_1 D_{1,0}}\Gamma_1\Gamma_2 \tag{55b}$$

$$D_{\omega,\lambda}^* = (1 - \omega_\lambda) D_s^* \tag{55c}$$

In Equation 55a, the local hydraulic conductivity at zero electric current, $k_{h,\lambda}$, is dependent on the concentration of the FC and the ions in solution. However, for plausible values of the parameters normally found in clay membranes, such dependency is generally negligible and, as a first approximation, may be assumed as follows:

$$k_{h,\lambda} \cong n d_h \gamma_w = k_h = L_h \gamma_w \left(\frac{q}{\Delta P}\right)_{\Delta\Pi=0, I_e=0}. \tag{56}$$

The global parameters, measured by means of experiments, are related to the local parameters as follows:

$$\omega = \left(\frac{\Delta P}{\Delta \Pi}\right)_{q=0, I_e=0} = \frac{1}{\Delta C_s}\int_{C_s''}^{C_s'} \omega_\lambda \, dC_s \tag{57a}$$

$$D_\omega^* = \frac{L_h}{n}\left(\frac{J_s}{\Delta C_s}\right)_{q=0, I_e=0} = \frac{1}{\Delta C_s}\int_{C_s''}^{C_s'} D_\omega^* \, dC_s = (1 - \omega) D_s^* \tag{57b}$$

Equation 57b provides the theoretical basis for Equation 41 that has been proposed on the basis of the experimental evidence.

Based on Equation 57a, the measured values of the chemico-osmotic efficiency coefficient depend on the salt concentration in the two reservoirs that are in contact with the sample, rather than on the simple average concentration, as assumed in Equation 37.

For a solution containing a 1:1 electrolyte (i.e. a salt consisting of mono-valence ions, such as NaCl or KCl), Dominijanni & Manassero (2005) developed the following analytical expression for $\omega$:

$$\omega = 1 + \frac{C_X}{2\Delta C_s}\left[Z_2 - Z_1 - (2t_1 - 1) \cdot \ln\left(\frac{Z_2 + 2t_1 - 1}{Z_1 + 2t_1 - 1}\right)\right] \tag{58}$$

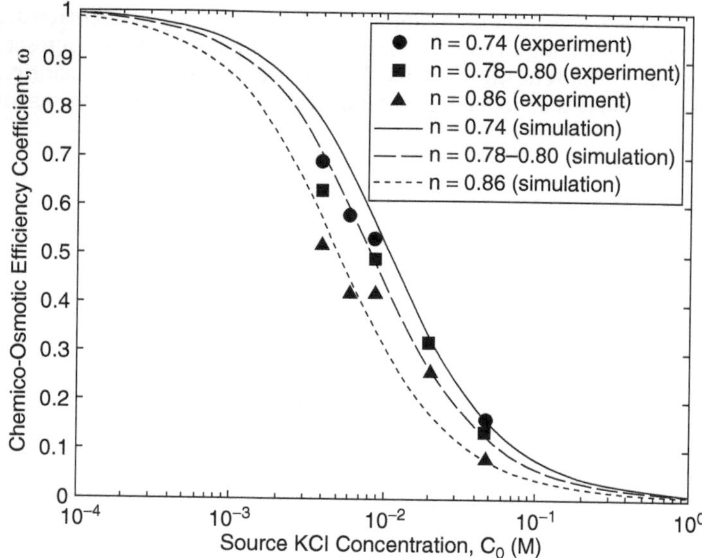


Figure 8. Interpretation of the experimental data of Malusis and Shackelford (2002a) assuming a constant fixed charge concentration. In the experiments $C_s' = C_0$ and $C_s'' = 0$.

where

$$Z_1 = \sqrt{1 + (2C_s'/C_X)^2}; \quad Z_2 = \sqrt{1 + (2C_s''/C_X)^2}$$

and $t_1 = D_{1,0}/(D_{1,0} + D_{2,0})$ is the cation transport number.

Using Equation 58 to interpret the experimental data of Malusis & Shackelford (2002a) for GCL specimens, Dominijanni & Manassero (2005) determined a fixed charge coefficient, $\phi_X$, equal to 0.04 (see Fig. 8).

Equation 55b describes the dependency of the local chemico-osmotic efficiency coefficient on the ion concentrations, the FC concentration, the ion diffusion coefficients and the ion valences. On the basis of Eq. 55b, $\omega_\lambda$ can assume negative values in the presence of low values of the transport number of the cation, $t_1 = D_{0,1}/(D_{0,1} + D_{0,2})$ (when the anions have a mobility much greater than cations) for 1:1 electrolytes. For example, in Figure 9, the dependency of $\omega_\lambda$ on the value of the transport number of the cation and the relative salt concentration, $\chi = C_s/C_X$, is illustrated for a 1:1 electrolyte. When $t_1$ is smaller than 0.5, $\omega_\lambda$ can assume negative values. When the salt concentration increases or the FC concentration decreases, the parameter $\omega_\lambda$ decreases.

When the solution contains a 2:1 electrolyte, the restrictive capacity of the membrane is reduced. This phenomenon is clearly illustrated in Figure 10, where $\omega_\lambda$ is plotted as a function of the relative salt concentration, $\chi$, for the case of KCl and $CaCl_2$ solution.

In order to evaluate the salt flux for steady state conditions, Equation 54b may be linearized, assuming as a first approximation that $\omega_\lambda \cong \omega$ and $D^*_{\omega,\lambda} \cong D^*_\omega$ for $C_s$ varying between $C_s'$ and $C_s''$. Making such an approximation, the steady state salt flux, $J_s$, is given by:

$$J_s \cong (1-\omega)q \frac{C_s' \exp\left[\dfrac{(1-\omega)qL}{nD^*_\omega}\right] - C_s''}{\exp\left[\dfrac{(1-\omega)qL}{nD^*_\omega}\right] - 1}. \tag{59}$$

Equation 59 allows a more accurate determination of the steady state salt flux than Equation 31.

The transient analysis of the transport of the solution should be based on the following coupled mass balances:

$$\frac{\partial(nC_w)}{\partial t} = -\frac{\partial(C_w q)}{\partial x} \tag{60a}$$

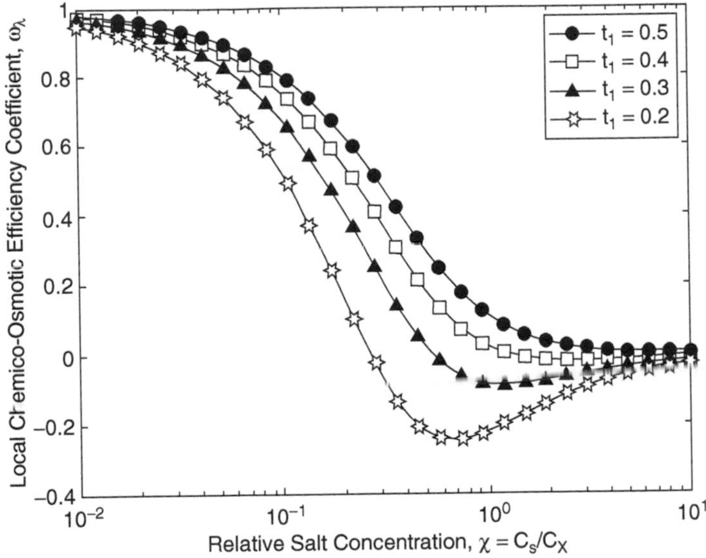


Figure 9. Local chemico-osmotic efficiency coefficient, $\omega_\lambda$, versus relative salt concentration, $\chi = C_s/C_X$, and the cation transport number, $t_1 = D_{0,1}/(D_{0,1} + D_{0,2})$, for a solution containing a (1:1) electrolyte.

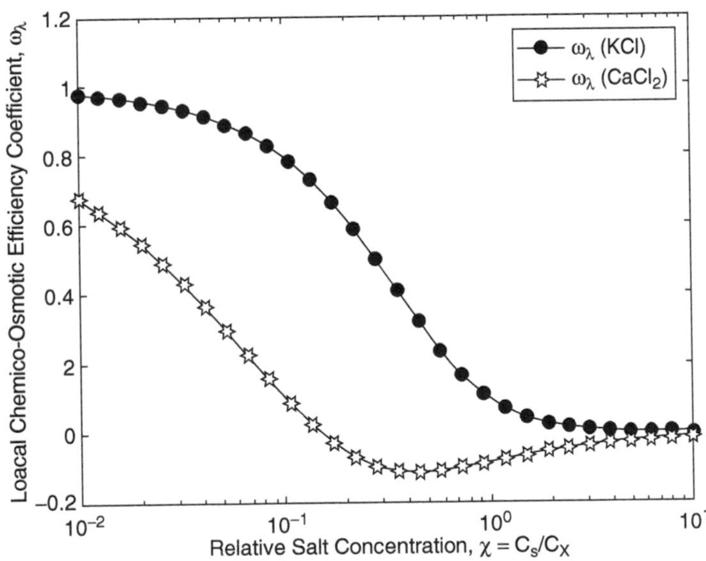


Figure 10. Local chemico-osmotic efficiency coefficient, $\omega_\lambda$, versus relative salt concentration, $\chi = C_s/C_X$, for solution containing cations with different valence ($z_1 = +1$ for $K^+$ and $z_1 = +2$ for $Ca^{2+}$).

$$\frac{\partial(n\Gamma_i C_i)}{\partial t} = -\frac{\partial(J_i)}{\partial x} i = 1,2 \tag{60b}$$

where $\rho_w$ = water density.

The analysis has been restricted to clays with only one cation and one anion in the pore solution. However, Equations 49 may be generalized in order to study systems containing a generic number of ions in solution. Also, the thermodynamic approach of Staverman (1952) may be generalized to a generic number of ions (Yaroshchuk, 1995) defining an appropriate set of phenomenological parameters.

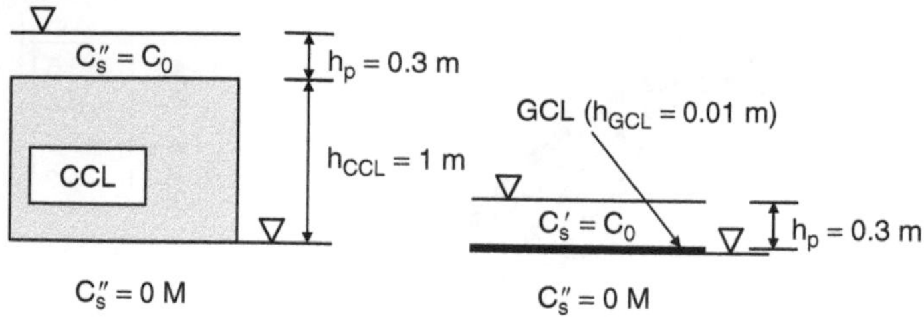


Figure 11. Example of a compacted clay liner (CCL) and a geosynthetic clay liner (GCL) used as containment contaminant barriers.

## 5 ROLE PLAYED BY OSMOTIC PHENOMENA IN CONTAMINANT TRANSPORT THROUGH GCLs

The primary objective of GCLs used as soil barriers for geoenvironmental applications is to minimize the migration of contaminants. With respect to this objective, the existence of membrane behaviour represents a potentially beneficial aspect that generally is not considered in the design and evaluation of GCLs (Malusis et al. 2003, Shackelford et al. 2003).

The membrane behaviour may induce two effects on the contaminant migration: (*i*) an advective transport, driven by the osmotic gradient and, normally, directed from lower concentration to higher concentration points; (*ii*) a restricted diffusive transport from higher concentration to lower concentration points.

The volumetric flow of the solution through a GCL at steady state conditions may be evaluated through Equation 28. The salt flux may be evaluated using Equation 31 or 59. In these last two equations, the osmotic effective diffusion coefficient, $D^*_\omega$, may be related to $\omega$ and $D^*_s$ through Equation 41 (or 57b).

The parameter that quantifies the membrane behaviour is the chemico-osmotic efficiency coefficient, $\omega$. Such a parameter should be measured by means of an experiment, in correspondence of the salt concentration range of interest. In the absence of experimental data, $\omega$ may be estimated through Equation 58, as a function of the FC concentration of the GCL. The FC concentration may be estimated from the CEC using Equation 48 and assuming, as a first approximation, the FC coefficient, $\phi_X$, varying between 0.01 and 0.1 for electrolytes characterized by monovalent ions.

The potential significance of membrane behaviour on the movement of contaminants through GCLs may be illustrated with reference to the scheme of Figure 11. The depth of the ponded liquid, $h_p$, is assumed to be 0.3 m. The exit boundary condition is a perfect-flushing boundary condition ($C_s'' = 0$). The geometry for this situation is analogous to the geometry associated with containment of miscible pollutants with engineered clay barrier materials (Shackelford 1997, Manassero et al. 2000). The perfect-flushing boundary condition has also been recommended as conservative approach for design of vertical barriers, such as vertical cutoff walls (Rabideau and Khandelwal, 1998).

A typical GCL is characterized by a hydraulic conductivity that is lower than that of a compacted clay liner (CCL). However, the hydraulic and concentration gradients across a GCL may be considerably higher due to its reduced thickness. The parameters characterizing a CCL and a GCL are reported in Table 2. The total salt flux at steady state conditions is generally higher for a GCL in comparison with a CCL. In Figure 12, for instance, the salt fluxes for a CCL and a GCL are reported as functions of the source KCl concentration. The data relative to the GCL are taken from the experimental results of Malusis & Shackelford (2002a). If the osmotic effects are not accounted for, the salt flux through the GCL is always higher than the salt flux through the CCL. However, if the osmotic effects are accounted for, the salt flux through the GCL is reduced and, for source concentrations lower than $10^{-3}$ M, is lower than the flux through the CCL.

The osmotic effects may reduce or even change the direction of the volumetric flux of the solution through the GCL. The salt diffusive flux is also reduced, with this effect being greater for low values of the source concentration.

In Figures 13 and 14, the salt mass flux through a CCL is compared with the flux through a composite barrier, constituted by a GCL placed over and in intimate contact with a CCL, having

Table 2. Parameters for the CCL and the GCL.

| | CCL | GCL |
|---|---|---|
| Thickness, $L_h$ (m) | 1 | 0.01 |
| Hydraulic conductivity, $k_h$ (m/s) | $10^{-9}$ | $1.63 \cdot 10^{-11}$ |
| Porosity, n | 0.4 | 0.79 |
| Tortuosity factor, $\tau_m$ | 0.4 | 0.14 |
| Fixed Charge Concentration, $C_X$ (mol/L) | 0 | 0.0123 |

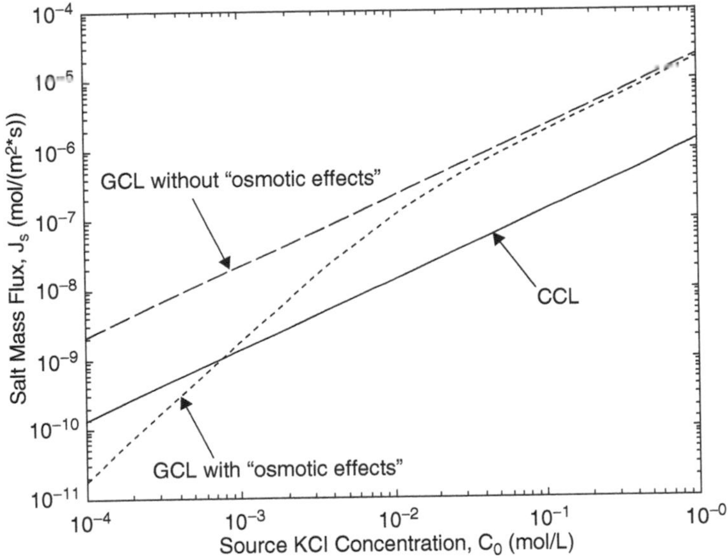


Figure 12. Salt mass flux versus source KCl concentration for a compacted clay liner (CCL) and a geosynthetic clay liner (GCL).

thicknesses equal to either 0.20 m (Fig. 13) or 0.40 m (Fig. 14). The composite barrier has a salt flux lower than that of the GCL only. If the osmotic effects are taken into account, the performances of the composite barrier are significantly improved.

The salt flux through a composite barrier may be obtained by imposing the conditions of equality of the volumetric and mass fluxes through all the layers of the barrier. The conditions of continuity of the fluxes allow the hydraulic pressures and salt concentrations to be evaluated at the boundaries between the layers. In the presence of osmotic phenomena, the solution of the flux continuity equations may be derived adopting iterative procedures.

In order to evaluate correctly the contaminant flux through GCLs, the fabric of the solid skeleton should be also taken into account. Clay particles, in fact, interact through long-range repulsive forces related to the electric field produced by the charge of the solid skeleton and the ions of the diffuse double layer. Such repulsive forces are depending on the distance between clay particles and the ion partition ability of the clay. The repulsive forces decrease when the salt concentration and the valence of the cations in the pore solution increase and the dielectric constant of the solvent decreases. Moreover, repulsive forces are combined with van der Waals attraction forces that tend to dominate in correspondence of high salt concentrations.

An increase of salt concentration or the replacement of monovalent cations (e.g. $Na^+$) with higher valence cations may induce flocculation of clay particles due to the reduction of the electric repulsive forces. As a consequence of this process, the void space available for the migration of the solution increases. The changes in the micro-fabric of the bentonite produce the formation of preferential flow paths for the permeating solution and cause increases in mechanical permeability, $d_h$, and possibly in the tortuosity factor, $\tau_m$, as well. The concentration of the FC, $C_X$, is reduced

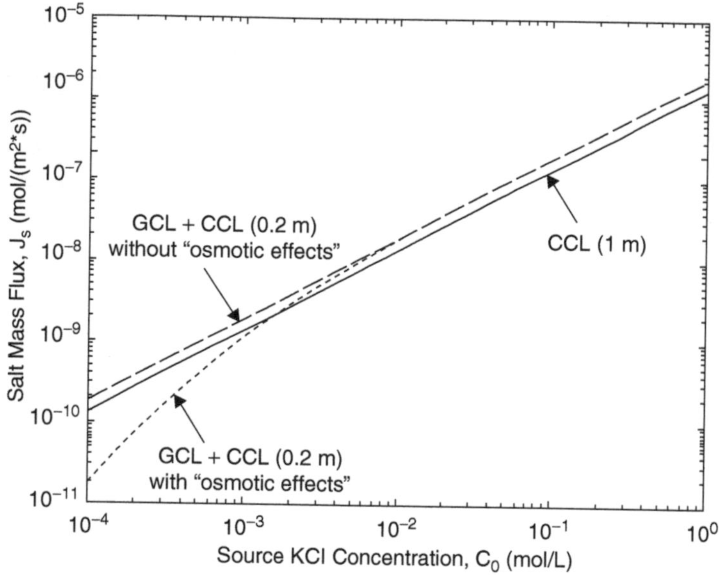


Figure 13. Salt mass flux versus source KCl concentration for a compacted clay liner (CCL) and a composite barrier constituted by a CCL, having thickness equal to 0.20 m, and a geosynthetic clay liner (GCL).

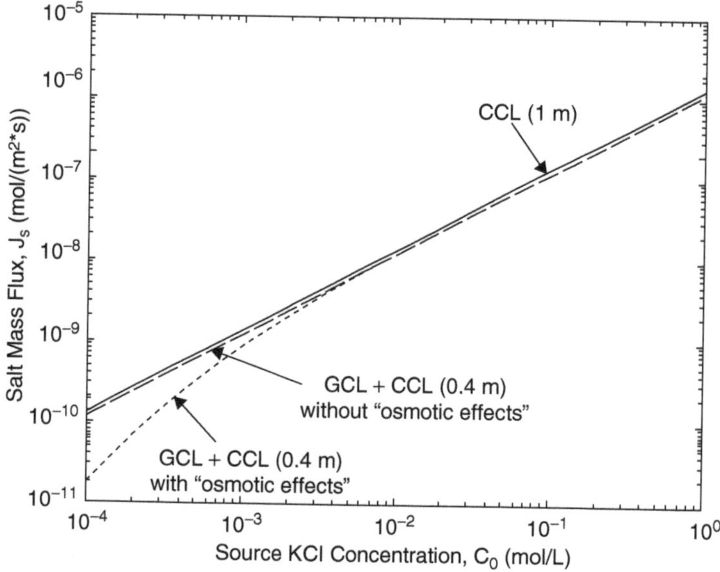


Figure 14. Salt mass flux versus source KCl concentration for a compacted clay liner (CCL) and a composite barrier constituted by a CCL, having thickness equal to 0.40 m, and a geosynthetic clay liner (GCL).

by flocculation, so that all the "osmotic effects" are destroyed. Shackelford & Lee (2003), for instance, found that diffusion of a 0.005 M $CaCl_2$ source concentration through a GCL specimen was sufficient to cause the complete destruction of an initially observed membrane behaviour. This observed sharp variation may be attributed to both a reduction of ion partition and a change of micro-fabric. As a result, typical analyses based on the simplifying assumption of unchangeable solid skeleton may overestimate the real osmotic properties and, consequently, the contaminant containment performances of GCLs when the salt concentration increases or monovalent cations in the pore solution are substituted by divalent cations. Changes of micro-fabric are also detected by large variations of mechanical permeability that induce large increases in hydraulic conductivity. Petrov & Rowe (1997), Shackelford et al. (2000) and Jo et al. (2005) showed that permeation with either strong ($\geq$0.1 M) solutions or solutions containing a large fraction of divalent cations can cause the hydraulic conductivity of GCLs to increase 1 order of magnitude or more. In correspondence of such large increases in hydraulic conductivity, the membrane behaviour is destroyed and the performances of GCLs as contaminant barriers are not directly comparable with CCLs.

Therefore, the final design of a barrier including a GCL should be based on the results obtained from experiments with solutions representative of the in-situ leachate. The laboratory tests should evaluate:

(1) the hydraulic conductivity, $k_h$, in correspondence of the average salt concentration of interest;
(2) the chemico-osmotic efficiency coefficient, $\omega$, and the osmotic effective diffusion coefficient, $D^*_\omega$, in correspondence of the boundary concentrations ($C'_s$ and $C''_s$) of interest.

For all the laboratory experiments, the porosity of the GCL should be representative of the porosity expected in-situ. If the permeating solutions contain only monovalent ions, the FC concentration may be assumed constant with respect to salt concentration changes only for salt concentrations lower than 0.1 M. However the FC concentration is a function of the porosity, the CEC and the type of cations that are present in the pore solution.

## 6 CONCLUSIONS

Due to the electric charge of bentonite particles, the transport of electrolyte solutions through GCLs may be significantly affected by osmotic phenomena. The formalism developed by Staverman (1952) in the context of the Thermodynamics of Irreversible Processes provides for a general framework for modelling the movement of electrolyte solutions through semipermeable membranes.

Accordingly, a set of phenomenological equations has been proposed based on the test conditions of the apparatus developed by Malusis et al. (2001) for solutions containing only one salt. The parameters that have to be measured to evaluate the volumetric and solute mass fluxes are the hydraulic conductivity, $k_h$, the chemico-osmotic efficiency coefficient, $\omega$, and the osmotic effective diffusion coefficient, $D^*_\omega$. The dependency of such coefficients on the salt concentration, the ion type and valence, the bentonite porosity and the cation exchange capacity (CEC) may be investigated by means of a series of tests and modelled using fitting functions. However, this approach doesn't provide any physical interpretation of the phenomena involved and is based on the simplifying assumption that the membrane may be considered as "discontinuous" between two compartments containing solutions having the same temperature, but different electro-chemical potential.

A physical explanation of the observed phenomena may be obtained from a model describing the friction and partition mechanisms governing the movement of electrolyte solutions within charged porous media. Based on such a physical interpretation, the chemico-osmotic phenomena are due to the transfer of momentum between the components of the pore solutions in the presence of an ion partition mechanism associated to the electric charge of the solid skeleton. Furthermore, in evaluating the fluxes through clay membrane, the condition of null electric current has to be added to the flux equations. The resulting macroscopic model is based on (i) the mechanical permeability, accounting for the energy dissipated by the friction between the liquid and the solid, (ii) the ion partition coefficients, accounting for the different distribution of ions within the pores, and (iii) the effective diffusion coefficient, accounting for the dissipation of energy associated to the friction between ions and solvent. The assumption of constancy of the effective diffusion coefficients is related to the hypothesis that all the effects due to the microscopic variations in concentration, velocity and electric potential (i.e. the "dispersive" effects) are negligible.

In the proposed theoretical approach the ion partition coefficient may be related to the CEC, the bentonite porosity and solid density through the Donnan equations and the electro-neutrality condition modified to account for the presence of the FC. In such a way, a theoretical expression for the chemico-osmotic efficiency as a function of the salt concentration of the bulk solutions in contact with GCLs and the physical and chemical properties of the bentonite and the permeating solution may be obtained. Moreover, the osmotic effective diffusion coefficient can be related linearly to the complement to unity of the chemico-osmotic efficiency coefficient, ω (i.e., 1 − ω). The validity of such theoretical conclusions has been verified through the experimental data obtained by Malusis & Shackelford (2002a, b) using GCL specimens.

Based on the experimental data, barriers that include GCLs have been compared to traditional compacted clay liners (CCLs). Accounting for the presence of osmotic effects, the contaminant flux through GCLs is considerably reduced and the performances of barriers including GCLs are improved. However, the design of barrier with GCL should also take into account the possibility of changes in micro-fabric of bentonite. In the presence of high salt concentrations, low dielectric constant solvents and when monovalent cations are replaced by poly-valent cations, flocculation of clay particles may induce the formation of preferential flow paths within the bentonite, causing an increase of the hydraulic conductivity and the tortuosity factor and the partial or complete destruction of the membrane behaviour. As a result, the final evaluation of the contaminant transport through GCLs should be always referred to experimental parameters, measured in correspondence of boundary conditions, permeating solutions and stress levels representative of the real in-situ conditions.

## REFERENCES

De Groot, S.R. & Mazur, P. 1962. *Non-equilibrium Thermodynamics*. London: Pergamon Press.

Dominijanni, A. & Manassero, M. 2005. Modelling osmosis and solute transport through clay membrane barriers. *Geotechnical Special Publication* (130–142): 3437–3448.

Dominijanni, A. 2005. *Osmotic properties of clay soils*. PhD Thesis. Torino: Politecnico di Torino.

Dormieux, L., Barboux, P., Coussy, O. & Dangla, P. 1995. A macroscopic model of the swelling phenomenon of a saturated clay. *European Journal of Mechanics, A/Solids* 14(6): 981–1004.

Groenevelt, P.H. & Bolt, G.H. 1969. Non-equilibrium thermodynamics of the soil-water system. *Journal of Hydrology* 7: 358–388.

Gu, W.Y., Lai, W.M. & Mow, V.C. 1999. Transport of multi-electrolytes in charged hydrated biological soft tissues. *Transport in Porous Media* 34(1–3): 143–157.

Hanshaw, B.B. & Coplen, T. B. 1973. Ultrafiltration by a compacted clay membrane-II. Sodium ion exclusion at various ionic strengths. *Geochimica et Cosmochimica Acta* 37: 2311–2327.

Jo, H.Y., Benson, C.H., Shackelford, C.D., Lee, J.-M. & Edil, T.B. 2005. Long-term hydraulic conductivity of a geosynthetic clay liner permeated with inorganic salt solutions. *Journal of Geotechnical and Geoenvironmental Engineering* 131 (4): 405–417.

Katchalsky, A. & Curran, P.F. 1965. *Nonequilibrium Thermodynamics in biophysics*. Cambridge: Harvard University Press.

Kemper, W.D. & Quirk, J.P. 1972. Ion mobilities and electric charge of external clay surfaces inferred from potential differences and osmotic flow. *Soil Science Society of America Proceedings* 36(3): 426–433.

Kemper, W.D. & Rollins, J.B. 1966. Osmotic efficiency coefficients across compacted clays. *Soil Science Society of America Proceedings* 30: 529–534.

Kemper, W.D. 1960. Water and ion movement in thin films as influenced by the electrostatic charge and diffuse layer of cations associated with clay mineral surfaces. *Soil Science Society of America Proceedings* 24: 10–16.

Kharaka, Y.K. & Berry, F.A.F. 1973. Simultaneous flow of water and solutes through geological membranes – I. Experimental results. *Geochimica et Cosmochimica Acta* 37: 2577–2603.

Malusis, M.A. & Shackelford, C.D. 2002a. Chemico-osmotic efficiency of a geosynthetic clay liner. *Journal of Geotechnical and Geoenvironmental Engineering* 128(2): 97–106.

Malusis, M.A. & Shackelford, C.D. 2002b. Coupling effects during steady-state solute diffusion through a semi-permeable clay membrane. *Environmental Science and Technology* 36: 1312–1319.

Malusis, M.A., Shackelford, C.D. & Olsen, H.W. 2001. A laboratory apparatus to measure chemico-osmotic efficiency coefficients for clay soils. *Geotechnical Testing Journal* 24(3): 229–242.

Malusis, M.A., Shackelford, C.D. & Olsen, H.W. 2003. Flow and transport through clay membrane barriers. *Engineering Geology* 70: 235–248.

Manassero, M. & Dominijanni, A. 2003. Modelling the osmosis effect on solute migration through porous media. *Geotechnique* 53(5): 481–492.

Manassero, M., Benson, C.H. & Bouazza, A. 2000. Solid waste containment systems. In *GeoEng2000. An International Conference on Geotechnical & Geological Engineering, 19–24 November 2000 Melbourne, Australia*. Lancaster: Technomic Publishing Company.

Meyer, K.H. & Sievers, J.F. 1936a. La perméabilité des membranes I, Théorie de la perméabilité ionique. *Helvetica Chimica Acta* 19(1): 649–664.

Meyer, K.H. & Sievers, J.F. 1936b. La perméabilité des membranes II. Essais avec des membranes sélectives artificielles. *Helvetica Chimica Acta* 19(1): 665–677.

Mitchell, J. K. and Soga, K. 2005. Fundamentals of soil behaviour (3rd edn). New York: John Wiley & Sons, Inc.

Onsager, L. 1931a. Reciprocal relations in irreversible processes I. *Physical Review* 37(4): 405–426.

Onsager, L. 1931b. Reciprocal relations in irreversible processes II. *Physical Review* 38(12): 2265–2279.

Petrov, R.J. & Rowe, R.K. 1997. Geosynthetic Clay Liner (GCL) Chemical compatibility by hydraulic conductivity testing and factors impacting its performance. *Canadian Geotechnical Journal* 34(6): 863–885.

Rabideau, A. & Khandelwal, A. 1998. Boundary conditions for modeling transport in vertical barriers. *Journal of Environmental Engineering* 124(11): 1135–1139.

Shackelford, C.D. & Lee, J.-M. 2003. The destructive role of diffusion on clay membrane behaviour. *Clays and Clay Minerals* 51(2), 186–196.

Shackelford, C.D. 1997. Modeling in environmental geotechnics: applications and limitations. In: Kamon, M. (ed.), *Second International Congress on Environmental Geotechnics, IS-Osaka '96, Osaka, Japan, Nov. 5–8, 1996, vol. 3*. Rotterdam: Balkema.

Shackelford, C.D., Benson, C.H., Katsumi, T., Edil, T.B. & Lin, L. 2000. Evaluating the hydraulic conductivity of GCLs permeated with non-standard liquids. *Geotextiles and Geomembranes* 18(2–4): 133–161.

Spiegler, K.S. 1958. Transport processes in ionic membranes. *Transactions of the Faraday Society* 54: 1408–1428.

Staverman, A.J. 1952. Non-equilibrium thermodynamics of membrane processes. *Transactions of the Faraday Society* 48(2): 176–185.

Teorell, T. 1935. An Attempt to formulate a quantitative theory of membrane permeability. *Proceedings of the Society for Experimental Biology and Medicine* 33: 282–283.

Teorell, T. 1937. The properties and functions of membranes, natural and artificial, II Artificial membranes, General discussion. *Transactions of the Faraday Society* 33: 1053, 1086–1088.

Yaroshchuk, A.E. 1995. Osmosis and reverse osmosis in fine-charged diaphragms and membranes. *Advances in Colloid and Interface Science* 60: 1–93.

# CHAPTER 7

# Gas permeability of geosynthetic clay liners

A. Bouazza
*Monash University, Melbourne, Victoria, Australia*

## 1 INTRODUCTION

In containment facilities for municipal solid waste (landfills), significant quantities of gases (methane, carbon dioxide, etc.) are generated as a result of anaerobic decomposition of organic materials. These gases may migrate through the engineered barrier system in cover liners and can cause significant threat to human health and the environment (greenhouse effect). In this respect, the potential impact of gas migration on the performance of the individual resistive barrier and ultimately, on the long term behaviour of the liners needs to be assessed to reduce safety and health risks.

Covers over solid waste landfills are typically multi-component systems that are constructed directly on top of the waste shortly after a specific cell has been filled to capacity. A conventional approach to cover system design is to construct a "resistive barrier" that utilises a liner with a low saturated hydraulic conductivity to reduce the water ingress into the landfill and to control biogas escape to the atmosphere. Geosynthetic clay liners (GCLs) are now widely used in landfill covers as the resistive barrier as an alternative to soil barriers. The application of GCLs in cover systems stems from the fact that they were found to be very effective as hydraulic barriers, easy to install, and could withstand distortion and distress while maintaining their low hydraulic conductivity (Bouazza 2002). As part of the evaluation process, GCLs hydraulic properties were considered to be the prime factors. However their gas performance has recently come under a growing scrutiny due to their increased use in capping.

## 2 BACKGROUND

The movement of gases in porous media such as soil or GCLs occurs by two major transport mechanisms: advective flow and diffusive flow. This chapter focuses on advective flow only; diffusive flow will be discussed in Chapter 11. In advective flow, the gas moves in response to a gradient in total pressure. To equalize pressure, a mass of gas travels from a region of higher pressure to a lower one. In the context of landfills, the primary driving force for gas migration, especially through cover systems, is pressure differentials due to natural fluctuations in atmospheric pressure (barometric pumping). Falling pressures tend to draw gas out of the landfill, increasing the gas concentration near the surface layers. Conversely, high or increasing barometric pressure tends to force atmospheric air into the landfill, diluting the near surface soil-gas and driving gas deeper into the landfill (this is a possible explanation on how VOCs can find their way into groundwater). A change in the leachate/water table or temperature can also give rise to pressure differences and lead to gas migration. It is important to note that the presence of a gas recovery system (i.e gas collection system) will have an impact on the magnitude of gas pressures build up. Prosser and Janachek (1995) reported that gas pressures can reach 400 kPa at the bottom of a landfill in the absence of a gas collection system. However, McBean et al. (1995) indicated that the build up of gas pressure under a cover system is unlikely to be higher than 10 kPa.

A number of events have brought the hazards associated with landfill produced methane very much into public view. The best known of these are the Loscoe, U.K, (Williams and Aitkenhead 1991); Skellingsted, Denmark, (Kjeldsen and Fisher 1995) and Masserano, Italy (Jarre et al. 1997) incidents, which resulted in extensive property damage and loss of lives. The Loscoe explosion in the United Kingdom for example, took place after atmospheric pressure dropped by 29 mbars in approximately 7 hours. The same phenomenon caused the Skellingsted and Masserano explosions. Another area of concern is the presence of a geomembrane in a cover system, where even nominal amounts of gas can be troublesome. One concern, amongst many others, is the possibility of landfill gas accumulation, which can gradually increase positive pressure in the landfill. This may induce geomembrane extrusion in landfill composite covers at points of inadequate overburden (Sherman 2000). In addition, the positive gas pressure under the barrier layers may induce the reduction of interface shear strength between the geomembrane and the underlying layer due to the insufficient normal stress acting on the barrier layer which in this case can contribute to a slope failure (Koerner and Daniel 1997, Richardson et al., 2008). Therefore, it is not surprising that nowadays gas pressures have been recognized as a design issue for landfill covers and a methodology has been put forward to address this issue (Thiel 1999).

In addition to posing a safety and health risk, the migration of gases from landfills poses a potentially serious problem by creating vegetation stresses or diebacks and contamination of surface waters. Haskell and Cochrane (2001) reported on a case study where gas migration underneath a composite cover contributed to contamination of surface waters. The review of the cover design indicated that the low confining load acting on the geomembrane limited the intimate contact of the geomembrane with the underlying layer, allowing gas condensate to flow to the underside of the cover and migrate upward to the perimeter drainage ditch due to capillary forces. A forensic investigation conducted by Peggs and McLaren (2002) on vegetation diebacks observed at the surface of a landfill has shown that it was caused by gas leaks through punctures in the geomembrane cover, as small as 2 mm, which were quite far from the location of maximum gas concentration above the cover soil. A 4 year study carried out in the U.K on migration of landfill gas through mineral and geomembrane liners has shown that despite the fact that every care has been taken in the installation of a geomembrane, small gas leaks still occurred. It was concluded that it was unrealistic to assume that a geomembrane will remain completely gas tight (Environment Agency, 2000).

## 3 GAS TRANSPORT DUE TO ADVECTION

Gas flow measurement performed by Alzaydi and Moore (1978) showed that Darcy's law could provide a fair approximation of gas flow in a low permeability material. Furthermore, Izadi and Stephenson (1992) confirmed that in contrary to coarse grain soils, the gas slippage flow through low permeable soils decreased as the degree of saturation decreased. This indicates that the magnitude of slip flow is very small relative to viscous flow. The velocity of gas flow at the pore walls cannot generally be assumed to be zero. The nonzero flow velocity at the pore walls is termed "slip flow" or "drift flow". Brusseau (1991) also indicated that slip flows are not observed when the pressure difference is lower than 20 kPa (i.e. in the range of pressures typically encountered in landfill covers) and can, on this basis, be excluded from the modeling process for gas advective transport conditions. The same study also stressed the fact that for low pressure differences the assumption of incompressible flow of gas in porous media is valid. Thus models developed for water flow can be used for gas flow. Massmann (1989) indicated that groundwater flow model provided good approximation for gas transport up to a differential pressure of 50 kPa.

Gas advection in porous media is generally analysed using Darcy's law which states that gas flow is directly proportional to the gas pressure gradient and the gas permeability. Darcy's law for one dimensional volumetric flow (Q) can be written as:

$$Q = -k\frac{K_r}{\mu}A\frac{dP}{dx} \tag{1}$$

where $k$ is the intrinsic permeability of the porous material, $K_r$ is the relative permeability for the permeant gas, $\mu$ is the dynamic viscosity of the gas, $A$ is the cross section of the porous material, and $dP/dx$ is the pressure gradient. It is assumed that the intrinsic permeability is a function only of the properties of the porous material, not the permeating gas. To avoid the complex calculation of $K_r$, Darcy's law can be formulated using the gas permeability $K_g$ as follow:

$$Q = -\frac{K_g}{\rho g}A\frac{dP}{dx} \tag{2}$$

where $g$ is the acceleration due to gravity and $K_g$ is given by:

$$K_g = \frac{\rho g}{\mu}kK_r \tag{3}$$

For gases, the rate of flow changes from one point to another point as the pressure decreases due to their compressibility. However, it can be assumed that landfill gases behave like ideal gases and the continuity equation of ideal gas can be written as:

$$\frac{\rho_0 T_0}{P_0} = \frac{\rho T}{P} \tag{4}$$

where $\rho_0$ is the gas density at standard pressure $P_0$ and standard temperature $T_0$, and $\rho$ is the gas density at pressure $P$ and temperature $T$. Assuming the rate of mass flow ($\rho Q$) is constant and the law of mass conservation is applied. A steady state flow ($d(\rho Q)/dt = 0$) of gas can be written as:

$$\frac{d}{dx}(\rho Q) = 0 \tag{5}$$

From equations 2, 4, and 5, a linear differential equation for the one-dimensional steady state flow in an isotropic homogeneous porous medium under isothermal conditions is obtained:

$$\frac{d^2}{dx^2}(P^2) = 0 \tag{6}$$

For a sample of thickness $L$, the solution to equation 6 is subject to the boundary conditions, $P = P_1$ at $x = 0$ and $P = P_2$ at $x = L$, hence:

$$P^2 = P_1^2 + \left(\frac{P_2^2 - P_1^2}{L}\right)x \tag{7}$$

From equation 2 and 7 the volumetric flow of gas at distance $x$ can be obtained from the following equation:

$$Q_x = -\frac{K_g A}{2\rho g L}\frac{(P_2^2 - P_1^2)}{\sqrt{P_1^2 + \dfrac{(P_2^2 - P_1^2)}{L}x}} \tag{8}$$

Considering the volumetric flow of gas at a distance $L$, equation 8 becomes:

$$Q_L = -\left(\frac{K_g}{\rho g}\right)A\,\frac{(P_2^2 - P_1^2)}{2LP_2} \tag{9}$$

Based on Equation (9), the gas permeability can be written as:

$$K_g = \frac{2\,Q_L\,\rho\, g\, L\, P_2}{A(P_1^2 - P_2^2)} \tag{10}$$

The application of Darcy's equation to the case of gas flow through porous media shows that the flow rate is not proportional to the differential pressure $\Delta P$ across the sample but rather to the pressure squared ($P_1^2 - P_2^2$).

It is known that the application of Darcy's law is only valid in a restricted domain, i.e. when the flow is laminar. The Reynolds number ($R_e$), a dimensionless number expressing the ratio of inertial to viscous forces, is generally used as a criterion to distinguish between laminar flow occurring at low velocities and turbulent flow. The flow rate at which the flow begins to deviate from a Darcy's law behaviour is observed when the Reynolds number exceeds some value between 1 and 10 (Bear 1972). For flow through porous media, the Reynolds number is defined as:

$$R_e = \frac{\mathrm{v}\, d}{\nu} \tag{11}$$

where v is the Darcy velocity, $d$ designates an average grain diameter of the porous matrix, and $\nu$ denotes the kinematic viscosity of the gas.

Bouazza and Vangpaisal (2003) and Vangpaisal and Bouazza (2004) showed that the Reynolds number for GCLs was well below the limit given by Bear (1972) and that laminar flow should

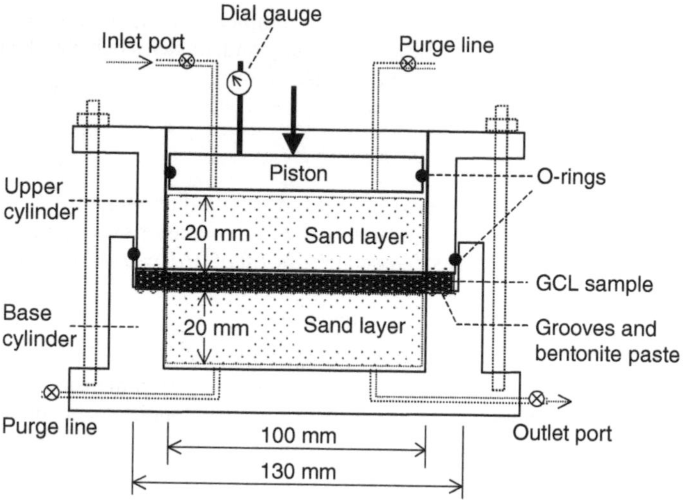


Figure 1. Cross section of gas permeability cell (after Bouazza and Vangpaisal, 2003).

prevail. Furthermore, gas flow is generally characterised by a relatively small velocity (pressure gradient and/or low flow) and as such, flow through fine porous media should be laminar. McBean et al. (1995) reported that, as a general rule, for flow rates typical of landfill gas migration and recovery, Darcy's law will apply if the characteristic grain sizes of the porous media are smaller than 2.0 mm.

## 4 GAS PERMEAMETER

Gas advective properties of GCLs are best determined using a gas permeability cell which can simulate the actual mechanical field conditions of landfill cover system (i.e., GCL is subjected to a confining stress and the effects of immediate supporting layers). Such cell, developed by Bouazza and Vangpaisal (2003), is shown in Figure 1 and consists of two separate parts: 1) a base cylinder, and 2) an upper cylinder with piston. The two parts are held together with retaining threaded rods. A piston situated in the upper cylinder is used to transmit the applied confining stress to the GCL sample. O-rings are used to seal the connections of the upper and the base cylinders, and the piston. The base cylinder has two different inside diameters, a diameter of 130 mm at the upper part and a diameter of 100 mm at the lower part, creating a shoulder on its wall, which is used to accommodate the GCL sample and the upper cylinder. Grooves are made on the inner shoulder of the base cylinder and on the bottom edge of the upper chamber intending to increase the sealing capability around the GCL perimeter when assembled. Dual gas inlet and outlet ports are provided on both the piston and the base chamber for gas inflow/outflow and cell purging before testing. Spaces inside the chambers are used to accommodate the supporting sandy material (medium sand, effective size 0.3 mm). A dial gauge can be attached to the piston to monitor the settlement of the sample during the test. A full description of the cell and testing procedures are given in Bouazza and Vangpaisal (2003).

## 5 GAS PERMEABILITY

GCLs are widely used in landfill applications and have been subject to considerable recent research. In this respect, there is a wide body of work available on their hydraulic performance and the measurements of their permeability to fluids are well documented in literature (see chapters 4, 5 and 6). However, experimental measurements of their permeability to gases are less widely available. Only recently, that information on their gas advective flow performance has become

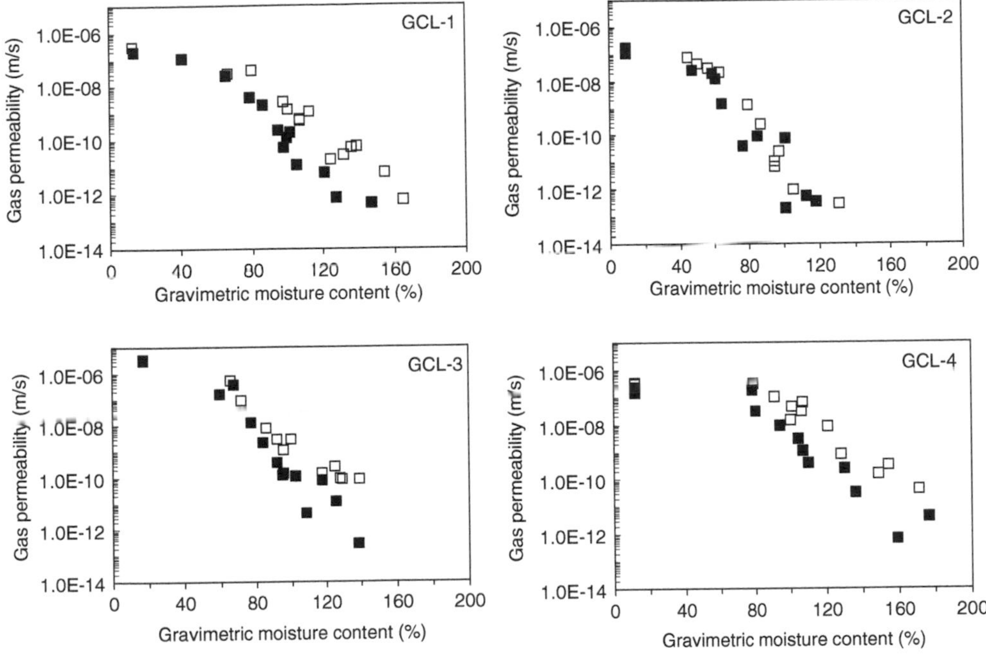


Figure 2. Relation between gas permeability and gravimetric moisture content for 4 different types of GCLs (Legends; □ = free swell hydration, ■ = confined hydration.).

available in the context of landfill capping (Didier et al., 2000; Bouazza and Vangpaisal, 2003, 2004, 2007; Vangpaisal and Bouazza, 2004; Bouazza et al., 2006). The published data show that the gas permeability or gas advective flux of GCLs is dependant on moisture content (gravimetric and volumetric), manufacturing process and operational conditions.

### 5.1 *Effect of moisture content*

Typical gas permeability results versus gravimetric moisture content for needle punched and stitch bonded GCLs are shown on Figure 2. Needle punched GCLs consisting of essentially dry bentonite [powder in GCL1 and GCL2 (with bentonite impregnated into the cover non-woven geotextile layer), and granular in GCL3] sandwiched between polypropylene geotextile layers and a stitch bonded GCL (GCL-4), were used in the investigations, reported by the above authors.

The results show that the decrease of gas permeability is associated with the increase in gravimetric moisture content. For the range of gravimetric moisture contents investigated, a decrease of around 5 to 7 and 4 to 6 orders of magnitude in the gas permeability was observed for confined hydration and free swell hydration, respectively. It appears from Figure 2 that the gas permeability of GCL-3 is more sensitive to moisture variation than other GCLs. In the case of confined hydration, a variation of 7 orders of magnitude in the gas permeability of GCL-3 was obtained for a gravimetric moisture content varying from 18% to 138%. This was due to the large difference in the bentonite form, which changed from coarse bentonite grains with large interconnected voids in the dry state to soft continuous bentonite gel at high gravimetric moisture content. Figure 2 also shows that the gas permeability of the GCLs varies according to the mode of sample hydration. The GCLs exposed to a surcharge during hydration tended to have lower gas permeability than the GCLs hydrated under zero confinement, particularly at medium to high gravimetric moisture content (>80%). This can be attributed to the fact that the application of a surcharge limited the swelling of hydrated bentonite and induced a more uniform distribution of moisture content throughout the GCL specimens. As a result, pore size and the interconnected voids in the bentonite component were likely to reduce, therefore, the lower gas permeability. This implies that the GCL should be subjected to confinement at time of installation or hydration. Interestingly, the modes of sample

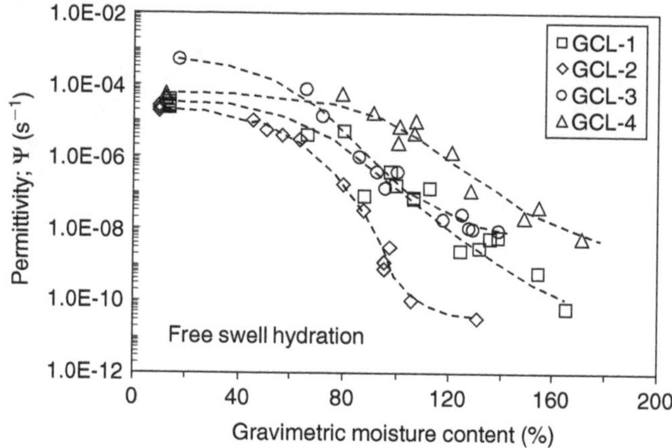


Figure 3. Variations of gas permittivity with gravimetric moisture content for GCLs under confined hydration.

hydration appeared to have no effect on the gas permeability of GCL-2. This was probably due to the effect of bentonite impregnation into the non-woven geotextile in GCL-2. The presence of the surcharge during hydration had less effect on the gas permeability at lower gravimetric moisture content. A common value of gas permeability was attained in the dry state (gravimetric moisture content as received from manufacturer).

The ability of GCLs to allow the flow of gas can also be expressed in terms of gas permittivity ($\Psi$). The permittivity is defined as the cross plane permeability ($K$) divided by the GCL thickness ($L$), $\Psi = K/L$. The variation of the permittivity against gravimetric moisture content is plotted in Figures 3 and 4 for the confined hydration and free swell hydration, respectively. The variation of permittivity followed the same trend for both hydration conditions. The higher permittivity values were obtained at lower gravimetric moisture content and lower permittivity values were obtained at higher gravimetric moisture content.

The effect of the bonding mechanism of GCLs on the gas permeability can be seen in the GCLs containing powdered bentonite (GCL-1, 2 and 4), in which GCL-4 tended to have higher permittivity. This difference is linked to the way that the GCLs are held together as a composite material. Stitch bonding is used in GCL-4, whereas needle punching is used in GCL-1 and GCL-2. As bentonite hydrates and swells GCL-4 tends to form pillow like shapes. This results in zones (at the stitch bonding level) with less bentonite to mitigate gas flow. In contrast, the bentonite tends to swell uniformly in the needle punched GCLs. In the case of dry GCLs, the effect of the holding mechanisms is insignificant because the large interconnected air voids in dry bentonite overrides the effect of needle punching and stitch bonding.

At the same level of gravimetric moisture content, GCL-2 tends to have lower permittivity than other GCLs for both hydration conditions. This is a result of the impregnation of bentonite in the non-woven geotextile, which induces an additional form of confinement, from the non-woven fibres, to the hydrated bentonite. The contribution of bentonite impregnation in lowering the gas permittivity was significant, particularly in the case of free swell hydration.

The variation of gas permittivity between different types of GCLs is more than one order of magnitude, and varies up to 3 orders of magnitude in the case of free swell hydration. The permittivity is also levelling off at higher gravimetric moisture contents, suggesting that gas advection becomes less significant (the measured flow rate approaches zero flow) and gas diffusion probably becomes the governing transport mechanism. The boundary of gravimetric moisture content, which was attained for zero advective flow measurement depended on the types of GCLs, i.e. gravimetric moisture content higher than 120% for GCL-2, and higher than 180% for GCL-4. At this level of gravimetric moisture content, the measured flow rate approaches the lower limit of the measuring technique and the advective flow is assumed as approaching a zero flow rate.

As expected, GCL-3 had an exceptionally high permittivity in the dry condition. This is because gas can easily flow through the large pore spaces of dry granular bentonite. The effect of bentonite form on the gas permeability of GCLs is clearer when the gas permittivity is plotted against

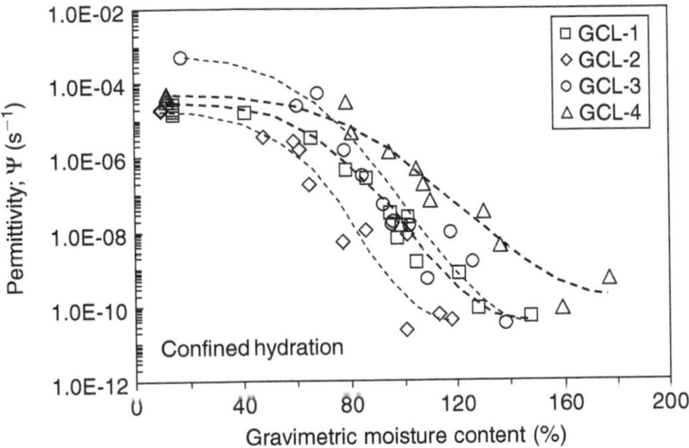


Figure 4. Variations of gas permittivity with gravimetric moisture content for GCLs under free swell hydration.

volumetric water content as shown in Figures 5 and 6. The variations of gas permittivity with volumetric water content follow a similar trend as the plot with moisture content, in other words the permittivity decreases as the volumetric water content increases.

Interestingly, among the needle punched GCLs (GCL-1, 2 and 3) investigated, GCL-3 (bentonite in granular form) tends to have higher permittivity than GCL-1 and 2. This is due to the large difference in the nature of bentonite form. The hydrated granular bentonite is stiffer than the hydrated powdered bentonite, and it is clearly visible as soft grains particularly at the lower level of volumetric water content. This indicates the presence of larger inter-granular pore spaces, which provide preferential gas flow paths. As indicated earlier in Chapter 3, each bentonite grain is slowly hydrated from the surface and as the volumetric water content increases the bentonite forms a gel surface and becomes softer, hence, the interconnected voids are decreased, and as a result the difference in the permittivity of GCL-3 to GCL-1 and GCL-2 is lower.

For the conditions tested, the effect of stitch bonding (in GCL-4) and the form of bentonite (in GCL-3) on gas permittivity are comparable at volumetric water content greater than 50%. However, the effect of the GCL structures and bentonite forms on gas permittivity tends to decrease as the volumetric water content increases when the GCLs are hydrated under a confining stress. In this case the effect of the surcharge overrides the effect of the differences between the GCLs, and a common permittivity value can be obtained, for all types of GCLs tested, at volumetric water content greater than 70%. At this level of volumetric water content, the advective flow is approaching a so-called zero advective flow condition and diffusive flow probably becomes significant. On the other hand, the gas permittivity of GCLs approaches the zero advective flow condition at different volumetric water content levels for the free swell hydration.

## 5.2 *Effect of hydration and desiccation*

The "resistive" barrier is normally required to maintain a low hydraulic conductivity during the lifetime of the cover system. While this is a relatively simple approach, engineers may encounter a number of practical problems with this barrier. For example, its integrity may be severely compromised if it undergoes cracking due to desiccation caused by environmental drying. Seasonal variations in precipitation and evaporation can lead to severe desiccation of the resistive barrier, depending on the materials used, and may result in high gas flux rates through the cover soil. Desiccation cracking can also occur due to moisture loss to atmosphere in final cover applications, which feature conditions very different to those encountered in liner applications. Desiccation can occur due to drying of the GCL as water evaporates to the atmosphere, but this process is also significantly influenced by heat received by the landfill cover (e.g., Henken-Mellies et al., 2002, Melchior, 2002).

The variation of gas permeability with moisture content for needle punched GCL samples (GCL1) tested after hydration (wetting) and after desiccation (drying) is presented in Figure 7.

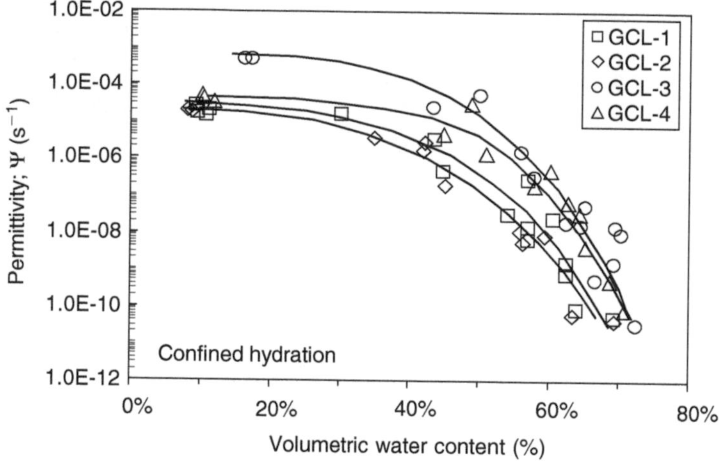


Figure 5. Variations of gas permittivity with volumetric water content for GCLs under confined hydration.

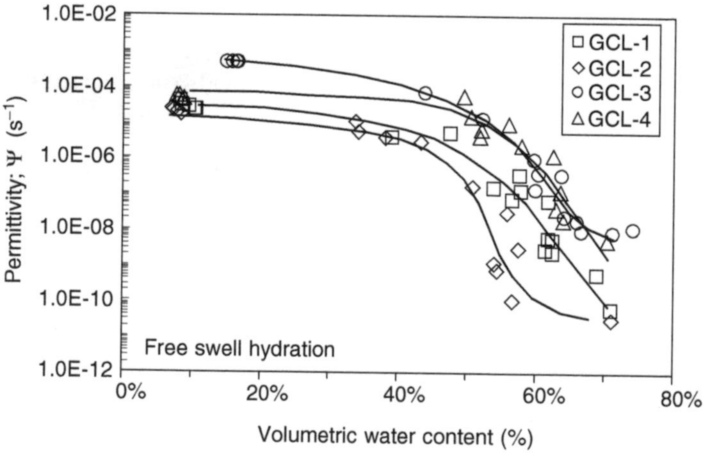


Figure 6. Variations of gas permittivity with volumetric water content for GCLs under free swell hydration.

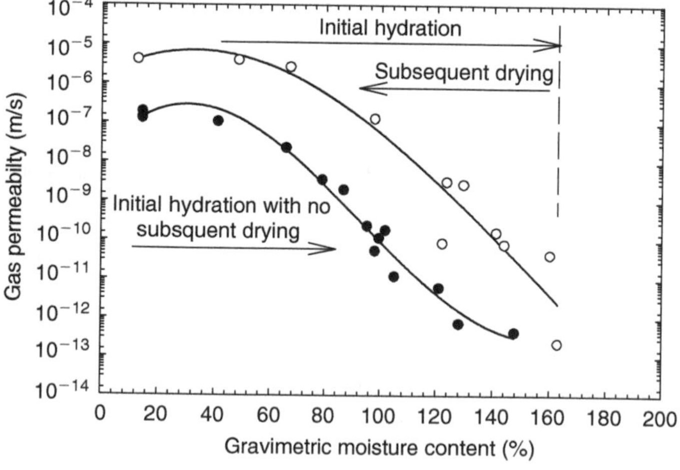


Figure 7. Variations of gas permeability versus moisture content under wetting and drying conditions.

The filled circle symbols represent GCL samples hydrated to achieve a specific moisture content, whereas the open circle symbols represent GCL samples hydrated to a moisture content of around 160% (i.e. a degree of saturation more than 80%) and then allowed to dry to given moisture contents. The results show that gas permeability decreases as the moisture content increases. However, at the same level of moisture content, the GCLs subjected to drying conditions have gas permeabilities of up to 2 orders of magnitude greater than the GCLs subjected to wetting.

Due to the very large moisture absorption and swelling capacity of bentonite, the hydrated GCLs tend to have lower gas permeability as more water is absorbed and the interconnected pore spaces in the hydrated bentonite component are decreased. In contrast, if the GCLs start to desiccate and lose the absorbed water, the hydrated bentonite in the GCL starts to shrink, which leads to a formation of interconnected gas flow paths across the bentonite layer. As a result, desiccated GCLs tend to have higher gas permeabilities. In the completely dry state (at residual shrinkage), the gas permeability of desiccated GCLs is around 1.5 orders of magnitude higher than that of the original, as supplied (non-hydrated) GCL. This difference in gas permeability is due to the fact that the bentonite component after desiccation is clearly different (see Figure 8b) from that for the original form (powder form) which does not contain any cracks.

In the present context, the desiccation of the bentonite is caused primarily by moisture loss to the atmosphere. As the desiccation process takes place (lowering in moisture content), the desiccation cracks of the hydrated bentonite component may be formed as micro cracks; visible fine cracks in the bentonite can be observed in Figure 8a at a moisture content of around 80%. Desiccation leading to cracks can be assumed to result from shear failure at the particle level and variation in suction pressures caused by the change in moisture contents. As shown in Figure 8b, an additional reduction in moisture content from 80% to 13% results in more prevalent cracking and wider cracks than are evident at a moisture content of 80%.

The larger interconnected pore spaces provide very high gas permeability compared with the finer interconnected pore spaces in the dry powdered bentonite. The variation in the gas permeability value is very small at moisture contents less than 80%, as the desiccation cracks are large enough to accommodate very high gas flow rate, rendering the effects of increase in crack size insignificant. As shown in Figure 8a, visible desiccation cracks in the bentonite were observed at a moisture content of around 80%, which is similar to a field investigation by Egloffstein (2000), where macroscopically visible desiccation cracks were observed at moisture contents ranging from 50% to 100% in GCLs exhumed from landfill covers. Also as shown in Figure 8b, the desiccation cracks are distributed more uniformly across the bentonite layer. The presence of needle punched fibres likely helped the development of uniform desiccation cracks by providing restraints against shrinkage of the bentonite as it dried. In any case, these desiccation cracks likely will close during rehydration provided the process is not impaired by reduction in swelling as a result of ion exchange or high ionic strength of the re-hydrating liquid.

For the conditions examined, and at comparable moisture contents, the desiccated GCLs tended to have higher gas permeability than the hydrated GCLs, because desiccation leads to the shrinkage of the bentonite component and to the formation of desiccation cracks that provide preferential gas flow paths. The results imply that a hydrated GCL in a cover system must be properly protected from desiccation, as there is a strong possibility that gas may escape if the GCL starts to desiccate.

### 5.3 *Effect of wet-dry cycles and ion exchange*

In the field, particularly in landfill cover system applications, a geosynthetic clay liner may be exposed to inorganic cations such as calcium, magnesium and aluminium which can alter the performance of the GCL bentonite component especially if accompanied by drying and re-wetting as a result of seasonal changes in temperature and rainfall.

Although it is documented that the self-healing capacity of GCLs is high, experimental evidence shows that this capacity can be impaired if the self-healing process is coupled with cation exchange. For example, if divalent cations such as calcium or magnesium ($Ca^{2+}$, $Mg^{2+}$) or trivalent cations are present in the infiltrating water, there can be an exchange of these cations for the monovalent sodium cation ($Na^{+}$) initially present on the bentonite of the GCL. This can cause irreversible

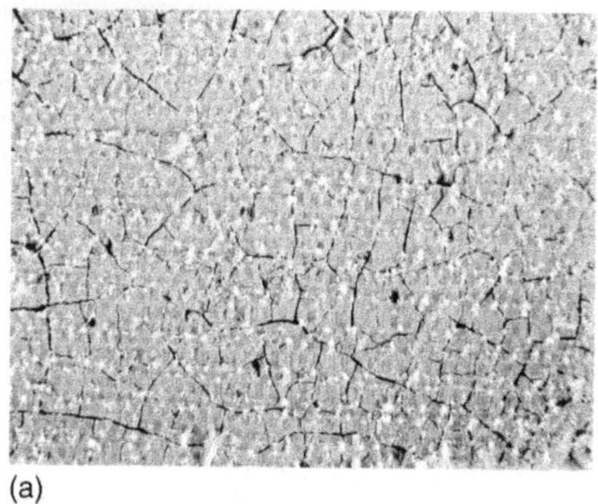
(a)

Figure 8(a). Visible macro-cracks for bentonite (moisture content = 80%), after 6 days of desiccation in the laboratory environment.

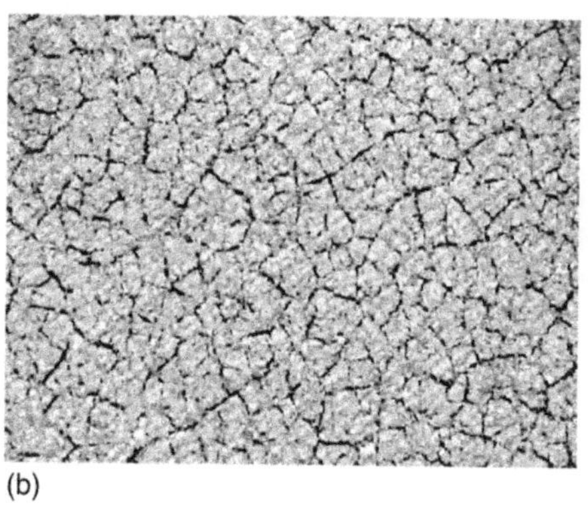
(b)

Figure 8(b). Desiccation cracks in bentonite (moisture content = 13%), after 10 days of desiccation in the laboratory environment. (Note: figure not at the same scale as Figure 5a).

damage to the bentonite resulting in a functional failure of the GCL depending on the type and concentration of cations in the pore water, the quantity of infiltrating water, the number of wet/dry cycles and the type and mass per unit area of bentonite (Dobras and Elzeas, 1993, James et al., 1997, Melchior, 1997, Lin and Benson, 2000, Bouazza et al., 2006, Meer and Benson, 2007, Benson and Meer, 2009). A detailed review on ion exchange in geosynthetic clay liners can be found in chapter 3.

Bouazza et al. (2006, 2007) conducted a large series of gas permeability tests on needle punched GCL specimens (see Table 1) which were subjected to a coupled effect of wet-dry cycles and ion exchange. Results of the gas permeability tests are summarized in Table 1. The GCL gas permeability varied with type of hydrating liquid, number of wet-dry cycles and the moisture content at the time of test. Determination of the gas permeability of the GCL hydrated with the stronger calcium chloride solution (0.125 M) was not possible since no valid relationship between differential gas pressure and the gas flow rate was obtained due to breakthrough flow during the tests. The GCLs hydrated with the calcium chloride solutions tended to have lower moisture contents than those hydrated with deionized water. This indicates, as would be expected, the reduction of water absorption capacity of the GCL after being repeatedly exposed to divalent cation solution.

As discussed earlier, sodium bentonite has a large swelling capacity when hydrated with deionized water, and is vulnerable to cracks due to shrinkage when desiccated. Without cation exchange reactions, the desiccated bentonite should be able to swell and self-heal the cracks when re-hydrated. For the conditions tested, wet-dry cycles seem to have no effect on the gas permeability of GCLs hydrated with deionized water. The gas permeability of the GCL, hydrated with deionized water and undergoing up to 10 cycles of wetting and drying, is shown in Figure 9. The GCL was able to maintain low gas permeability over these 10 cycles. As long as the cation exchange reaction between $Na^+$ ions in the bentonite and $Ca^{2+}$ ions in the hydrating water can be minimized, the gas permeability of the GCL subjected to wet-dry cycles can be comparable to that of the hydrated GCL (the baseline values). The bentonite component was able to regain its swelling capacity when re-hydrated, and all cracks that developed during desiccation were healed. The rearrangement of bentonite particles during multiple wetting and drying may also have contributed to very low gas permeability. On the other hand, cation exchange combined with wet-dry cycles had an effect on the gas permeability of the GCLs due to the change of swelling and self-healing properties of the bentonite component.

For the GCL hydrated with the weaker calcium chloride solution (0.0125 M), the bentonite component of the GCL tended to become less plastic and less swollen during the wet-dry cycles. A reduction of GCL performance as a barrier material was evident when cracks in the bentonite component were formed during desiccation. For the specimens subjected to wet-dry cycles, the bentonite component eventually lost its swelling ability and the desiccation cracks were not fully healed when re-hydrated. Preferential gas flow paths developed and the gas permeability increased. For the conditions examined, the gas permeability of the GCL hydrated with 0.0125M calcium

Table 1. Summary of gas permeability of GCLs hydrated with different solutions and subjected to wet-dry cycles.

| Sample number | Hydrating liquid | Number of cycles | Final gravimetric moisture content (%) | Gas permeability (m/s) |
|---|---|---|---|---|
| 1 | DI water | 10 | 174 | $1.22 \times 10^{-13}$ |
| 2 | DI water | 10 | 190 | No flow |
| 3 | 0.0125M $CaCl_2$ | 2 | 73 | $2.55 \times 10^{-7}$ |
| 4 | 0.0125M $CaCl_2$ | 2 | 101 | $2.55 \times 10^{-9}$ |
| 5 | 0.0125M $CaCl_2$ | 4 | 55 | $1.15 \times 10^{-6}$ |
| 6 | 0.0125M $CaCl_2$ | 4 | 85 | $3.20 \times 10^{-9}$ |
| 7 | 0.0125M $CaCl_2$ | 4 | 103 | $9.80 \times 10^{-9}$ |
| 8 | 0.0125M $CaCl_2$ | 6 | 74 | $2.35 \times 10^{-7}$ |
| 9 | 0.0125M $CaCl_2$ | 6 | 76 | NA |
| 10 | 0.0125M $CaCl_2$ | 8 | 67 | $1.05 \times 10^{-7}$ |
| 11 | 0.125 M $CaCl_2$ | 1 | 14* | $3.20 \times 10^{-3}$ |
| 12 | 0.125 M $CaCl_2$ | 2 | 70 | NA |
| 13 | 0.125 M $CaCl_2$ | 2 | 73 | NA |
| 14 | 0.125 M $CaCl_2$ | 3 | 69 | NA |
| 15 | 0.125 M $CaCl_2$ | 4 | 65 | NA |
| 16 | 0.125 M $CaCl_2$ | 4 | 68 | NA |
| 17 | 0.125 M $CaCl_2$ | 6 | 73 | NA |

*The specimen was not rehydrated before tested.
NA = Not applicable (breakthrough of gas flow occurred before gas permeability could be measured due to cracking).

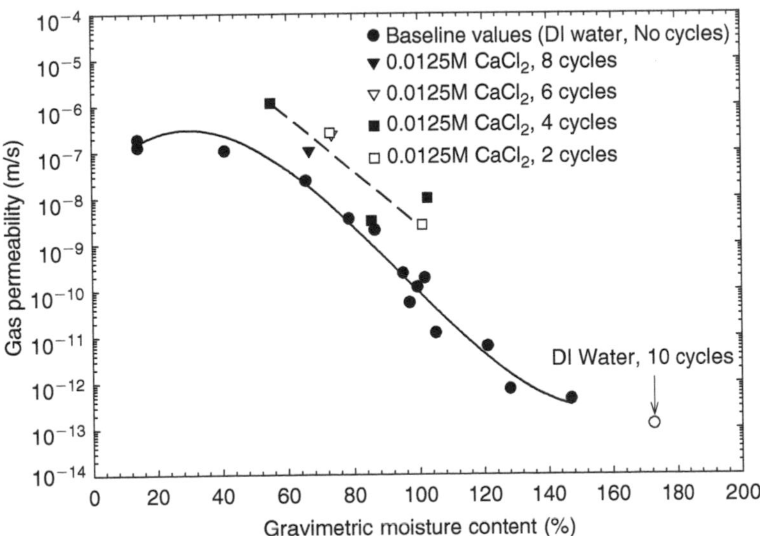


Figure 9. Effects of cation exchange and wet-dry cycles on the variation of gas permeability GCL hydrated with 0.0125 M $CaCl_2$.

chloride solution was approximately one to 1.5 orders of magnitude higher than the GCL hydrated with deionized water. This increase in gas permeability is not large due to the fact that the GCL was always dried to a constant moisture content of 50% to 60% that is usually encountered on site. The breakthrough flow observed for sample number 9 (Table 1) possibly indicated the existence of dry spots, which are known to lead to very high gas permeability (Bouazza and Vangpaisal, 2003). Figure 10 shows desiccation cracks in the bentonite component of a GCL sample hydrated with 0.0125 M calcium chloride solution after 2 wet-dry cycles. The average final moisture content

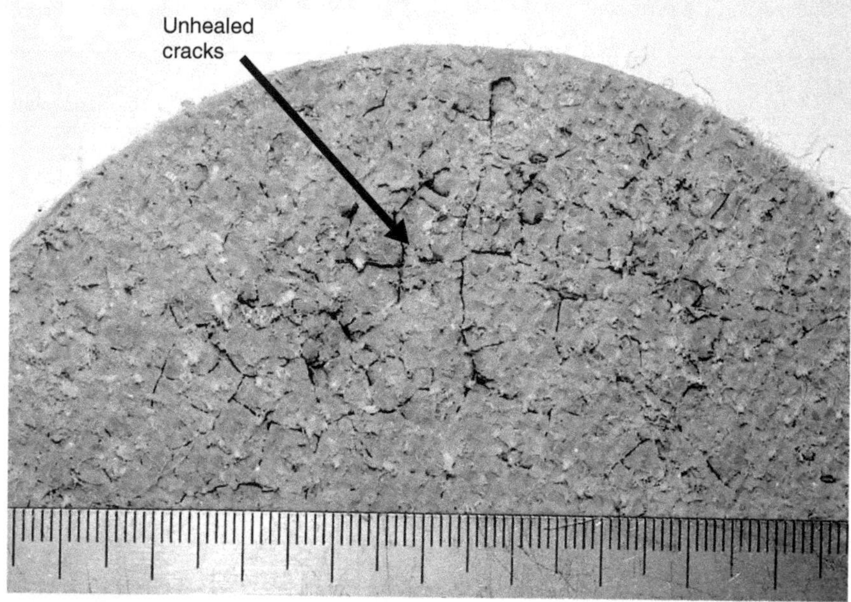


Figure 10. Unhealed desiccation cracks in the bentonite component of the GCL hydrated with 0.0125 M $CaCl_2$ solution after 2 wet-dry cycles, Moisture content = 73%.

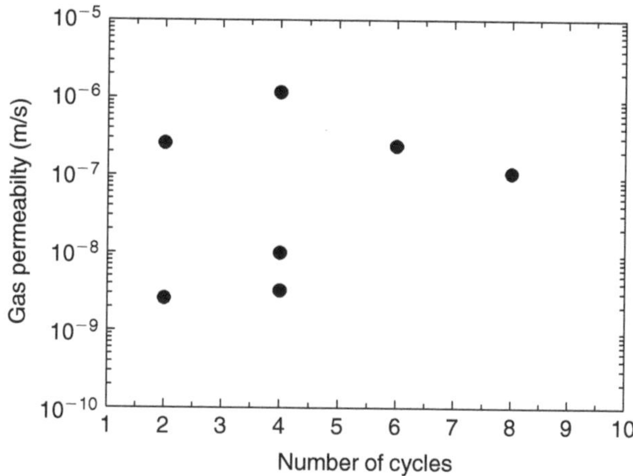


Figure 11. Variation of gas permeability versus number of cycles.

was 73%. Desiccation cracks in the bentonite were not completely healed and became preferential paths for gas migration.

There was no clear relationship between the number of wet-dry cycles (and hence the degree of cation exchange) and the reduction in gas permeability of the GCL (Figure 11). This may be due to the variation of the final moisture content of the tested GCL sample and the size of the unhealed cracks in each sample. This aspect may lead to large variations in the measured gas flow rate, as shown in Figure 9 for GCL samples hydrated with 0.0125 M calcium chloride solution, subjected to 4 wet-dry cycles and tested at moisture contents of 85% and 55%, respectively. The area covered by unhealed cracks was larger in the lower moisture content sample (55%) than in the higher moisture content sample (85%), as shown in Figure 12. This resulted in a difference of up to 2 orders of magnitude in the gas permeability between the two samples.

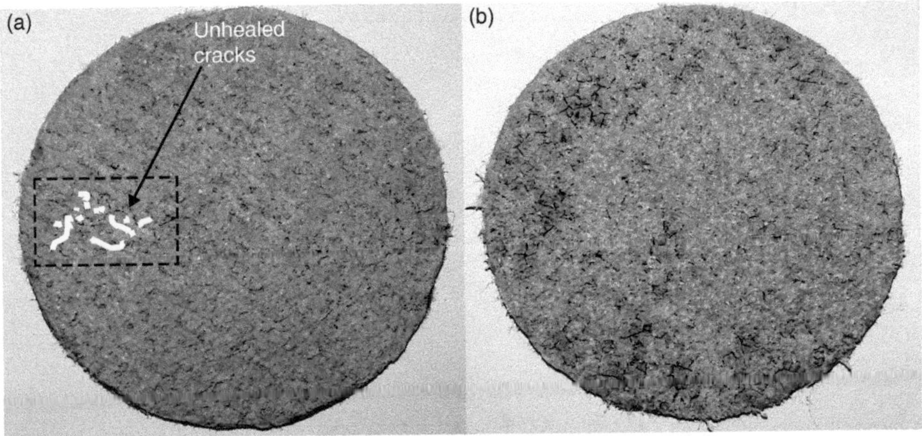


Figure 12. Unhealed desiccation cracks in the bentonite component of the GCL hydrated with 0.0125 M $CaCl_2$ solution after 4 wet-dry cycles; (a) Moisture content = 85%; (b) Moisture content = 55%. Note: cracks were highlighted in (a) due to poor contrast.

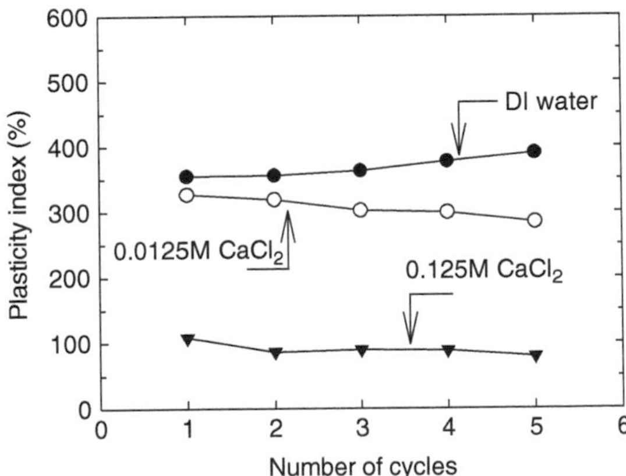


Figure 13. Plasticity index versus number of cycles.

For the GCL hydrated with a strong concentration solution (0.125 M $CaCl_2$), cracks were formed in the bentonite component of the GCL after desiccation. When re-hydrated with the same solution, the GCL was unable to regain its original swelling capacity due to the ion exchange reactions of bentonite and the compression of the diffuse double layer of bentonite by high ionic concentration of the pore water. As a result, desiccation cracks in the bentonite were not healed. Open cracks in the bentonite layer possibly were not completely healed after the first drying cycle. Although the cracks in the bentonite layer were not markedly observable because all pore spaces were filled with the hydrating water, the change of the bentonite properties was clear. The bentonite component became much softer and less plastic than in the case of hydration with deionized water as evidenced in Figure 13.

The existence of the unhealed cracks is supported by observations during the gas permeability tests. An abrupt increase in gas flow, which progressively developed, occurred for the GCL hydrated with the 0.125 M calcium chloride solution as shown in Figure 14. The existence of cracks can be explained by the reduction in the swelling and water absorption capacities of the bentonite component. The swelling on rehydration was insufficient to heal the desiccation cracks in the bentonite (Vangpaisal, 2002). However, the void spaces (fine open cracks) would have been filled with the hydrating liquid. As a result, no gas flow was observed at very low differential gas

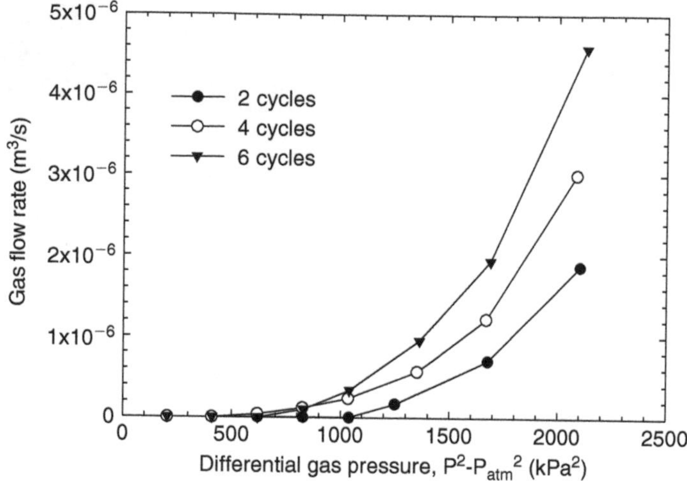


Figure 14. Variation of gas flow rate with differential gas pressure for GCL hydrated with 0.125 M $CaCl_2$ solution, indicating the non validity of Darcy's law.

pressure since all the void spaces were filled with liquid. As the differential gas pressure was increased and eventually exceeded the capillary entry pressure of the cracks, the excess hydrating liquid was expelled from the cracks and an increase in moisture content of geotextile covers and the supporting sand adjacent to the specimen was observed. Once the liquid expulsion was initiated, preferential gas flow paths were formed and resulted in a progressive increase of the gas flow rate. At higher differential gas pressures, more pore water was expelled resulting in larger preferential gas paths. The mechanism of gas flow was changed. As shown in Figure 14, the relationship of the gas flow rate to the differential gas pressure became non-linear, indicating that Darcy's law was no longer valid for the determination of gas permeability.

Although desiccation of GCLs will take much longer to occur in the field due to the presence of soil covers, and very high cation concentrations are unlikely to be present in natural soils, the test results show the potentially damaging effects of cation exchange in conjunction with wet-dry cycles on GCLs under partially saturated conditions as may exist in cover systems. Long-term GCL exposure to cation exchange and wet-dry cycles can be a major area of concern, as this can eventually change the swelling and healing ability of the bentonite component. A similar effect may be observed in cover systems where the GCL is not properly protected from the wet-dry effects (for example, through insufficient top soil cover) or from cation exchange (for example, high concentrations of leachable divalent or trivalent cations in the cover soil). If each year a GCL is hydrated by pore water containing natural divalent cations during the wet season, followed by desiccation in the dry season, then the bentonite component of the GCL will gradually become less plastic and more brittle and will swell less when re-hydrated. As a result, the GCL when partially hydrated will eventually become more permeable to gas due to larger effective pore spaces caused by the reduced swelling of the bentonite and hence the presence of unhealed desiccation cracks.

A sufficiently thick soil layer should be placed above the GCL in cover systems to avoid excessive desiccation as well as to provide the overburden stress required to close desiccation cracks in the bentonite after rehydration – and indeed to inhibit their formation by promoting vertical shrinkage rather than lateral shrinkage which leads to cracking. Natural cover soils with low divalent or trivalent cation contents leachable by ground water or acid rain water may be used to minimize the potential for cation exchange. However, before such soils are sought, assessments of the potential for cation exchange should be performed to confirm that the proposed soils can produce sufficient benefit – most soils will contain some leachable multi-valent cations and benefits may be modest. The GCL may also be covered with a geomembrane, as in a composite barrier, to protect the GCL from infiltrating water and to minimize desiccation – provided that as a result of the reduction in water input, localized desiccation does not occur due to thermally driven moisture migration.

## 6 CONCLUSIONS

Unlike the fully saturated hydraulic conductivity, the gas permeability of partially hydrated GCLs depends on the bentonite component as well as the GCL structures. The differences in GCL structures (i.e. bentonite impregnation, needle punching and stitch bonding) and the form of bentonite (granular and powdered) have a significant effect on the variation of gas permeability. The needle punched GCLs tend to have lower gas permeability than the stitch bonded GCLs, and the GCLs containing granular bentonite tend to have higher gas permeability than the GCLs containing powdered bentonite (mostly in the low range of gravimetric/volumetric water content). The bentonite impregnation of the non-woven geotextile also contributes to lower gas permeability. For comparable conditions, these effects resulted in a reduction of up to 3 orders of magnitude of gas permittivity from one GCL to another. However, the effect of the differences between the GCLs on gas permeability at high volumetric water content (>70%) was overridden by the presence of the overburden pressure during hydration. Furthermore, the overburden pressure also had an important role in the reduction of gas permeability, which implies that the GCL should be subjected to confinement at time of installation or hydration in order to obtain a low gas permeability.

The effects of cation exchange and wet-dry cycles on the gas permeability of a needle punched GCL are similar to the effects of the same factors on the hydraulic conductivity, as previously presented in the literature. However, their impacts on gas permeability are more significant than those previously reported for hydraulic conductivity, presumably due to the greater fluidity (lower viscosity) attributed to gases relative to liquids.

It is interesting to note that wet-dry cycles have no measurable effect on the gas permeability of a GCL hydrated with deionized water. On the other hand, the gas permeability of a GCL hydrated with 0.0125 M calcium chloride solution is approximately one order of magnitude higher than that of the GCL hydrated with deionized water. The apparent exchange of $Na^+$ ions in the bentonite by $Ca^{2+}$ ions in the hydrating liquid decreases the swelling and self-healing capacities of the bentonite component. Desiccation cracks formed in the bentonite component during the drying stages may not have been fully healed when re-hydrated. As a result, the gas permeability of the desiccated GCL hydrated with solutions containing divalent calcium ions are higher than that of the GCL hydrated with deionized water. The GCL hydrated with 0.125 M calcium chloride solution likely lost most of its swelling capacity. Desiccation cracks, which were formed in the bentonite component, were probably not fully healed when re-hydrated. Gas breakthrough flow is observed for all but the first wetting cycle.

## REFERENCES


Alzaydi, A.A. and Moore, C.A.1978. Combined pressure and diffusional transition region flow of gases in porous media. *American Institute of Chemical Engineering Journal*, 24(1): 35–43.

Bear, J. 1972. *Dynamics of fluids in porous media*. Dover publications Inc., 764 pp.

Benson, C., Meer, S. 2009. Relative abundance of monovalent and divalent cations and the impact of desiccation on geosynthetic clay liners, *Journal of Geotechnical and Geoenvironmental Engineering,* 135 (3): 349–358.

Bouazza, A. 2002. Geosynthetic clay liners. *Geotextiles & Geomembranes*, 20 (1): 3–17.

Bouazza, A., Vangpaisal, T. 2003. An apparatus to measure gas permeability of geosynthetic clay liners. *Geotextiles and Geomembranes,* 21(2): 85–101.

Bouazza, A., Vangpaisal, T. 2004. Effect of straining on gas advective flow of a needle punched GCL. *Geosynthetics International*, 11(4): 287–295.

Bouazza, A., Vangpaisal, T. 2007. Gas permeability of GCLs: Effect of poor distribution of needle punched fibres. *Geosynthetics International*, 14 (4): 248–252.

Bouazza, A., Vangpaisal, T., Jefferis, S. 2006. Effect of wet dry cycles and cation exchange on gas permeability of geosynthetic clay liners. *Journal of Geotechnical and Geoenvironmental Engineering*, 132 (8): 1011–1018.

Bouazza, A., Jefferis, S., Vangpaisal, T. 2007. Analysis of degree of ion exchange on atterberg limits and swelling of geosynthetic clay liners. *Geotextiles and Geomembranes*, 25 (3): 170–185.

Bouazza, A., Vangpaisal, T, Abuel-Naga, H.M., Kodikara, J. 2008. Analytical modelling of gas leakage rate through a geosynthetic clay liner-geomembrane composite liner due to a circular defect in the geomembrane. *Geotextiles and Geomembranes,* 26 (2): 109–204.

Brusseau, M.L. 1991. Transport of organic chemicals by gas advection in structured or heterogeneous porous media: Development of a model and application to column experiments. *Water Resources Research*, 27(12): 3189–3199.

Didier, G., Bouazza, A., Cazaux, D., 2000. Gas permeability of geosynthetic clay liners. *Geotextiles and Geomembranes*, 18(2): 235–250.

Dobras, T.N., Elzea, J.M. 1993. In-situ soda ash treatment for contaminated geosynthetic clay liners. *Proceedings Geosynthetics 93*, Vancouver, Canada, pp. 1145–1159.

Egloffstein, T.A. 2000. Natural bentonites - influence of the ion exchange and partial desiccation on permeability and self healing capacity of bentonites used in GCLs. *Proceedings 14th GRI conference: Hot Topics in Geosynthetics - I.*, Las Vegas, USA, 164–188.

Environment Agency, 2000. The effectiveness of liners in inhibiting the migration of landfill gas. R&D Technical Report No. 256. Swindon, U.K.

Haskell, K., Cochrane, D. 2001. Case history of the performance of a passive gas venting system in a landfill closure. *Proceedings 8th International Waste Management and Landfill Symposium*, Cagliari, 663–670.

Henken-Mellies, W.U., Zanzinger, H., Gartung, E. 2002. Long term field tests of a clay geosynthetic barrier in a landfill cover system. *Proceedings International Symposium on Clay Geosynthetics Barrier*, Nuremberg, 303–309.

Izadi, M.T., Stephenson, R.W. 1992. Measurement of gas permeability through clay soils. *Current Practices in Ground Water and Vadose Zone Investigations; ASTM STP 1118*, Philadelphia, 3–20.

James, A.N., Fullerton, D., Drake, R. 1997. Field performance of GCL under ion exchange conditions. *Journal of Geotechnical and Geoenvironmental Engineering*, 123 (10): 897–901.

Jarre, P., Mezzalama, R., Luridiana, A. 1997. "Lessons to be learned from a fatal landfill gas Explosion." *Proceedings 6th International Landfill Symposium*, Cagliari, Italy, 497–506.

Kjeldsen, P., Fisher, E.V.1995. Landfill gas migration – field investigations at Skellingsted landfill, Denmark. *Journal of Waste Management and Research*, 13: 467–484.

Koerner, R.M., Daniel, D.E. 1997. *Final covers for solid waste landfills and abandoned dumps*. ASCE Press, 256 pp.

Lin, L.C., Benson, C.H., 2000. Effect of wet-dry cycling on swelling and hydraulic conductivity of GCLs. *Journal of Geotechnical and Geoenvironmental Engineering,* 126 (1): 40–49.

Massmann, J.W. 1989. Applying groundwater flow models in vapor extraction system design. *Journal of Environmental Engineering*, ASCE, 115(1): 129–149.

McBean, E.A., Rovers, F.A., Farquhar, G.J. 1995. *Solid waste landfill: Engineering and design*. Prentice Hall PTR, New Jersey, 521 pp.

Meer, S. R., Benson, C. G. 2007. Hydraulic conductivity of geosynthetic clay liners exhumed from landfill final covers. *Journal of Geotechnical and Geoenvironmental Engineering*, 133(5):550–563.

Melchior, S. 1997. In-Situ studies on the performance of landfill caps. *Proceedings 1st International Conference on Containment Technology*, St. Petersburg, FL, pp. 365–373.

Melchior, S. 2002. Field studies and excavations of geosynthetic clay barriers in landfill covers. Proceedings International Symposium on Clay Geosynthetics Barrier, Nuremberg, 321–330.

Peggs, I.D., McLaren, S. 2002. Portable infrared spectroscopy for the rapid monitoring of leaks in geomembrane lined landfill caps. *Proceedings 7th International Conference on Geosynthetics*, Nice, vol. 2, 775–778.

Prosser, R., Janachek, A. 1995. Landfill gas and groundwater contamination. *Proceedings Landfill Closure: Environmental Protection and Land Recovery*, Geotechnical Special Publication No53, ASCE, 258–271.

Richardson, G.N., Smith,S.A., Scheer, P.K. 2008. Active LFG control: an unreliable aid to veneer stability. *Proceedings 1st Pan American Geosynthetics Conference*, Cancun, Mexico, 817–825.

Sherman, V.W. 2000. GCL durability and lifetime – A state's perception. *Proceedings 14th GRI conference: Hot Topics in Geosynthetics – I.,* Las Vegas, USA, 238–241.

Thiel, R. 1999. Design of a gas pressure relief layer below a geomembrane cover to improve slope stability. *Proceedings Geosynthetics' 99*, Boston, USA, 235–251.

Vangpaisal, T. 2002. Gas permeability of geosynthetic clay liners under various conditions. Department of Civil Engineering, Monash University.

Vangpaisal, T., Bouazza, A. 2004. "Gas permeability of partially hydrated geosynthetic clay liners." *Journal of Geotechnical and Geoenvironmental Engineering,* 130 (1):93–102.

Williams, G.M., Aitkenhead, N. 1991. Lessons from Loscoe: The uncontrolled migration of Landfill gas. *Quarterly Journal Engineering Geology*, 24: 191–207.

# CHAPTER 8

# Internal and interface shear strength of geosynthetic clay liners

J.G. Zornberg
*The University of Texas at Austin, Austin, TX, USA*

J.S. McCartney
*University of Colorado at Boulder, Boulder, CO, USA*

## 1 INTRODUCTION

Geosynthetic clay liners (GCLs) with geomembranes (GMs) placed on slopes as part of composite liner systems may be subject to a complex, time-dependent state of stresses. Stability is a major concern for side slopes in bottom liner or cover systems that include GCLs and GMs because of the wide range of commercially available GCL products, the change in behavior with exposure to water, variability in the quality of internal GCL reinforcement and GM texturing, and the low shear strength of hydrated sodium bentonite. Accordingly, proper project- and product-specific shear strength characterization is needed for the different materials and interfaces in composite liner systems.

A major concern when GCLs are placed in contact with GMs on steep slopes is the interface friction, which must be sufficiently high to transmit shear stresses generated during the lifetime of the facility. Shear stresses are typically generated in the field from static or seismic loads and waste decomposition. The need for a careful design of GCL-GM interfaces has been stressed by the failures generated by slip surfaces along liner interfaces, such as at the Kettleman Hills landfill (Byrne et al. 1992; Gilbert et al. 1998) and in the EPA test plots in Cincinnati, Ohio (Daniel et al. 1998). Another concern is the possibility of internal failure of GCLs (i.e. failure through the bentonite core), although failures have only been observed in unreinforced GCLs in the field, such as at the Mahoning landfill (Stark et al. 1998). The internal shear strength of GCLs should be characterized due to variations in shear strength due to moisture effects and manufacturing quality control. In addition, the use of GCLs in high normal stress applications such as heap-leach pads requires the identification of the internal and interface shear strength of GCLs.

Several studies have focused on experimental investigations of the different factors affecting the internal shear resistance of GCLs (Gilbert et al. 1996, 1997, Stark et al. 1996, Eid and Stark 1997, Fox et al. 1998, Eid et al. 1999; Chiu and Fox 2005; Zornberg et al. 2005) and the shear resistance of GCL-GM interfaces (Gilbert et al. 1996, 1997; Hewitt et al. 1997; Triplett and Fox 2001, McCartney et al. 2009). Some of these studies were used to guide the development of an ASTM standard for GCL internal and interface shear strength testing (ASTM D6243), which has been in effect since 1998. Several comprehensive reviews have already been compiled, including reviews on GCL internal and GCL-GM interface shear strength testing methods and representative shear strength values (Frobel 1996; Swan et al. 1999; Marr 2001; Bouazza et al. 2002; Fox and Stark 2004; McCartney et al. 2009), as well as evaluations of databases including internal and interface GCL shear strength results assembled from commercial testing laboratories (Chiu and Fox 2004; Zornberg et al. 2005; Koerner and Narejo 2005, McCartney et al. 2009). Recent research has focused on the shear strength of GCLs under dynamic loading (Fox and Olsta 2005). As a result of these studies, significant progress has been made in understanding and measuring GCL internal strength and GCL-GM interface strengths.

This chapter is geared toward providing practicing engineers a basic understanding of the variables that affect the GCL internal and GCL-GM interface shear strength determined via laboratory testing using a direct shear device with static loading. Specifically, the effects of GCL shear strength testing equipment and procedures, GCL reinforcement type, GM texturing and polymer type, normal stress, moisture conditioning, and shear displacement rate on the GCL internal and GCL-GM interface shear strength are discussed. This chapter also includes a discussion of GCL shear strength

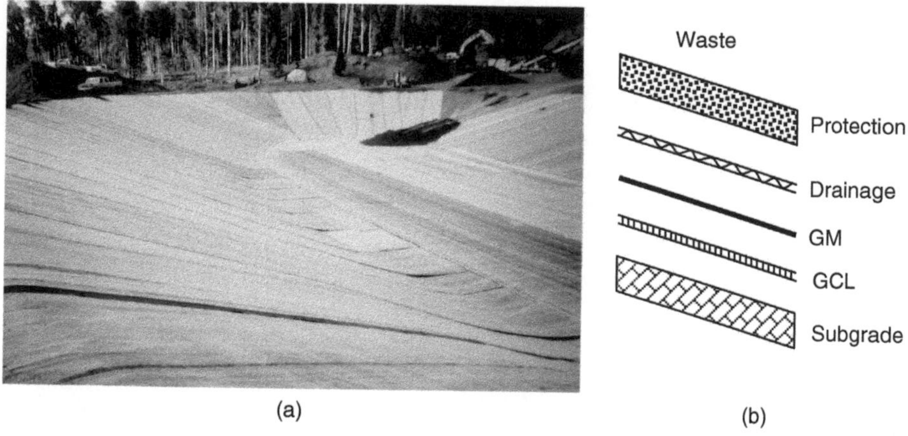


Figure 1. Composite liner system including a GCL: (a) Picture of GCL installation; (b) Components in a single composite liner system.

variability, which may be significant compared with the variability in other engineering materials. Finally, a discussion of the relationship between field and laboratory shear strength values is included. Shear strength values from the literature and from a database of 414 GCL internal and 534 GCL-GM large-scale direct shear tests presented by Zornberg et al. (2005, 2006) and McCartney et al. (2009) are used to guide the discussion. The database was assembled from tests that were performed for commercial purposes between 1992 and 2003 by the Soil-Geosynthetic Interaction Laboratory of GeoSyntec Consultants, currently operated by SGI Testing Services (SGI). Test conditions reported for each series in the GCLSS database include specimen preparation and conditioning procedures, hydration time ($t_h$), consolidation time ($t_c$), normal stress during hydration ($\sigma_h$), normal stress during shearing ($\sigma_n$), and shear displacement rate (*SDR*).

## 2 MATERIALS

### 2.1 *GCL reinforcement types*

Several unique GCL products have been proposed to offer a compromise between the hydraulic conductivity and shear strength requirements of containment projects. These products can be broadly categorized into unreinforced and reinforced GCLs. Unreinforced GCLs typically consist of a layer of sodium bentonite that may be mixed with an adhesive and then affixed to geotextile or geomembrane backing components with additional adhesives (Bouazza 2002). The geotextile or geomembrane backing components of a GCL are typically referred to as the "carrier" geosynthetics. If hydrated, the strength of unreinforced GCLs is similar to that of the bentonite component (Gilbert et al. 1996). However, they are still useful for applications where slope stability is not a serious concern. For applications that require higher shear strength, reinforced GCLs transmit shear stresses to internal fiber reinforcements as tensile forces. The two predominant methods of GCL reinforcement are stitch-bonding and needle-punching (Bouazza 2002). Stitch-bonded GCLs consist of a layer of bentonite between two carrier geotextiles, sewn together with continuous fibers in parallel rows. Needle-punched GCLs consist of a layer of bentonite between two carrier geotextiles (woven or nonwoven), reinforced by pulling fibers from the nonwoven geotextile through the bentonite and woven geotextile using a needling board. The fiber reinforcements are typically left entangled on the surface of the top carrier geotextile. Since pullout of the needle-punched fibers from the top carrier geotextile may occur during shearing (Gilbert et al. 1996), some needle-punched GCL products are thermal-locked to minimize fiber pullout. Thermal-locking involves heating the GCL surface to induce bonding between individual reinforcing fibers as well as between the fibers and the carrier geotextiles (Lake and Rowe 2000). For simplicity, thermal-locked needle-punched GCLs are typically referred to simply as thermal-locked GCLs.

Nonwoven or woven carrier geotextiles are used in fiber reinforced GCLs to achieve different purposes. Nonwoven carrier geotextiles provide puncture protection to the bentonite layer of the

GCL, allow in-plane drainage and filtration, and provide interlocking capabilities with internal fiber reinforcements and textured geomembrane interfaces (McCartney et al. 2005). Woven carrier geotextiles provide tensile resistance to the GCL and allow bentonite migration, referred to as extrusion, which leads to improved hydraulic contact (Stark and Eid 1996). However, bentonite extrusion may lead to lubrication of interfaces between the GCL and the adjacent geomembrane, lowering the shear strength (Triplett and Fox 2001; McCartney et al. 2009).

### 2.2 *Geomembrane texturing types*

Geomembranes are flexible, polymeric sheets that have low hydraulic conductivity and are typically used as water or vapor barriers. Geomembranes come in a variety of polymer types, interface characteristics and thicknesses. High Density Polyethylene (HDPE) is the polymer predominantly used in geomembranes for landfill applications due to high chemical resistance and long-term durability. However, HDPE is relatively rigid, so more flexible polymers such as Low-Linear Density Polyethylene (LLDPE), Very-Low Density Polyethylene (VLDPE), and Polyvinyl Chloride (PVC) are often used for situations in which differential deformations are expected. During shearing, the flexibility of the geomembrane is related to the formation of ridges in the direction of shearing (plowing) which may help to increase the shear strength of geomembrane interfaces (Dove and Frost, 1999).

Geomembranes may have a smooth finish or textured finish. The textured features, typically referred to as asperities, are formed either by passing nitrogen gas through the polymer during formation (coextrusion), spraying of particles onto the geomembrane during formation (impingement), or by a physical structuring process (Koerner 2005). Asperities allow greater connection between the GM and GCL, which implies that the particular type of GCL fiber reinforcement can influence the interface strength. For instance, the interface of a needle-punched GCL with entangled fibers on the surface will likely have different behavior than a needle-punched GCL that has been thermal-locked. The effect of GM thickness on the shear strength of a GCL interface has not been investigated in detail, although McCartney et al. (2009) indicate that it may not be a significant factor.

## 3 GCL SHEAR STRENGTH TESTING EQUIPMENT

### 3.1 *Shear strength testing alternatives*

Different aspects of shear testing conditions of GCLs must be thoroughly understood in order to reproduce representative field conditions in the laboratory. A number of devices have been developed to investigate the variables affecting GCL internal and interface shear strength, including the large-scale direct shear device, the ring shear device, and the tilt table (Marr 2001). The particular mechanisms of normal and shear load application, specimen size, and specimen confining method for a given device may have a significant impact on the GCL internal or GCL-GM interface shear strength.

The large-scale direct shear device is the testing approach most often used in industry, and it is recommended by the current testing standard for GCL internal and interface shear strength, ASTM D6243. The large-scale direct shear box is conceptually similar to the conventional direct shear test for soils (ASTM D3080) in that a horizontal, translational force is applied to a specimen to induce failure on a horizontal plane. However, larger shear boxes are used for GCL testing (300 mm by 300 mm in plan view) to reduce boundary effects from specimen confinement and allow a representative amount of internal reinforcements within the specimen. Figure 2 shows the picture of a direct shear device. Frobel (1996), Swan et al. (1999), and Marr (2001) provide comprehensive summaries of issues pertinent to GCL internal and GCL-GM interface shear strength testing using a large-scale direct shear device. These studies identify the practical nature of using direct shear devices regarding ease of specimen preparation and availability.

Ring shear devices have been typically used for investigation of GCL shear strength at large displacements. In this device, ring shaped specimens of GCL are cut, and the top and bottom carrier geosynthetics are clamped onto ring-shaped platens. The top platen is then rotated about a central axis with respect to the bottom platen, inducing a shear stress in the specimen. This device is capable of applying large displacements in a single direction making it suitable for investigations of

Figure 2. Large scale direct shear device.

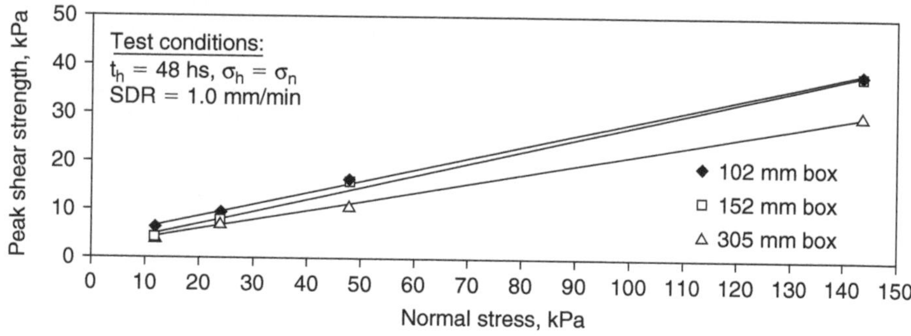


Figure 3. Effect of shear box size on GCL internal peak shear strength.

residual shear strength (Eid and Stark 1997). Additional benefits are that the contact area remains constant during shearing and the normal load moves with the top rotating rigid substrate. However, the rotational shearing mechanism may not be the same as that mobilized in the field. Ring shear devices have limitations related to specimen confinement and edge effects, testing difficulty, as well as possible lateral bentonite migration during loading (Eid and Stark 1997).

The tilt table has been employed by some laboratories to test large specimens under field loading conditions (Marr 2001). In this test, a dead weight is placed above the specimen, and the entire system is tilted slowly from one side to induce a shear force. The angle of tilt and the displacements are measured. This test is a stress controlled test (*i.e.*, a constant shear force is applied throughout the specimen due to the inclination of the specimen and normal load), so it is not capable of measuring a post-peak shear strength loss. The tilt table can only be used for a limited range of normal stresses due to safety reasons.

### 3.2 *Specimen size for the direct shear device*

The standard large-scale direct shear device for GCLs consists of separate upper and lower boxes (or blocks) which have plan dimensions of at least 300 mm by 300 mm. This size has been selected as it provides a balance between the limitations of small boxes and large boxes. Small boxes typically give higher shear strength due to boundary effects from specimen confinement, and larger boxes have problems with uneven stress distribution and specimen hydration (Pavlik 1997). The results in Figure 3 show that testing of a 305 mm square specimen result in substantially lower shear strength than testing of a 102 mm square specimen.

Large-scale direct shear devices typically allow shear displacements of 50 to 75 mm, although the area during shearing may not be constant with shear displacement. For interface testing, the bottom box in the large-scale direct shear device is occasionally 50 to 60 mm longer (in the direction of shearing) than the top box, which provides a constant area during shearing. A longer bottom

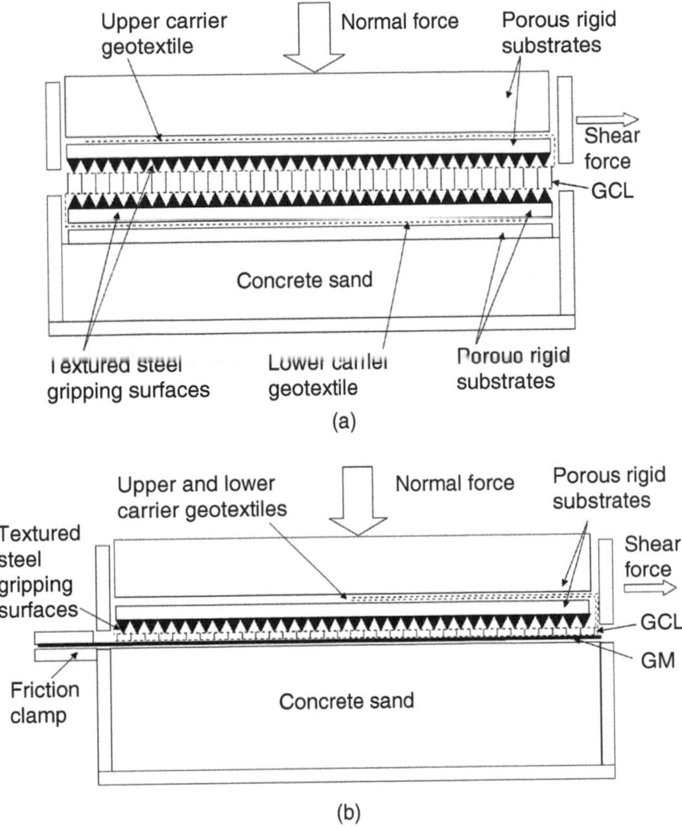


Figure 4. Large-scale direct shear device: (a) GCL internal detail; (b) GCL-GM interface detail.

shear box is not recommended when shearing GCLs internally, as previously undisturbed and unconsolidated bentonite may enter the shear plane with further displacement. To further decrease the influence of the boundaries and allow greater shear displacements, Fox et al. (1997b) developed a larger direct shear device with dimensions of 405 mm by 1067 mm and maximum shear displacement of 203 mm.

### 3.3 *GCL specimen confinement for the direct shear device*

The direct shear device configurations for internal GCL and GCL-GM interfaces are similar, differing only in specimen confinement. Swan et al. (1999) provides a review of issues relevant to the impact of specimen confinement on GCL internal and GCL-GM interface shear strength. Figure 4(a) shows the schematic view of a GCL confined within a direct shear box for internal shearing, while Figure 4(b) shows a GCL and GM confined within a direct shear box for interface shearing.

According to ASTM D6243, the GCL specimen should be confined between two porous rigid substrates (usually plywood, porous stone or porous metal) using textured steel gripping teeth, which are placed between the upper and lower boxes. The steel gripping teeth allow the shear force applied to the box to be transferred completely to the inner GCL interface. Slippage between the rigid substrate and the GCL should be minimized during shearing, as this may cause shear stress concentrations or tensile rupture of the carrier geosynthetics, which are not representative of field failure conditions. Figure 5 shows the picture of a specimen being manually trimmed to the dimensions of the rigid substrate. To speed the time of trimming GCL specimens for commercial testing, cutting dies have been developed to trim the GCL to the correct dimensions with minimal loss of bentonite. The steel gripping teeth shown in Figure 5 were constructed by adhering steel rasps

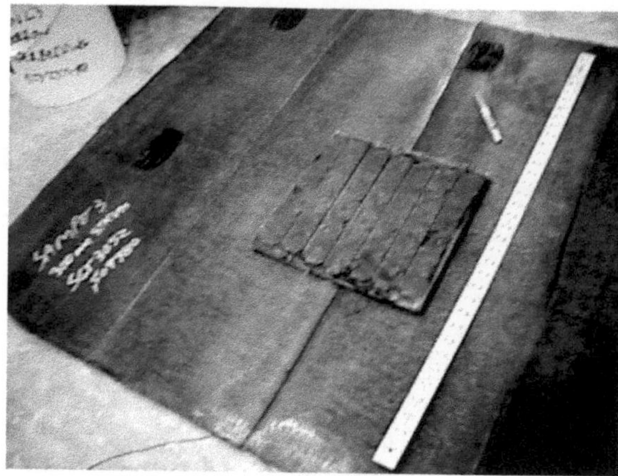

Figure 5. Specimen being trimmed to the dimension of the textured rigid substrate.

Figure 6. Specimen confinement for GCL (top) and geotextile (bottom) interface testing.

(truss plates) to the rigid substrate. More intense gripping surfaces have recently been proposed to eliminate slippage problems under low confining pressures, which involve sharp teeth that enter into the carrier geotextile of the GCLs. The Geosynthetics Research Institute (GRI) is developing a new standard (GRI GCL3) that includes recommendations for GCL confinement.

The top or bottom carrier geosynthetics may be clamped into position by wrapping a flap of the GCL around the rigid substrate, and placing another rigid substrate onto the flap. Figure 6 shows a picture of a GCL-geotextile interface in which the specimens are wrapped around the rigid substrates. To provide confinement, a second rigid substrate is placed atop this assembly, which effectively clamps the top portion of the GCL into place. Gluing should not be used, as it may affect the behavior of the internal reinforcements.

For GCL internal shear strength testing, the GCL is positioned in the direct shear device so that the top box is attached to the top carrier geosynthetic, and the bottom box is attached to the bottom carrier geosynthetic. The confined GCL specimen and the rigid substrates are placed atop a foundation of concrete sand, plywood or PVC plates in the bottom box, and the top box is placed so that it is in line with the top rigid substrate. The width of the gap between the boxes is adjusted to the mid-plane of the GCL.

For GCL-GM interface testing, the GCL is typically attached to the rigid substrate by wrapping extensions of the carrier geosynthetics around the textured rigid substrate in the direction of shear, then placing another rigid substrate above this to provide a frictional connection. As a GM is stiffer than a GCL and cannot be wrapped around the rigid substrate, it is typically placed atop a foundation of concrete sand (flush with the top of the bottom box). The GM is connected with a frictional clamp to the bottom box, on the opposite side to that of the shearing direction. As local slippage has been observed between the sand and GM in low normal stress tests, rigid substrates with gripping teeth have been proposed to enhance contact between the GM and the bottom box. The gap is then set to the height of the GCL-GM interface. Although the grip system forces failure at the GCL-GM interface, slack should remain in the carrier geosynthetics so that the initial shear stress distribution is not influenced by the grips.

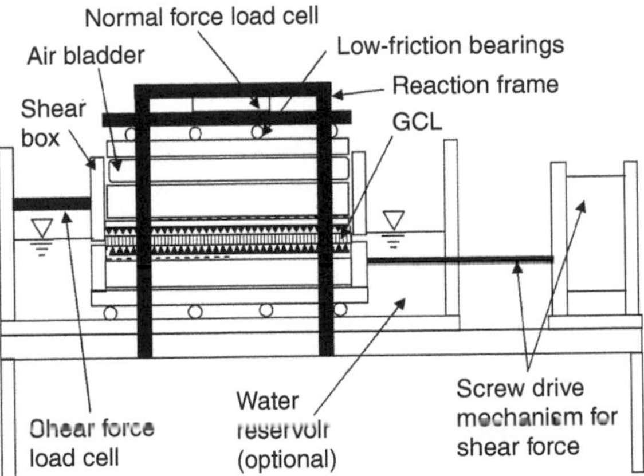


Figure 7. Load application configuration in the direct shear box (shearing occurs to the left).

The specimens may also be gripped so that failure occurs along the weakest interface (*e.g.*, either internally within the GCL or at the GCL-GM interface), in which case the extension of the lower GCL carrier geosynthetic is not wrapped around the rigid substrate. The gap setting should be wide enough to allow failure on the weakest plane.

GCL specimen sampling from different sections of the roll is not specifically addressed by ASTM D6243, although it is stated that specimens should not be chosen from near the edge of the GCL roll (a minimum distance of 1/10 the total width of the GCL roll). A specimen with a width of 305 mm and a length of twice the shear box (610 mm) should be trimmed from a bulk GCL specimen.

### 3.4 *Normal and shear load application for the direct shear device*

Figure 7 shows a schematic view of the load application configuration for a large-scale direct shear test. Direct shear devices can typically apply normal forces to the specimen ranging from 450 N (4.8 kPa for a 305 mm square specimen) to 140,000 N (1915 kPa for a 305 mm square specimen). Dead weights can be placed over the specimen in tests conducted under normal forces less than 500 kg, while an air bladder or hydraulic cylinder are used to apply higher forces between the specimen and a reaction frame, as shown in Figure 7. Dead weights are typically discouraged due to rotation of the top box during shear, which leads to an uneven normal stress distribution. The normal force is typically measured using a load cell placed under the cylinder, or using a system of load cells placed between a load distribution plate and the top rigid substrate. The latter option allows definition of the stress distribution during shearing. Although the top box remains stationary during testing, rollers are typically placed between the reaction frame and the load distribution plate to prevent moments induced during shearing, which may affect the shear force reading on the load cell. Most direct shear devices include an optional water bath for submerged testing. Uniform application of the normal stress over the area of the GCL is critical, as lateral migration of hydrated bentonite may occur (Stark 1998).

For strain-controlled tests, a constant shear displacement rate is typically applied to the bottom box. As the normal load is applied to the stationary top box, translating the bottom box prevents the normal load from translating across the GCL specimen during shear. A guiding system of low friction bearings must be used to ensure that the movement between the boxes is in a single direction. The industry standard large scale direct shear device uses a mechanical screw drive mechanism to apply the constant shear displacement. The screw drive mechanism permits shear displacement rates ranging from 1.0 mm/min (time to failure of 75 minutes) to 0.0015 mm/min (time to failure of 35 days). The displacement rate is typically measured using linearly variable displacement transformers (LVDTs). For stress-controlled tests, a constant shear force is applied to the bottom box, and the resulting displacements are measured. The shear force is typically applied using a system of pulleys connected to a dead weight, or using an electric screw drive mechanism with a feedback system that regulates a constant pressure as displacement occurs. The shear force is increased incrementally to reach failure.

## 4 TESTING PROCEDURES

### 4.1 *Moisture conditioning*

The bentonite clay in GCLs is initially in a powdered or granular form, with initial moisture content of approximately 12%. Moisture conditioning of GCL specimens for shear strength testing involves hydration and (in some cases) subsequent consolidation of the bentonite. During hydration, the bentonite absorbs water and increases in volume. Daniel et al. (1993) indicated that GCLs are typically expected to reach full hydration in the field unless encapsulated between two geomembranes. GCLs should be hydrated with a liquid that is representative of liquids found in the field. GCLs used in landfill liners may become hydrated before waste placement or after construction with waste leachate, while landfill covers may become hydrated from percolation through the vegetated cover or moisture in landfill gases (Gilbert et al. 1997; Bouazza 2002).

The hydration process typically used for GCLs is described by Fox et al. (1998). The specimen and rigid substrates are placed under a specified hydration normal stress ($\sigma_h$) outside the direct shear device and simultaneously submerged in a specified liquid, and allowed to hydrate during a specified hydration time ($t_h$). This assembly is then transferred to the direct shear device. Shearing commences immediately for specimens hydrated under the normal stress used during shearing. To simulate hydration of the GCL before loading occurs (*i.e.*, before waste placement or cover construction), the GCL is allowed to hydrated under a lower normal stress than that used during shearing. After this point, the GCL may or may not be consolidated before shearing. The GCL may not be consolidated to simulate situations where normal stress increases quickly in the field and drainage doesn't occur. When consolidating GCLs, the normal stress is increased in stages during a specified consolidation time ($t_c$), or until vertical displacement ceases.

Figure 8(a) shows the average moisture content of GCLs during hydration under different normal stresses. For low normal stresses, a significant increase in moisture content is observed, with constant moisture content observed after approximately 4 days. However, Gilbert et al. (1997) and Stark and Eid (1996) indicate that complete hydration typically requires 2 weeks. A percentage change in vertical swell of less than 5% can be obtained for unconfined GCLs in a period of 10 to 20 days, indicating full hydration (Gilbert et al. 1997). As the hydration normal stress increases, the moisture content only increases to 70%, and does not change significantly in water content with time.

Figure 8(b) shows the spatial distribution in moisture content in a needle-punched GCL specimen after different times of hydration. The specimen had an initial gravimetric moisture content of 12% and was hydrated under a normal stress of 9.6 kPa. Significant hydration occurs during 24 hs (moisture content change of 123%), although hydration continues after this time. The edge of the specimen has a greater increase in moisture content than the center, likely due to a rigid substrate with poor drainage. Pavlik (1997) and Fox et al. (1998) found that increased specimen size results in incomplete hydration at the center of GCL specimens as there is little lateral movement of water through the bentonite and carrier geotextiles. ASTM D6243 recommends using rigid substrates that are porous or have grooves to channel water during testing along with a time of hydration greater than 24 hs. Porous, textured, rigid substrates allow even hydration and better dissipation of pore water pressures during shear.

Figure 9(a) shows the vertical displacement during hydration for 4 needle-punched GCLs under different hydration normal stresses. Swelling occurs for GCLs with low hydration normal stress, and compression occurs during hydration for higher normal stress. The trend in vertical displacement with hydration normal stress is indicative of the swell pressure of the bentonite, which is defined as the hydration normal stress at which the GCL does not swell beyond its initial thickness. Petrov et al. (1997) reported swell pressures ranging from 100 to 160 kPa for a thermal-locked GCL, while lower values were reported by Stark (1997) for a needle-punched GCL. Zornberg et al. (2005) interpreted the internal peak shear strength of GCLs with different normal stresses, and shear displacement rates to infer that GCLs sheared internally will likely change in behavior above and below the swell pressure.

Figure 9(b) shows the vertical displacement of a GCL during hydration under a low normal stress and subsequently consolidated to a normal stress of 1000 kPa in stages over nearly two years. The

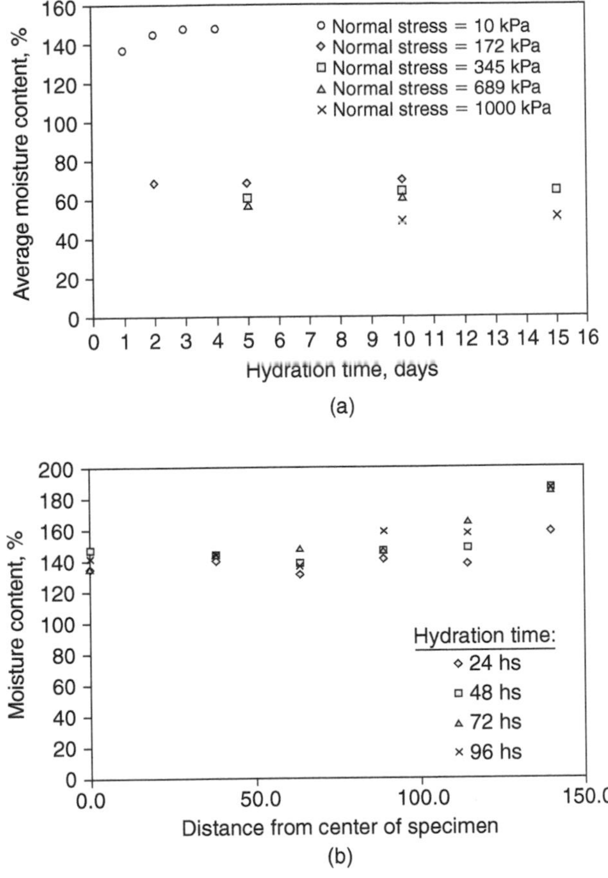


Figure 8. (a) Variation in bentonite moisture content with time during hydration under increasing $\sigma_h$; (b) Spatial variation in moisture content with time in a needle-punched GCL specimen.

specimen had an initial thickness of approximately 20 mm before hydration. Approximately 4 mm of vertical displacement was observed due to swelling during hydration, followed by 15 mm of settlement during consolidation. In practice, the normal load is applied in a single increment. This can lead to significant lateral movement of bentonite within the GCL, as well as extrusion through the carrier geotextiles. Lake and Rowe (2000) and Triplett and Fox (2001) observed extrusion of bentonite during moisture conditioning as well as during shearing. Bentonite extrusion may lead to lubrication of the GCL-GM interface, and may prevent drainage of water through the carrier geotextiles.

### 4.2 *Shearing procedures*

Shearing is conducted after GCL conditioning by applying the shear load under a constant shear displacement rate. ASTM D6243 recommends using a shear displacement rate (SDR) consistent with the conditions expected in the field application. As it is not likely for pore pressures to be generated in the field for drained conditions such as those during staged construction of landfills (Gilbert et al. 1997; Marr 2001), an adequately low shear displacement rate should be used to allow dissipation of shear-induced pore water pressure. The approach used by Gibson and Henkel (1954) is typically used to define the shear displacement rate, as follows:

$$SDR = \frac{d_f}{50 t_{50} \eta} \tag{1}$$

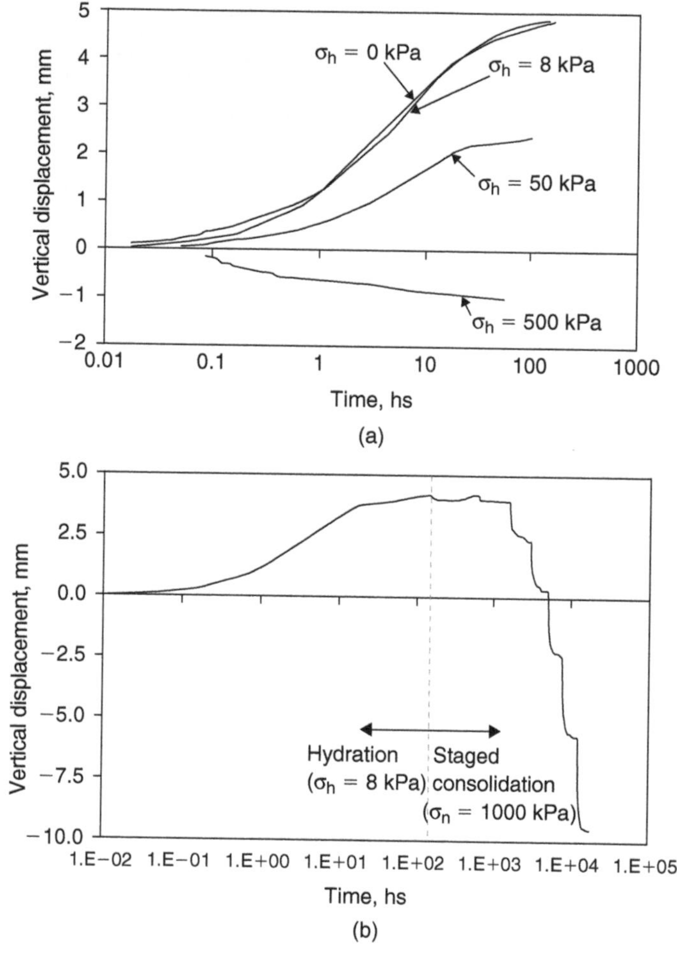


Figure 9. Conditioning of needle-punched GCLs: (a) Effect of hydration normal stress on vertical displacement during hydration; (b) Vertical displacement during hydration and consolidation.

where $d_f$ is the estimated horizontal displacement, $t_{50}$ is the time required to reach 50% consolidation assuming drainage from the top and bottom of the specimen, and $\eta$ is a factor used to account for drainage conditions. An appropriate value of $t_{50}$ can be obtained using the vertical displacement measurements during hydration and consolidation for the GCL, shown in Figure 9(a). However, the determination of $t_{50}$ may take several days, which may be prohibitive for commercial testing programs. In practice, engineers typically prescribe faster shear displacement rates to prevent long testing times and increased costs, despite the possible effects of shear-induced pore water pressure generation on the shear strength results. While relatively fast for guaranteeing drained conditions anticipated in the field, a *SDR* of 1.0 mm/min is typically used in engineering practice because of time and cost considerations. For $d_f$, ASTM D6243 requires a minimum of 50 mm of displacement when reporting the large-displacement shear strength of a GCL. ASTM D6243 defines the possible values of $\eta$ to be 1 for internal shearing, 4 for shearing between the GCL and an impermeable interface, and 0.002 for shearing between the GCL and a permeable interface.

The shear displacement is typically measured using an LVDT or dial gauge, and the shear force measured using a load cell, are used to define the shear stress-displacement curve. Peak shear strength is reported as the maximum shear stress experienced by the interface. The large-displacement shear strength is reported as the shear stress when there is constant deformation with no further change in shear stress, or the shear stress at a displacement of 50 to 70 mm. The

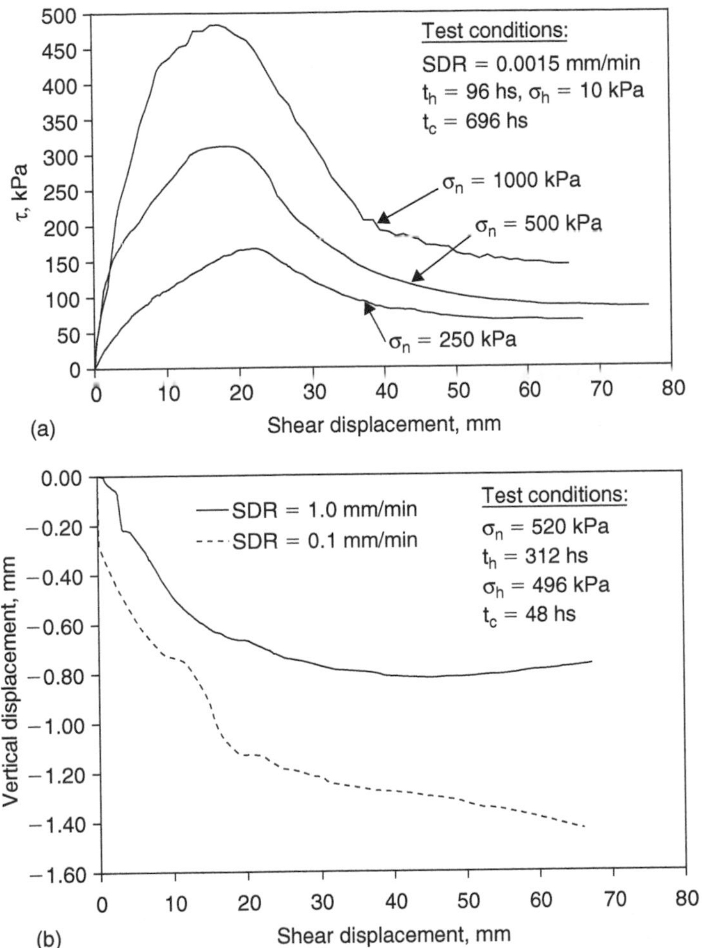


Figure 10. (a) Shear stress – shear displacement curves for needle-punched GCL sheared internally under different normal stresses; (b) Vertical displacement – Shear displacement curves.

large-displacement shear strength is the shear stress that remains after all fiber reinforcements (if any) in the failure plane rupture and the soil particles in the shear zone align into the direction of shear. Large-displacement shear strength is typically reported instead of residual shear strength because the shear displacement capability of the direct shear device is often not sufficient to mobilize GCL residual shear strength. The residual shear strength of a GCL may not be reached until displacements as large as 700 mm (Fox et al. 1998), although some testing facilities have observed residual conditions after approximately 300 mm.

## 5 GCL INTERNAL SHEAR STRENGTH

### 5.1 *Shear stress-displacement behavior*

Figure 10(a) shows a typical set of shear stress-displacement curves for a needle-punched GCL tested under a wide range of normal stresses. A prominent peak value is observed at a shear displacement less than 25 mm, followed by a post-peak drop in shear strength. As normal stress increases, the secant modulus at 10 mm of displacement increases significantly, and the displacement at peak increases. For similar normal stress and conditioning procedures, Hewitt et al. (1997), Fox et al. (1998) and Zornberg et al. (2005) found that the needle-punched and thermal-locked GCLs have similar shaped shear-stress displacement curves with peak strength occurring at shear

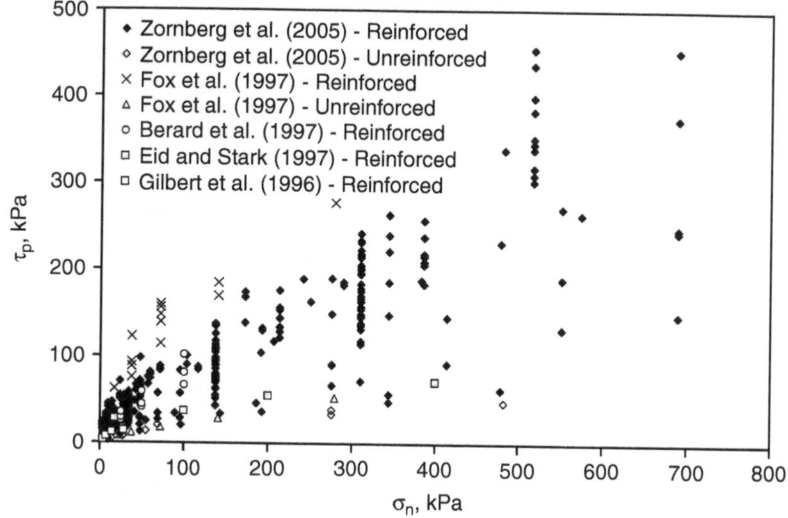


Figure 11. Overview of GCL internal peak shear strength values.

displacements ranging from 10 to 30 mm, while stitch-bonded GCLs have lower peak shear strength values occurring at a shear displacement ranging from 40 to 70 mm. Due to the greater amount of reinforcement in needle-punched GCLs than in stitch-bonded GCLs, needle-punched GCLs act in a more brittle manner than stitch-bonded.

Figure 10(b) shows typical vertical displacement vs. shear displacement curves for GCLs sheared internally under the same normal stress and conditioning procedures, but different shear displacement rates. The data in this figure indicate compressive deformations during shearing, which indicate consolidation of the GCL. Consolidation is likely occurring in these tests due to ongoing dissipation of excess pore water pressures from the increase in normal stress from 500 to 520 kPa, as well as due to dissipation of shear-induced positive pore water pressures. It should be noted that vertical displacement measurements in direct shear tests only measure the overall vertical displacement of the top box, which may be caused by GCL expansion/contraction during shear, continued consolidation of the GCL, and tilting of the top box from moments caused by the increase in horizontal stress. These changes may not be the same as those on the failure surface.

Inspection of the shear displacement curves may indicate the quality of the test. For example, testing problems such as slippage between the gripping system and GCL or plowing of the grip system into the GCL may manifest in the shear displacement curves. Slippage is typically indicated by a very rough shear displacement curve with erratic changes in shear load with increased displacement. Rapid application of the normal stress during consolidation may lead to excessive bentonite extrusion from the GCL, which may in turn cause interaction between the upper and lower grips during shear. Interaction between the grips is typically indicated by a constantly increasing shear stress, or a peak shear strength value that occurs at a displacement greater than 25 mm. A quality shear strength test will have a shear displacement curve with a peak occurring at less than 25 mm and a smooth post-peak reduction to a large-displacement shear strength resulting in a large-displacement friction angle of 8 to 13 degrees.

## 5.2 *Preliminary shear strength overview*

Figure 11 shows GCL internal peak shear strength values $\tau_p$ reported in the literature. The wide range in shear strength reported by the different studies can be explained by differences in GCL reinforcement types, moisture conditioning procedures, shear displacement rates, as well as testing procedures and equipment used in the various studies, although significant variability is still apparent. Generally, reinforced GCLs show higher shear strength and greater variability than unreinforced GCLs.

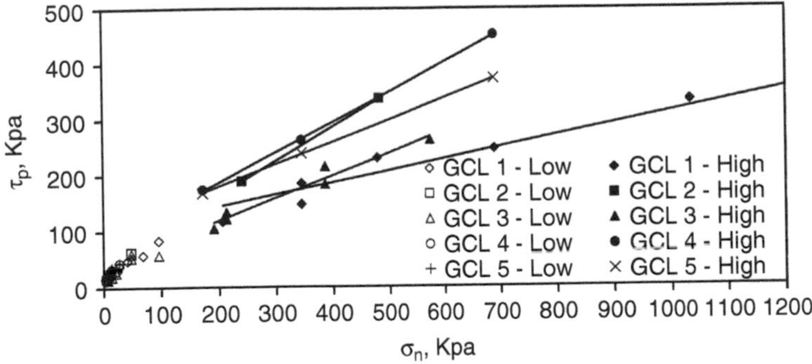


Figure 12. Normal stress effects on needle-punched GCL internal shear strength.

## 5.3 *Variables affecting GCL internal shear strength*

### 5.3.1 *Effect of normal stress*

GCLs are frictional materials, so their shear strength increases with normal stress. Also, the internal reinforcements give the GCL strength at low normal stress. Accordingly, the GCL internal peak shear strength for a set of tests with the same conditioning procedures and shear displacement rate is typically reported using the Mohr-Coulomb failure envelope, given by:

$$\tau_p = c_p + \phi_p \tan \sigma_n \tag{2}$$

where $\tau_p$ is the peak shear strength, $c_p$ is the cohesion intercept, $\sigma_n$ is the normal stress and $\phi_p$ is the interface friction angle. ASTM D6243 requires a minimum of three points $[(\tau_{p1}, \sigma_{n1}), (\tau_{p2}, \sigma_{n2}), (\tau_{p3}, \sigma_{n3})]$ to define the peak or residual failure envelope for the given interface. Chiu and Fox (2004) and Zornberg et al. (2005) provide a range of internal peak shear strength parameters for different GCLs, hydration procedures, ranges of normal stresses, and shear displacement rates. Zornberg et al. (2005) also found that large-displacement shear strength was represented well by a linear failure envelope, although the cohesion intercept was typically negligible.

Most previous studies on GCL shear strength (Gilbert et al. 1996; Daniel and Shan 1993; Stark and Eid 1996; and Eid and Stark 1997, 1999) were for tests under low levels of normal stresses (typically below 200 kPa). Fox et al. (1998), Chiu and Fox (2004), and Zornberg et al. (2005) report shear strength values for a wider range in normal stresses, and indicate that linear failure envelopes do not represent the change in shear strength with normal stress. Accordingly, multi-linear or nonlinear failure envelopes are recommended. Gilbert et al. (1996) and Fox et al. (1998) used the model presented by Duncan and Chang (1970) to represent nonlinear trends in shear strength with normal stress, as follows:

$$\tau_p = \sigma_n \left[ \phi_0 - \Delta\phi \log\left(\frac{\sigma_n}{P_a}\right) \right] \tag{3}$$

where $P_a$ is the atmospheric pressure, $\phi_0$ is the secant friction angle at atmospheric pressure, and $\Delta\phi$ is the change in secant friction angle with the log of the normal stress normalized by $P_a$.

Figure 12 shows the trend in GCL internal peak shear strength with normal stress for different needle-punched GCLs. The results within each data set were obtained from direct shear tests with the same conditioning procedures and shear displacement rate. For tests under low normal stress ($\sigma_n < 100$ kPa), a significant increase in peak shear strength is observed with normal stress, while for tests under high normal stress ($\sigma_n > 100$ kPa), a less prominent increase in peak shear strength is observed.

The data in Figure 12 suggests that bilinear failure envelopes represent the data well for normal stresses under low normal stresses (<100 kPa) and high normal stresses (>100 kPa). Zornberg et al. (2005) reported shear strength parameters for several bilinear failure envelopes. In addition, Chiu and Fox (2004) reported shear strength parameters for nonlinear failure envelopes. Both approaches tend to represent the shear strength of GCLs over a wide range in normal stress. The change in GCL shear strength behavior with normal stress is important to consider when specifying

Table 1. Shear strength parameters for GCL internal peak shear strength.

| GCL description | Peak envelope | | |
|---|---|---|---|
| | $c_p$ (kPa) | $\phi_p$ (Degrees) | $\tau_{50}$ (kPa) |
| Reinforced GCLs | 40.9 | 18.0 | 57 |
| Unreinforced GCLs | 5.0 | 5.7 | 10 |
| Needle-punched GCLs | 40.5 | 19.5 | 61 |
| Stitch-bonded GCLs | 28.5 | 5.6 | 33 |
| Thermal-locked GCLs | 33.2 | 22.7 | 54 |
| W-NW needle-punched GCLs | 19.1 | 40.9 | 58 |
| NW-NW needle-punched GCLs | 35.0 | 24.5 | 58 |

normal stresses to define a failure envelope, which should always be within the range expected in the application.

The normal stress applied to a GCL may affect the lateral transmissivity of the carrier geotextiles. Depending on if the carrier geotextile is woven or nonwoven and the carrier geotextile compressibility, high normal stresses may lead to decreased lateral transmissivity of the carrier geotextiles. Combined with bentonite extrusion, the carrier geotextiles may not aid the dissipation of shear-induced pore water pressures under high normal stresses. The normal stress may also affect the strength of GCL reinforcements. Gilbert et al. (1996) reported that the resistance of fiber reinforcements to pullout from the carrier geotextiles increased with normal stress because of the frictional nature of the connections.

### 5.3.2 *GCL Reinforcement*

Table 1 shows typical shear strength parameters for different sets of GCLs reported by Zornberg et al. (2005). The peak shear strength calculated at a normal stress of 50 kPa, $\tau_{50}$, is also shown in Table 1 for comparison purposes. The peak internal shear strength of reinforced GCLs is significantly higher than that of unreinforced GCLs. The reinforced GCLs have a substantial intercept, while the unreinforced GCLs have a relatively low cohesion intercept and friction angle. The data in Table 1 indicates that needle-punched GCLs and thermally-locked GCLs have similar shear strength, while stitch-bonded GCLs have lower shear strength.

The difference between the needle-punched and thermal-locked GCLs may be explained by the pullout of reinforcements from the woven geotextile of the thermal-locked GCL during hydration and shearing (Lake and Rowe 2000). The fiber reinforcements in needle-punched GCLs are typically left entangled on the surface of the woven carrier geotextile, so significant swelling or shear displacement is required for pullout of the fibers from the carrier geotextile. On the other hand, the fibers in thermal-locked GCLs are melted together at the surfaces of the woven carrier geotextile. Stitch-bonded GCLs have less fiber reinforcement per unit area (stitches are typically at a 3-inch spacing), but the fiber reinforcements are continuous throughout the length of the GCL. Fox et al. (1998) and Zornberg et al. (2005) observed that the continuous fiber reinforcements in GCL *B* did not break during shearing, but instead the woven carrier geotextile ruptured at large shear displacements. The lower reinforcement density and mechanism of failure influences in the direct shear device leads to the low shear strength of these GCLs. Stitch-bonded GCLs are typically not used in practice. Table 1 also indicates that GCLs with nonwoven carrier geotextiles have similar shear strength at a normal stress of 50 kPa to woven carrier geotextiles. However, the greater friction angle of GCLs with woven carrier geotextiles leads to higher shear strength at high normal stresses for these GCLs.

Fox et al. (1998) found that the type of fiber reinforcement used in GCLs (needle-punched or stitch-bonded) has minor effect on the residual shear strength of GCL, although Zornberg et al. (2005) found that the type of fiber reinforcement still has an effect on the shear strength at a displacement of 75 mm.

Many studies have been conducted to investigate whether the internal shear strength of a needle-punched GCLs vary with the amount of needle punching per unit area of the GCL. Needle-punched GCLs are manufactured using a production line assembly which employs several threaded needles

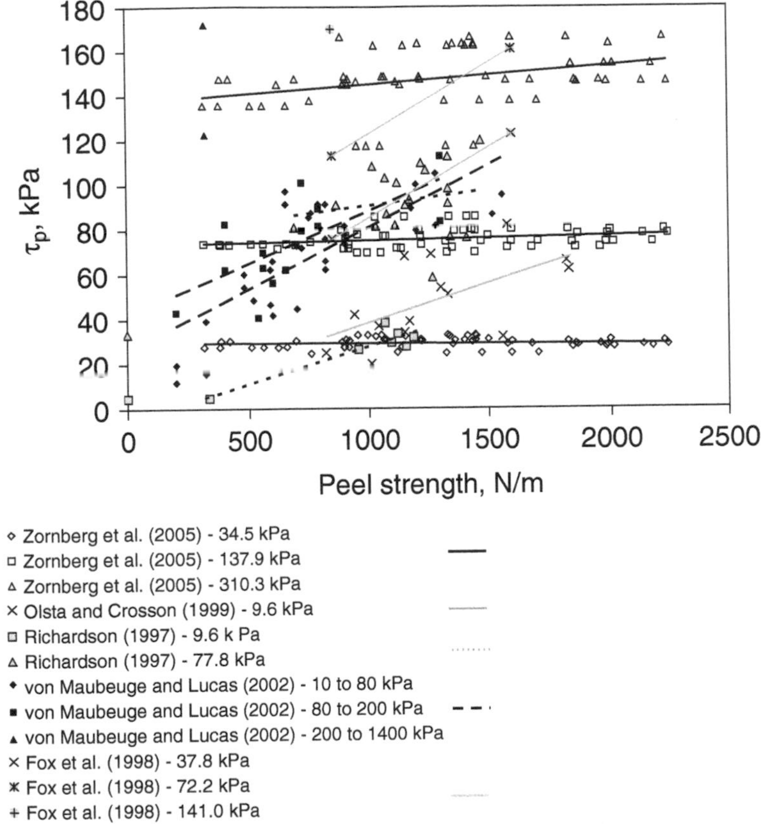


Figure 13. Relationship between GCL internal peak shear strength and GCL peel strength.

connected to a board (von Maubeuge and Ehrenberg 2000). As the lifetime of the needle-punching boards increases, more needles break and a lower density of fiber reinforcements may be apparent in the GCL with wear of the needle-punching board. The peel strength test (ASTM D6496) has been used as a manufacturing quality control test, as well as an index of the density (and possibly the contribution) of fiber reinforcements in needle-punched GCLs (Heerten et al. 1995, Eid and Stark 1999). Several studies have correlated the peak internal shear strength of needle-punched GCLs with peel strength (Berard 1997; Richardson 1997; Fox et al. 1998; Eid et al. 1999; Olsta and Crosson 1999; von Maubeuge and Lucas 2002; Zornberg et al. 2005). Figure 13 shows a comparison of the trends in peak shear strength of needle-punched GCLs with peel strength from several of these studies. An increase in peak shear strength is apparent in some of these data sets, although only Zornberg et al. (2005) used the same GCL, same conditioning procedures, and shear displacement rate. Overall, no trend is observed in this data.

Stark and Eid (1996) performed shear strength tests on reinforced GCLs with and without a sodium bentonite component (filled and unfilled, respectively) to find the effect of the reinforcement of the shear strength of reinforced GCLs. They found that the peak shear strength of unfilled GCLs was higher than that of filled GCLs, which indicates that the shear resistances of the sodium bentonite and the reinforcements are not additive. This trend may be due to pullout of the fiber reinforcements due to swelling of the bentonite during hydration. The presence of reinforcement may cause an adhesive component in the shear strength failure envelope of the GCL, because the fiber reinforcements provide tensile resistance to the bentonite clay. The tensile strength of the fiber reinforcements provides confinement of the sodium bentonite portion of the GCL. Lake and Rowe (2000) found that reinforced GCLs provide additional confinement to the bentonite, which may prevent swelling of the bentonite during hydration.

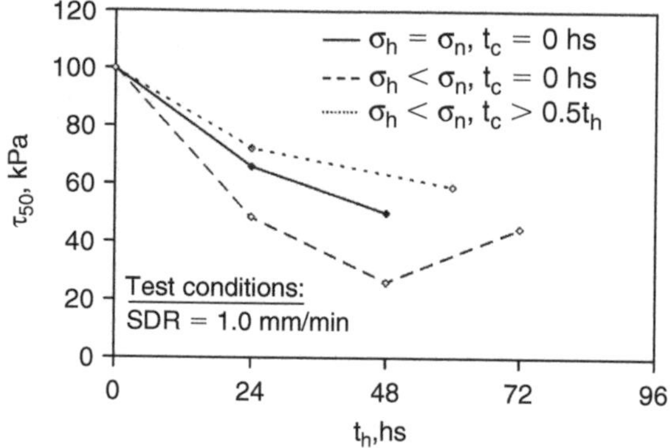


Figure 14. Effects of moisture conditioning on GCL internal peak shear strength.

### 5.3.3 *Moisture conditioning*

Hydration of the bentonite layer has been reported to result in a decrease in shear strength (Gilbert et al. 1997; McCartney et al. 2004). Figure 13 shows a comparison between the internal peak shear strength at a normal stress of 50 kPa for a needle-punched GCL. Hydration of the GCL under the normal stress used during shearing during 48 hs led to a decrease in shear strength of about 40% from unhydrated conditions. Hydration of the GCL under a normal stress less than that used during shearing without allowing time for consolidation led to an even greater decrease in shear strength. Conversely, GCLs that were consolidated after hydration have shear strength closer to that of unhydrated GCLs.

Stark and Eid (1996) found that hydration of GCLs from a water content of approximately 10 to 20% to a water content of approximately 150 to 200% during 250 hours led to a reduction in the peak and residual friction angles by about 40%. However, GCLs do not lose shear strength at a rate proportional to the increase in specimen water content. Daniel and Shan (1993) found that partially hydrated GCL specimens have similar shear strength as fully hydrated GCL specimens for normal stress below 100 kPa. Their analysis showed that GCLs with a water contents between 50 and 80% ($t_h$ < 24 hours) have similar shear strength to GCLs with water contents between 180 and 200% ($t_h$ > 24 hours).

Hydration of GCLs under low hydration normal stresses leads to swelling of the sodium bentonite, which leads to an increase in the bentonite void ratio. Further, bentonite swelling leads to stretching or pullout of the fiber reinforcements. Conversely, high hydration normal stresses do not allow swelling of the bentonite. As consolidation occurs, the void ratio decreases during consolidation, leading to an increase in bentonite shear strength. However, consolidation does not lead to shear strength recovery in the case that fiber reinforcement pullout occurs during hydration. Further, the effective stress in the partially hydrated GCLs is higher due to negative pore pressures present before shearing occurs, while the effective stress in the GCLs hydrated under low normal stress and sheared at a higher stress without allowing consolidation is higher due to positive pore pressures before shearing occurs.

The liquid used in hydrating GCL test specimens may also yield varying shear strength results. Daniel and Shan (1993) tested the shear strength of GCLs with several different hydration liquids, including simulated leachate, and found that the use of distilled water leads to the lowest shear strength. Gilbert et al. (1997) reported that the magnitude of sodium bentonite swell also depends on the hydration fluid, where distilled water induces the largest amount of swell, and inorganic and organic fluids cause the least. The data shown in Figure 14 is for GCLs hydrated with tap water, which has similar cation content to groundwater. Tap water provides a compromise between the effects of leachate and distilled water.

### 5.3.4 *Shear displacement rate*

The effect of SDR on the peak and large-displacement shear strength has been reported by Stark and Eid (1996), Gilbert et al. (1997), Eid and Stark (1997), Fox et al. (1998) and Eid et al. (1999).

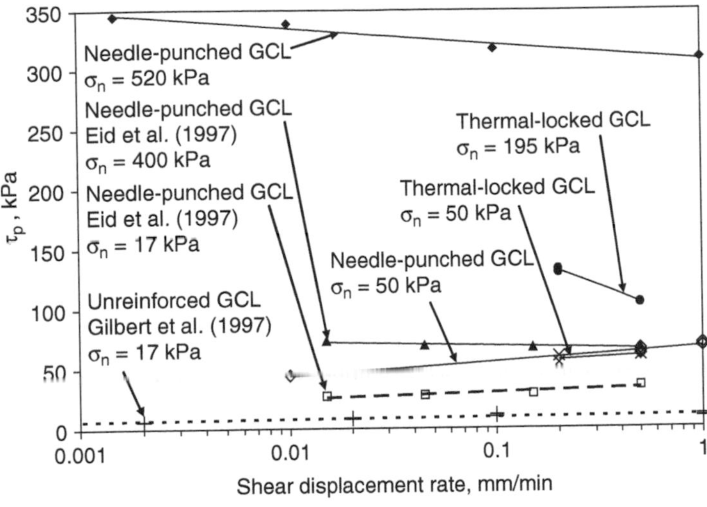


Figure 15. Effects of shear displacement rate on GCL internal peak shear strength.

These studies, which primarily focused on the response of tests conducted under relatively low $\sigma_n$, reported increasing peak shear strength with increasing shear displacement rate. Gilbert et al. (1997) conducted direct shear tests at shear displacement rates ranging between 0.0005 and 1.0 mm/min on unreinforced GCLs with normal stresses of 17 kPa and 170 kPa. Eid et al. (1999) conducted ring shear tests at shear displacement rates ranging between 0.015 mm/min and 36.5 mm/min for needle-punched GCLs sheared at normal stresses between 17 and 400 kPa. Figure 15 shows the effect of shear displacement rate on the internal peak shear strength of reinforced and unreinforced GCLs sheared under a range of normal stresses. The data reported by Zornberg et al. (2005) and McCartney et al. (2002) included specimens that were trimmed from the center of the same roll to prevent variations in reinforcement density that may occur with the width. The data in Figure 15 follows an increasing trend in peak shear strength with shear displacement rate for tests conducted under low normal stresses (<100 kPa), and a decrease in shear strength for tests conducted under high normal stresses (>100 kPa).

Explanations proposed to justify the trend of increasing peak shear strength with increasing *SDR* observed in other studies, conducted under relatively low $\sigma_n$, have included shear-induced pore water pressures, secondary creep, undrained frictional resistance of bentonite at low water content, and rate-dependent pullout of fiber reinforcements during shearing. However, the results obtained from tests conducted under both low and high $\sigma_n$ suggest that the observed trends are consistent with generation of shear-induced pore water pressures. Further, longer tests allow additional time for hydration and consolidation of the GCL, which may have an additional effect on the shear strength of the GCL. The GCLs tested under low normal stress may tend to swell during hydration, so longer testing times may lead to lower shear strength. Conversely, the GCLs tested under high normal stress may tend to consolidate during hydration, so longer testing times will lead to specimens with lower void ratio and higher shear strength.

The shear-induced pore pressure response of the sodium bentonite may indicate its positive or negative contribution to the shear strength of the GCL. Shear-induced pore water pressures are expected to be negative in tests conducted under low $\sigma_n$ (*i.e.*, below the swell pressure of GCLs). Consequently, increasing the shear displacement rate will lead to increasingly negative pore water pressures and higher peak shear strength. On the other hand, shear-induced pore water pressures are expected to be positive in tests conducted under high $\sigma_n$ (*i.e.*, above the swell pressure of GCLs). In this case, increasing SDR will lead to increasingly positive pore water pressures and thus lower peak shear strength.

Gilbert et al. (1997) reported a pore water pressure dissipation time analysis using a model proposed by Gibson and Henkel (1954). This model predicts that shear displacement rates less than 0.001 mm/min result in constant peak shear strength for unreinforced GCLs. Gilbert et al. (1997) suggested that slow shear displacement rates could result in creep, as it is a rate-dependent mechanism. This may explain the decreasing trend in shear strength with decreasing shear displacement rate.

Kovacevic Zelic et al. (2002) reported that the peak and large-displacement shear strength of an unreinforced GCLs increase with increasing shear displacement rates for normal stresses of between 50 and 200 kPa. This was postulated to be a result of changing effective stress (*i.e.*, generation of positive pore water pressure) on the failure plane or rate effects such as creep. However, when the vertical displacement during shearing was measured, inconsistent findings were apparent. Swelling occurred during shearing for slow shear displacement rates, and settlement occurred during shearing for fast shear displacement rates. Swelling is associated to negative pore water pressure generation, which should yield higher shear strength, which was the opposite observed. The authors were inconclusive with respect to the effect of the shear displacement rate on pore water pressure generation. Vertical displacements measured in the direct shear box are typically not representative of the displacements along the shear plane due to principal stress rotation during shear, tilting of the top platen, and uneven distribution of pore water pressures (or strains) throughout the bentonite thickness.

Stark and Eid (1996) observed that the peak shear strength increased with increasing shear displacement rates for GCLs with the sodium bentonite component removed (*i.e.*, the interface between two geotextiles, needle-punched together). This phenomenon was postulated to arise from tensile rupture of the fiber reinforcements at high shear displacement rates, and gradual pullout of the fiber reinforcements at slow shear displacement rates. For filled GCLs, the peak shear strength was constant below 0.4 mm/min, then increased with increasing shear rate up to about 1.5 mm/min, and then decreased for greater shear displacement rates. Stark and Eid (1996) hypothesized several mechanisms that explained these phenomena. Although pore water pressures may have been generated in the GCL at shear displacement rates between 0.4 mm/min and 1.5 mm/min, the increased shear strength was associated with rapid rupture of the fiber reinforcements. The decrease in strength associated with excess pore water pressures in the sodium bentonite dominated at shear displacement rates above 1.5 mm/min. At shear displacement rates below 0.4 mm/min, no excess pore water pressures generated, and gradual pullout failure of the fiber reinforcements may have led to the minimum peak shear strength. Stark and Eid (1996) recommended that shear displacement rates less that 0.4 mm/min be used for shearing of needle-punched GCLs. Eid and Stark (1999) reported a decrease in final water content at high normal stresses as a result of consolidation during shear.

Stark and Eid (1996) and McCartney et al. (2002) found that the large-displacement shear strength of reinforced GCLs remains constant with decreasing shear rate. The shear-induced pore pressures are expected to fully dissipate upon reaching residual shear strength.

### 5.4 *GCL internal shear strength variability*

The data shown in Figure 11 indicate that GCL internal shear strength is an inherently variable property. Zornberg et al. (2005) indicated that potential sources of GCL internal shear strength variability include: (*i*) differences in material types (type of GCL reinforcement, carrier geosynthetic), (*ii*) variation in test results from the same laboratory (repeatability), and (*iii*) overall material variability. In turn, the overall material variability includes more specific sources such as: (*iii-a*) inherent variability of fiber reinforcements, and (*iii-b*) inherent variability of the shear strength of sodium bentonite. This study found that the most significant source of shear strength variability was due to inherent material variability. Figure 16 shows the peak shear strength of GCLs with the same conditioning procedures and shear displacement rate. The spread in peak shear strength values about the mean trend line, represented by the probability density functions, increases with normal stress. In particular, a mean shear strength of 177 kPa is observed at a normal stress of 310 kPa, but the shear strength varies from approximately 115 kPa to 230 kPa. Chiu and Fox (2004) also provide measures of shear strength variability, and McCartney et al. (2004) presented an application of GCL shear strength variability in stability design.

## 6 GCL-GM INTERFACE SHEAR STRENGTH

### 6.1 *Shear stress-displacement behavior*

Figure 17 shows typical shear stress-displacement curves for interfaces between the woven carrier geotextile side of a needle-punched GCL-textured HDPE GM with the same conditioning

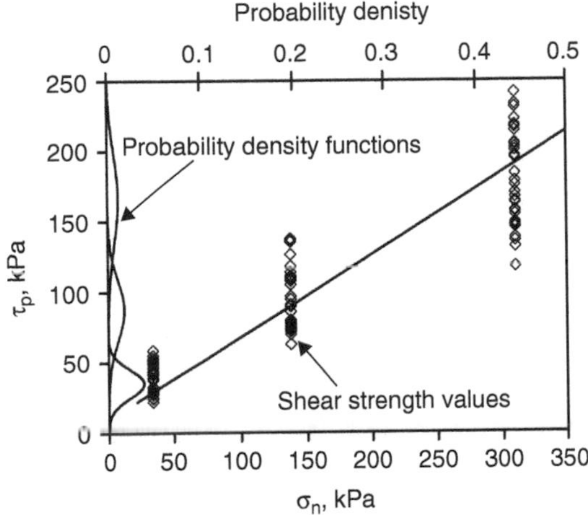


Figure 16. Variability in GCL internal peak shear strength for constant moisture conditioning ($t_h = 168$ hs, $\sigma_h = 20.7$ kPa, $t_c = 48$ hs, SDR = 1.0 mm/min) with mean failure envelope.

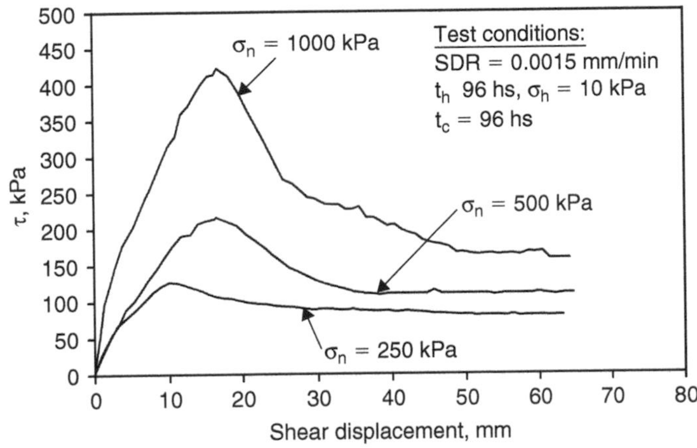


Figure 17. Shear stress – displacement curves for interface shearing of the woven carrier geotextile side of a needle-punched GCL and a textured GM under different normal stresses.

procedures and sheared under a slow shear displacement rate under different normal stresses. Similar to the GCL curves, the secant modulus tends to increase with normal stress, but the displacement at peak tends to increase with normal stress from 10 to 20 mm.

The relatively short displacement required to reach peak shear strength, combined with the large post-peak shear strength loss are important factors to consider when designing for static and dynamic design loads. The displacement at peak shear strength for the GCL-textured GM interface is slightly less than that measured for internal GCLs. The peak strength of textured GM interfaces is usually developed at 7 to 20 mm when the interlocking connections between the GCL and the textured geomembrane rupture. However, the peak shear strength of smooth geomembranes is typically developed at shear displacements less than 3 mm (Triplett and Fox 2001; McCartney et al. 2009). Hewitt et al. (1997) observed that the shear stress-displacement curves of different GCLs with different interfaces follow similar behavior to the GCL internal shear stress-displacement curves, especially for stitch-bonded GCLs. Little post-peak shear strength loss is observed for smooth geomembrane interfaces.

Triplett and Fox (2001) and McCartney et al. (2009) observed that textured HDPE geomembrane interfaces experience a greater post-peak shear strength loss than smooth HDPE geomembranes. In fact, smooth GM interfaces rarely experience any post-peak shear strength reduction. Accordingly,

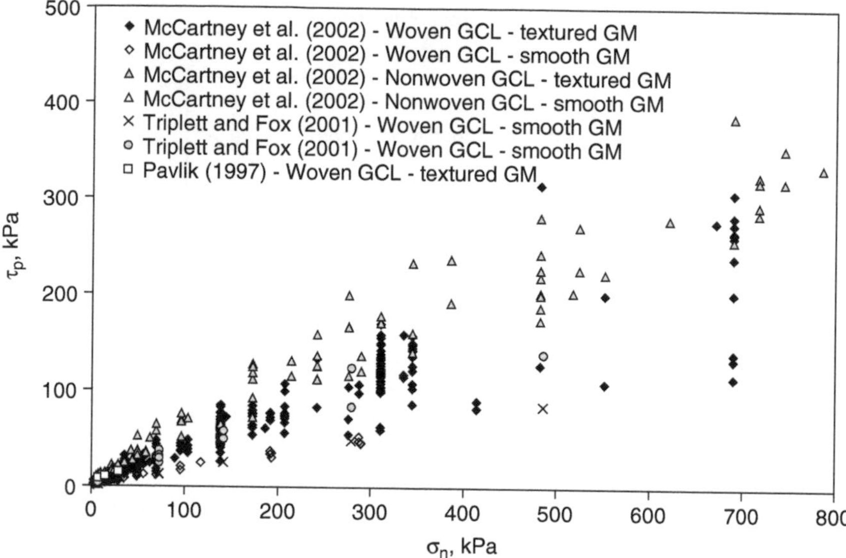


Figure 18. Overview of GCL-GM interface shear strength values.

the difference between peak and residual shear strengths was greater for textured GM interfaces. This is most likely due to the fact that at the peak shear strength, the interlocking capabilities of the geotextile and fiber reinforcements with the geomembrane asperities rupture, resulting in a large loss of strength.

## 6.2 *Preliminary GCL-GM interface shear strength overview*

Figure 18 shows the range of peak shear strength values obtained from several studies on GCL-GM interfaces. Good agreement is observed between the results from the different studies, although significant variability is observed. The textured GM interfaces are significantly stronger than the smooth GM interfaces. In general, the nonwoven carrier geotextile side of the GCL has higher interface shear strength than the woven carrier geotextile side. Different from the data for GCL internal shear strength shown in Figure 11, the GCL-GM interface shear strength does not show a strong cohesion intercept.

## 6.3 *Variables affecting GCL internal shear strength*

### 6.3.1 *Normal stress effects*

The GCL-GM interface behaves in a frictional manner with an increase in shear strength, but it does not have a significant cohesion intercept. Accordingly, the GCL-GM interface shear strength is well represented using a Mohr-Coulomb failure envelope with a zero intercept, or a nonlinear envelope. GCL-GM interface shear strength has less pronounced nonlinearity in shear strength with normal stress than GCL internal shear strength, but the peak shear strength is best represented by the Duncan and Chang (1970) model shown in Equation (3).

Eid and Stark (1997) found that for interface tests between a GM and an unreinforced GCL involving a layer of sodium bentonite adhered to a GM, an adhesive failure occurs between the two components at high normal stresses. At lower normal stresses, the GCL failed at the expected sodium bentonite-geomembrane interface. Eid and Stark (1997) also found that the peak shear strength failure envelope for this interface is slightly non-linear.

The critical interface in a layered system may change with normal stress due to the difference in the friction angles of the internal and interface GCL failure envelopes (Stark and Eid 1996). The failure envelopes may cross at a certain normal stress, above which the interface shear strength may be the critical interface (Gilbert et al. 1996). As the normal stress may vary along the length

Table 2. Shear strength parameters for GCL-GM interface peak shear strength.

| Interface description | Peak envelope $\phi_p$ (degrees) | $\Delta\phi_p$ (degrees) | $\tau_{50}$ (kPa) |
|---|---|---|---|
| Woven GCL – textured GM interfaces | 23.8 | 7.7 | 25 |
| Nonwoven GCL – textured GM interfaces | 33.1 | 11.4 | 37 |
| Woven GCL – smooth GM interfaces | 11.5 | 7.1 | 12 |
| Nonwoven GCL – smooth GM interfaces | 13.7 | 5.0 | 14 |
| GCL-textured HDPE GM interfaces | 23.4 | 6.2 | 24 |
| GCL-textured VLDPE GM interfaces | 22.3 | 19.5 | 27 |
| GCL-textured LLDPE GM interfaces | 28.0 | 4.2 | 28 |
| Needle-punched, woven GCL – textured HDPE GM interfaces | 24.2 | 5.9 | 24 |
| Stitch-bonded, woven GCL – textured HDPE GM interfaces | 16.0 | 10.3 | 17 |
| Thermal-locked, woven GCL – textured HDPE GM interfaces | 23.0 | 5.2 | 23 |

of a slope (*i.e.*, in a mounded layer of waste), the critical interface may change along the length of the slope if the failure envelopes cross at the normal stress on the interface.

### 6.3.2 *Effects of GCL reinforcement type and GM polymer*

Table 2 shows a summary of the shear strength parameters for different sets of GCL-GM shear strength data using the Duncan and Chang (1970) approach shown in Equation (3). Consistent with the data in Figure 18, the nonwoven carrier geotextiles interfaces of a GCL are typically stronger than the woven carrier geotextiles interfaces, for both textured and smooth GMs.

The data in Table 2 also indicates that GMs with more flexible polymers may have higher peak shear strength than stiffer GMs. The GMs with LLDPE and VLDPE polymers generally have higher shear strength than HDPE GMs. For smooth GMs, McCartney et al. (2009) indicated that the flexible PVC GMs tended to have the highest interface shear strength of the smooth GM interfaces. The data in Table 2 indicates that needle-punched GCL interfaces have slightly higher shear strength than thermal-locked GCL interfaces and greater shear strength than stitch-bonded GCL interfaces.

Triplett and Fox (2001), Stark and Eid (1998), Gilbert et al. (1996) and Hewitt et al. (1997) identified that interface shear strength of GCL-textured GM interface shear strength arises from: (*i*) frictional resistance to shearing between the un-textured portions of the GM and the woven carrier geotextile of the GCL, (*ii*) interlocking between the woven carrier geotextile and the textured GM asperities, and (*iii*) interlocking between the fiber reinforcements of the GCL on the surface of the woven carrier geotextile and the GM asperities. The difference between smooth and textured GM interface shear strength shown in Table 2 indicate the impact of the first two mechanisms. The difference between the shear strength of the different GCL types shown in Table 2 indicates the effect of the third mechanism. In needle-punched GCLs, the fiber reinforcements formed small bundles (not thermal-locked) or asperities (thermal-locked) on the surface of the woven carrier geotextile, which generally flattened during shearing (Gilbert et al. 1996).

The observations of Lake and Rowe (2000) indicate that the difference between the interface shear strength of GCLs with different reinforcement types may be related to moisture conditioning. Specifically, during hydration of the bentonite, the needle-punched fiber reinforcements typically pullout from the carrier geotextile while thermal-locked fiber reinforcements resist swelling of the bentonite. This implies that more bentonite extrusion will occur from thermal-locked and stitch-bonded GCLs as the bentonite is rigidly confined between the carrier geotextiles. Hewitt et al. (1997) confirmed that the least amount of bentonite extrusion occurs in needle-punched GCLs. Triplett and Fox (2001) found that less bentonite was extruded from the GCLs when a smooth geomembrane was used, likely due to less interaction between the GCL and GM during shearing.

### 6.3.3 *Effect of GM texturing*

GM texturing is often used to increase the peak shear strength of GCL-GM interfaces, and recent studies have found that it leads to an increase in the shear interaction between the GCL and the GM

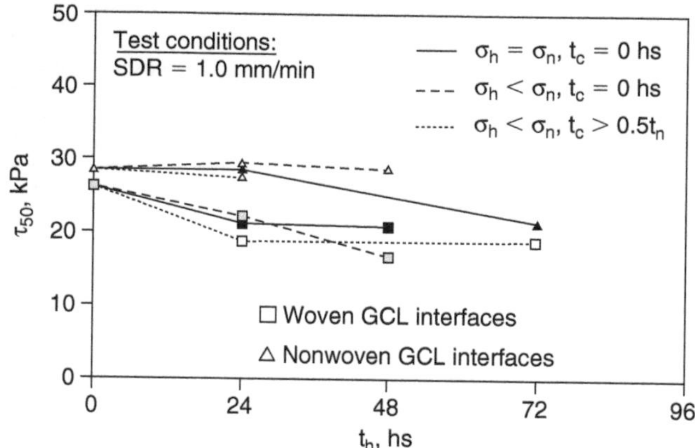


Figure 19. Effects of moisture conditioning on GCL-GM interface peak shear strength.

using asperity heights (Ivy 2003; McCartney et al. 2005) and post-failure examination (Triplett and Fox 2001). Triplett and Fox (2001) and McCartney et al. (2005) found that GM texturing leads to an improvement in the peak and large-displacement shear strength of GCL-GM interfaces. Different texturing approaches have not been shown to influence GCL-GM interface shear strength. Ivy (2003) and McCartney et al. (2005) found that GM asperity heights, despite being highly variable, were a good indicator of the peak shear strength of GCL-GM interfaces and clay-GM interfaces. McCartney et al. (2005) found that GM asperity height is an inconsistent indicator of GCL-GM large-displacement shear strength, most likely due to rupture of asperities-GCL connections during shearing.

#### 6.3.4 *Moisture conditioning*

Hydration of the GCL has been reported to result in extrusion of bentonite from the GCL, and impregnation of the carrier geotextiles with sodium bentonite. Figure 18 shows a comparison between the interface peak shear strength of the woven and nonwoven carrier geotextiles of a needle-punched GCL and a textured HDPE GM at a normal stress of 50 kPa. Hydration of the woven GCL interfaces under the normal stress used during shearing led to a decrease in shear strength of about 30% from unhydrated conditions, but little impact is observed for nonwoven GCL interfaces, except for a hydration time longer than 72 hs. Hydration of the nonwoven and woven GCL interfaces under a normal stress less than that used during shearing without allowing time for consolidation led to similar to slightly lower shear strength. Unlike the trend observed for GCL interfaces, both woven and nonwoven GCL interfaces consolidated after hydration under low normal stress have the lowest shear strength.

As mentioned, the sodium bentonite component of GCLs swells during hydration, which leads to extrusion and impregnation of bentonite in the carrier geotextiles. The extruded bentonite lubricates the connections between the fiber reinforcements and the woven or nonwoven carrier geotextiles. The nonwoven carrier geotextile of needle-punched GCLs is generally thicker than the woven carrier geotextile, so more extrusion is expected from the woven carrier geotextiles. The data shown in Figure 19 indicates that moisture conditioning does not have a significant impact on the shear strength of nonwoven GCL carrier geotextile interfaces. However, moisture conditioning has a greater effect on the woven GCL carrier geotextile interface. Consolidation of the interface after hydration does not remove the lubrication effect of the hydrated bentonite. Impregnation of woven carrier geotextiles with sodium bentonite has been shown to lead to lower shear strength than clean woven geotextile-GM interfaces (Lake and Rowe 2000).

#### 6.3.5 *Shear displacement rate*

The available literature on GCL-GM interface shear strength indicates that there is not a significant effect of SDR on the peak shear strength value (Triplett and Fox 2001; McCartney et al. 2009).

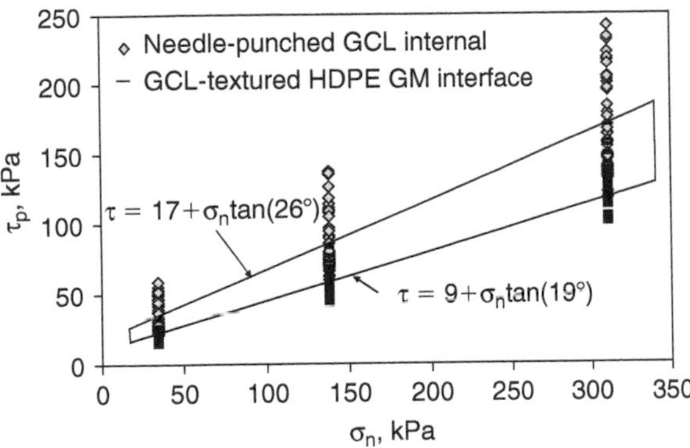


Figure 20. Internal and interface shear strength for tests with the same conditioning procedures.

Eid et al. (1999) found that the shear displacement required to reach peak strength conditions does not vary with shear displacement rate. This finding is consistent with results for geotextile-GM interfaces (Stark et al. 1996). The lack of a trend indicates that shear displacement effects for GCL internal shear strength are likely due to the contribution of the fiber reinforcements and bentonite. The lack of trend also implies that faster shear displacement rates may be used in practice to replicate field conditions.

### 6.4 *GCL internal shear strength variability*

The spread in the data shown in Figure 18 indicates that GCL-GM interface shear strength is as variable as GCL internal shear strength. The variability in GCL internal and GCL-GM interface shear strength values may have implications on the weakest plane in a liner system. Figure 20 shows a comparison between the variability in GCL internal and GCL-GM interface shear strength for tests with the same conditioning procedures and shear displacement rate. The mean failure envelopes for the sets of data indicate that the GCL internal shear strength is significantly above the GCL-GM interface shear strength. However, due to variability, a zone of overlap is observed. In this zone, the GCL internal shear strength may be less than the GCL-GM interface shear strength. The data shown in Figure 20 indicate that design of composite liner systems on a slope will be governed by GCL-GM interface shear strength except in the overlap zone delineated by the two failure envelopes. In this zone, either the GCL internal or GCL-GM interface shear strength may be the lower strength.

The GCL and GM specimens with shear strength shown in Figure 20 were obtained from the center of different manufacturing lots (*i.e.*, a set of rolls manufactured in a given batch). Zornberg et al. (2005) found that the variability in peak shear strength of GCLs sampled from the same manufacturing lot is less than the variability within different rolls due to manufacturing differences over time. Because the direct shear test provides the shear strength of a relatively small specimen with respect to the area of the roll, the variability in shear strength within a given roll is expected to average out over the length of a slope. However, McCartney et al. (2005) observed that variability in shear strength between rolls from different lots used at a given job site may not average out over the area of the landfill, leading to zones that have lower shear strength than others. This observation emphasizes the need to conduct project-specific and product-specific shear strength testing.

## 7 LABORATORY AND FIELD SHEAR STRENGTH COMPARISONS

Although analyses of slope failures involving GCLs are not widely published, the slope failures at the Mahoning landfill site (Stark et al. 1998) and the US Environmental Protection Agency (EPA) GCL test section in Cincinnati, Ohio (Daniel et al. 1998; Stark and Eid 1996) provide excellent opportunities to verify the results of different test devices for GCLs. Stark et al. (1998) observed that failure of the Mahoning landfill occurred along an unreinforced, geomembrane-backed GCL located in the landfill base liner. Further, settlement of the overlying waste caused down-drag on the liner system, resulting in large shear displacements. To simulate field conditions, the investigation included a study of the loading procedures and the hydration process for the GCL. The results of ring shear tests on hydrated GCL internal and GCL-GM interfaces were consistent with the shear strength of the soil back-calculated using two-dimensional limit equilibrium analysis.

Stark and Eid (1996) present a comparison of the results of a three-dimensional back-analysis of failures at the EPA GCL test sections with direct shear and a ring shear test results on the same GCLs and GMs. Daniel et al. (1998) observed failure at the interfaces between the woven geotextiles of a needle-punched GCL and a textured HDPE GM, as well as between a stitch-bonded GCL and a textured HDPE geomembrane lying on a slope of about 23.5° (Daniel et al. 1998). The failures occurred 20 and 50 days after construction, respectively. For the interface failures between the GCLs and GMs, the mobilized interface friction angle was approximately 21.5°. Ring and direct shear tests were conducted on the same materials under fully hydrated conditions, and it was found that the ring shear test obtained a peak friction angle of 22.5°, and the direct shear test obtained a peak friction angle of 23.8°. Both test methods obtained friction angles slightly above the actual back-calculated three dimensional friction angles, so the difference in actual and experimental friction angles may be attributed to three dimensional effects, such as the strength contribution of the vertical failure surfaces at the edges of the veneer failure or differences in scale. In addition, further differences in the shear strength measured in the laboratory and the shear strength back-calculated from forensic studies may arise from the tensile forces that may develop in the carrier geotextiles of the GCL. This phenomenon may increase the back-calculated shear strength of the GCL as the tensile forces provide additional resistance to down-slope deformations.

## 8 CONCLUSIONS

This chapter uses shear strength data from the literature and commercial databases to indicate the basic concepts behind geosynthetic clay liner (GCL) internal and interface shear strength values obtained from laboratory testing. The information in this chapter can be used guide the design of a site-specific shear strength testing program. The effects of different variables on the GCL and GCL-geomembrane interface shear strength are quantified. Specifically, the effects of normal stress, GCL reinforcement, geomembrane texturing and polymer type, moisture conditioning, shear displacement rate, and normal stress are assessed.

## REFERENCES


American Society of Testing and Materials, 1998. Standard Test Method for Determining the Internal and Interface Shear Resistance of Geosynthetic Clay Liner by the Direct Shear Method. *ASTM D6243*. West Conshohocken, Pennsylvania.

American Society of Testing and Materials, 1999. Standard Test Method for Determining Average Bonding Peel Strength between the Top and Bottom Layers of Needle-Punched Geosynthetic Clay Liners." *ASTM D6496*. West Conshohocken, Pennsylvania.

Berard, J. F. 1997. Evaluation of Needle-Punched Geosynthetic Clay Liners Internal Friction. *Geosyn. '97*, St. Paul, MI. pp. 351–362.

Bouazza, A., Zornberg, J.G., and Adam, D. 2002. Geosynthetics in Waste Containments: Recent Advances. *Proc. 7th Int. Conf. on Geosynthetics*, Nice, France. vol. 2, pp. 445–507.

Byrne, R.J., Kendall, J., and Brown, S. (992. Cause and Mechanism of Failure, Kettleman Hills landfill B-19, Unit IA. *Proc. ASCE Spec. Conf. On Performance and Stability of Slopes and Embankments*, Vol. 2, ASCE. pp. 1188–1215.

Daniel, D. E., and Shan, H. Y. 1991. "Results of Direct Shear Tests on Hydrated Bentonitic Blankets." Geotechnical Engineering Center, University of Texas at Austin, Austin, TX.

Daniel, D. E., Shan, H. Y., and Anderson, J. D. 1993. Effects of Partial Wetting on the Performance of the Bentonite Component of a Geosynthetic Clay Liner. *Geosyn. '97*, St. Paul, MN.

Daniel, D. E., Carson, D. A., Bonaparte, R., Koerner, R. M., and Scranton, H. B. 1998. Slope Stability of Geosynthetic Clay Liner Test Plots. *Journal of Geot.and Geoenv. Eng.*, ASCE, 124(7), 628–637.

Dove, J. E., and Frost, J. D. 1999. Peak Friction Behavior of Smooth GM-Particle Interfaces. *Journal of Geot. and Geoenv. Engineering*, ASCE, 125(7), 544–555.

Duncan, J. M. & Chang, C.-Y. 1970. Nonlinear Analysis of Stress and Strain in Soils. *Journal of the Soil Mechanics and Foundations Division*. ASCE. 96(SM5), 1629–1653.

Eid, H. T., and Stark, T. D. 1997. Shear Behavior of an Unreinforced Geosynthetic Clay Liner. *Geosyn. Int.* IFAI, 4(6), 645–659.

Eid, H. T., Stark, T. D., and Doerfler, C. K. 1999. Effect of Shear Displacement Rate on Internal Shear Strength of a Reinforced Geosynthetic Clay Liner. *Geosyn. Int.* IFAI, 6(3), 219–239.

Fox, P. J., Rowland, M. G., and Scheithe, J. R. 1998. Internal Shear Strength of Three Geosynthetic Clay Liners. *Journal of Geot. and Geoenv. Eng.*, ASCE, 124(10), 933–944.

Fox, P.J., Rowland, M.G., Scheithe, J.R., Davis, K.L., Supple, M.R., and Crow, C.C. 1997. Design and Evaluation of a Large Direct Shear Machine for Geosynthetic Clay Liners. *Geotechnical Testing Journal*. 10(3).

Fox, P.J., and Olsta, J. 2005. Current Research on Dynamic Shear Behavior of Geosynthetic Clay Liners. *Proceedings of Geo-Frontiers 2005*. ASCE.

Gibson, R.E., and D.J. Henkel. 1954. Influence of Duration of Tests at Constant Rate of Strain on Measured Drained Strength. *Geotechnique*. 4, 6–15.

Gilbert, R. B., Fernandez, F. F., and Horsfield, D. 1996. Shear Strength of a Reinforced Clay Liner. *Journal of Geot. and Geoenv. Engineering*, ASCE, 122(4), 259–266.

Gilbert, R. B., Scranton, H. B., and Daniel, D. E. 1997. Shear Strength Testing for Geosynthetic Clay Liners. *Testing and Acceptance Criteria for Geosynthetic Clay Liners*, L. Well, ed., American Society for Testing and Materials, Philadelphia, 121–138.

Heerten, G., Saathoff, F., Scheu, C., von Maubeuge, K. P. 1995. On the Long-Term Shear Behavior of Geosynthetic Clay Liners (GCLs) in Capping Sealing Systems. *Proceedings of the International Symposium "Geosynthetic Clay Liners"*. Nuremberg, 141–150.

Helsel, D.R. and Hirsh, R.M. 1991. *Statistical Methods in Water Resources*. United States Geologic Survey.

Hewitt, R. D., Soydemir, C., Stulgis, R. P., and Coombs, M. T. 1997. Effect of Normal Stress During Hydration and Shear on the Shear Strength of GCL/Textured Geomembrane Interfaces. *Testing and Acceptance Criteria for Geosynthetic Clay Liners*, L. Well, ed., American Society for Testing and Materials, Philadelphia, 55–71.

Koerner, G.R. and Narejo, D. Direct Shear Database of Geosynthetic-to-Geosynthetic and Geosynthetic-to-Soil Interfaces. GRI Report #30. Folsom, PA. 106 pp.

Kovacevic Zelic, B., Znidarcic, D., and Kovacic, D. 2002. Shear Strength Testing on Claymax 200R. *7th International Conference on Geosynthetics*. Nice, France.

Lake, C. G., and Rowe, R. K. 2000. Swelling Characteristics of Needle-Punched, Thermal Treated Geosynthetic Clay Liners. *Geotextiles and Geomembranes*, 18, 77–101.

Marr, W. A. 2001. Interface and Internal Shear Testing Procedures to Obtain Peak and Residual Values." *15th GRI Conference: Hot Topics in Geosynthetics (Peak/Residual; RECMs; Installation; Concerns)*. Houston, TX, pp. 1–27.

McCartney, J.S. and Zornberg, J.G. 2005. Effect of Geomembrane Texturing on Geosynthetic Clay Liner – Geomembrane Interface Shear Strength. *ASCE GeoFrontiers 2005*. Austin, TX January 27–29, 2005.

McCartney, J.S., Zornberg, J.G., Swan, Jr., R.H., and Gilbert, R.B. 2004. Reliability-Based Stability Analysis Considering GCL Shear Strength Variability. *Geosyn. Int.*. 11(3), 212–232.

McCartney, J.S. and Zornberg, J.G. 2004. Effect of Specimen Conditioning on GCL Shear Strength." *GeoAsia 2004: 3rd Asian Regional Conference on Geosynthetics*. Eds. Shim, J.G., Yoo, C., and Jeon, H-Y. Seoul, Korea. pp. 631–643

McCartney, J.S., Zornberg, J.G., and Swan, R. 2002. *Internal and Interface Shear Strength of Geosynthetic Clay Liners (GCLs)*. Geotechnical Research Report, Department of Civil, Environmental and Architectural Engineering, University of Colorado at Boulder, 471 p.

McCartney, J.S., Zornberg, J.G., and Swan, R.H. 2009. Analysis of a Large Database of GCL-Geomembrane Interface Shear Strength Results. *Journal of Geotechnical and Geotechnical Engineering, ASCE*, 134(2), 209–223.

Olsta, J. and L. Crosson 1999. Geosynthetic clay liner peel index test correlation to direct shear. *Proceedings Sardinia 99, Seventh International Waste Management and Landfill Symposium.* S. Margherita di Pula, Cagliari, Italy. 4–8 October 1999.

Pavlik, K. L. 1997. U.S. Army Corps of Engineers GCL Interface Testing Program. Geosynthetics '97, St. Paul, MN, 877–884.

Petrov, R. J., Rowe, R. K., and Quigley, R. M. 1997. Selected Factors Influencing GCL Hydraulic Conductivity. *Journal of Geot. and Geoenv. Eng. ASCE*, 123(8), 683–695.

Richardson, G. N. 1997. GCL Internal Shear Strength Requirements. *Geosyn. Fabric Report*, March, 20–25.

Stark, T. D., and Eid, H. T. 1996. Shear Behavior of a Reinforced Geosynthetic Clay Liner. *Geosyn. Int.* IFA. 3(6), 771–785.

Stark, T. D. 1997. Effect of Swell Pressure on GCL Cover Stability. *Testing and Acceptance Criteria for Geosynthetic Clay Liners*, ASTM STP 1308, L. W. Well, Ed., ASTM, 30–44.

Stark, T. D., Arellano, D., Evans, W. D., Wilson, V. L., and Gonda, J. M. 1998. Unreinforced Geosynthetic Clay Liner Case History. *Geosynthetics International*, IFAI, 5(5), 521–544.

Swan Jr., R. H., Yuan, Z., and Bachus, R. C. 1996. Factors Influencing Laboratory Measurement of the Internal and Interface Shear Strength of GCLs. *ASTM Symposium on Testing and Acceptance Criteria for Geosynthetic Clay Liners*, Atlanta, GA.

Swan Jr., R. H., Yuan, Z., and Bachus, R. C. 1999. Key Factors Influencing Laboratory Measurement of the Internal and Interface Shear Strength of GCLs. *ASTM on Grips, Clamps, Clamping Techniques and Strain Measurement for Testing Geosynthetics*, Memphis, TN.

Triplett, E. J., and Fox, P. J. 2001. Shear Strength of HDPE Geomembrane/Geosynthetic Clay Liner Interfaces. *Journal of Geot. and Geoenv. Engineering*, ASCE, 127(6), 543–552.

von Maubeuge, K. P., and Ehrenberg, H. 2000. Comparison of Peel Bond and Shear Tensile Test Methods for Needle-Punched Geosynthetic Clay Liners. *Geot. and Geomem.*, 18, 203–214.

von Maubeuge, K. P. and Lucas, S. N. 2002. Peel and shear test comparison and geosynthetic clay liner shear strength correlation. *Clay Geosynthetic Barriers*. H. Zanzinger, R. M. Koerner, and E. Gartung, eds. Swets & Zeitlinger. Lisse, the Netherlands. 105–110.

Zornberg, J.G., McCartney, J.S., and Swan, R.H. 2005. Analysis of a Large Database of GCL Internal Shear Strength Results. *Journal of Geotechnical and Geotechnical Engineering, ASCE*, 131(3), 367–380.

Zornberg, J.G., McCartney, J.S., and Swan, R.H. 2006. Analysis of a Large Database of GCL Internal Shear Strength Results. Closure, *Journal of Geotechnical and Geotechnical Engineering*, ASCE, 132(10), 1376–1379.

# CHAPTER 9

# Slope stability with geosynthetic clay liners

R.B. Gilbert & S.G. Wright
*The University of Texas at Austin, Austin, TX, USA*

## 1 INTRODUCTION

Geosynthetic clay liners provide a unique challenge in designing stable slopes for cover and liner systems. They are very weak compared to most if not all other materials in a landfill. They can exhibit a significant loss in shear strength with displacement. Their behavior in shear is complicated because they contain a variety of different components that interact with one another internally and that interact with other materials externally at interfaces on either side of the GCL.

The types of slope failures that involve GCLs in a cover or a liner system are shown schematically on Figure 1. A potential sliding or failure surface can include a GCL over either part or all of the surface; it can also extend into the waste, the subgrade and other materials in a cover or liner system. Furthermore, a potential failure surface involving a GCL can be within the GCL or at interfaces with the GCL. Figure 2 shows an example of a slope failure involving a GCL.

Designing a stable slope with a GCL consists of the following steps:

1. Define the geometry, loading conditions and consequences of a failure for the slope during construction, operation and after closure.
2. Select appropriate material properties for the GCL and all other materials in the slope. Consider rate of loading, deformations, normal stresses and fluid pressures in this selection.
3. Analyze and evaluate the stability.
4. Take measures to mitigate any concerns about stability of the slope.

In this chapter, these four steps are described, discussed and illustrated with examples for slopes with GCLs. It is expected that the reader has a basic knowledge of slope stability and shear strength for soils. The emphasis here is on practical methods and guidance. References are provided if the reader desires to explore topics in greater depth.

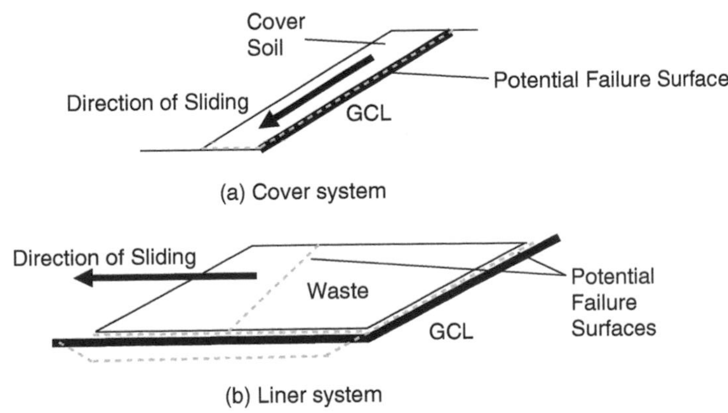


Figure 1. Types of slope failures that involve a GCL in a cover or liner system.

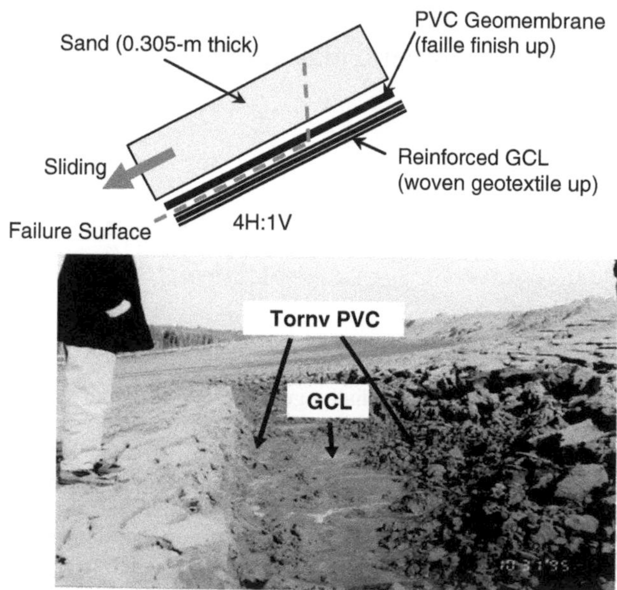


Figure 2. Illustration of a cover slope failure involving a GCL (failure described in Liu et al. 1997; picture courtesy of R. Thiel).

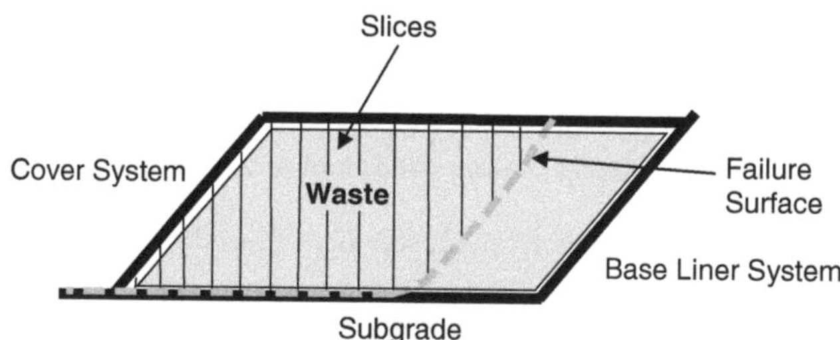


Figure 3. Schematic cross-section for method of slices.

## 2 METHODS OF ANALYSIS

The methods used to analyze the stability of a slope with a GCL are the same as the methods used to analyze the stability of any engineered or natural slope containing soils. Since these methods are well documented elsewhere (e.g., Duncan and Wright 2005), the emphasis in this chapter will be on particulars in how these methods are applied to slopes in cover and liner systems with GCLs.

The predominant method for analyzing the stability of a slope is a two-dimensional limit equilibrium analysis. The forces required to maintain force and moment equilibrium are compared to the forces driving instability. The analysis is performed on one or more two-dimensional cross-sections that represent the actual three-dimensional geometry of the slope. The slope is divided into a series of slices in order to account for variations in the geometry and the material properties throughout the cross-section (Fig. 3). A variety of possible failure surfaces, with different locations and shapes, can be analyzed and those providing the greatest potential for failure can be identified.

This method requires assumptions about how internal forces are distributed within the sliding mass, i.e., the orientation and magnitude of forces between slices. Spencer's procedure (Spencer 1967) is the only practical procedure that makes assumptions about inter-slice forces that satisfy complete static equilibrium for each slice. Therefore, Spencer's procedure is the recommended procedure for conducting a limit equilibrium analysis. Computer programs are convenient for implementing and analyzing the results; commonly available and used programs include PC STABL (Achilleous 1988), SLOPE/w (Geoslope International 2004), UTEXAS (Wright 1991), and XSTABL (Sharma 1994). Since these programs offer a variety of options regarding assumptions about inter-slice forces, it is important to check and not assume that Spencer's procedure is being used in the analysis.

### 2.1 *Simplified wedge procedures*

There are a variety of simplified procedures to conduct a limit equilibrium analysis where the slope is divided into a small number of slices or wedges (e.g., Giroud and Beech 1989, Koerner and Hwu 1991, Giroud et al. 1995 and Corps of Engineers 2003). The advantage of these procedures over Spencer's procedure is that they can be solved easily by hand. Therefore, they provide a useful means to check a computer analysis and to conduct sensitivity analyses. However, these procedures impose assumptions on the analysis that are more restrictive than Spencer's procedure and not necessarily realistic. They are best suited for long veneer slopes with planar failure surfaces, like in a cover system. Results from a simplified analysis should always be checked with those from Spencer's procedure, and vice versa.

### 2.2 *Possible failure surfaces*

Liner and cover systems will generally contain numerous possible surfaces, at interfaces between materials and within materials upon, which a failure surface can form. With the exception of very small slopes, the tensile capacities of the geomembranes, non-woven geotextiles and geonets used in liners or covers are essentially negligible. Therefore, the failure surfaces can readily jump from one material or interface to another within the liner or cover, the waste, and the subgrade. Failure surfaces can be highly non-linear. The material or interface providing the minimum shear resistance at any location along a possible failure surface will depend on the effective normal stress, which will depend on the total normal stress and fluid pressures at that location. Therefore, numerous possibilities need to be considered in identifying potential failure surfaces.

Great care is required in using computer programs that search automatically for potential failure surfaces to make sure that the entire range of possibilities is considered. Generally, different failure surfaces will need to be used as the starting point for an automated search in order to consider all of the possibilities in a liner or cover system (e.g., Fig. 1b).

### 2.3 *Representative cross-sections and three-dimensional geometries*

An analysis on a two-dimensional cross-section is intended to represent a three-dimensional geometry. A representative cross-section is one in which the relative driving and resisting forces on a failure surface are representative of those in the three-dimensional slope. A practical approach to selecting a representative cross-section is to hypothesize a direction of sliding and then select a cross-section that is parallel to this direction and contains the largest mass of material above the potential failure surface. An example of a representative cross-section is shown on Figure 4. Conducting sensitivity analyses with multiple cross-sections and possible sliding directions is always helpful.

One challenge specifically related to liner systems is that there may be sharp breaks in a potential failure surface, such as at the toe of the side slope for a failure surface located along the liner system in Figure 4b. Spencer's procedure assumes that all of the inter-slice forces are oriented at the same angle to the horizontal. This assumption can be slightly conservative when there is a sudden change in the orientation of the failure surface (e.g., Gilbert et al. 1998). A deformation analysis would be required in order to develop a more realistic assumption for side force inclinations, which then could be implemented in using a procedure such as Morgenstern and Price (1965).

Conducting a truly three-dimensional analysis is generally not necessary nor is it practical. A limit equilibrium analysis conducted on a representative two-dimensional cross-section will provide a reasonable if slightly conservative assessment of the stability. Loehr et al. (1998) provide a simplified means to account for three-dimensions using a series of two-dimensional cross-sections. Lam and Fredlund (1993) propose a procedure to conduct a three-dimensional analysis using three-dimensional columns versus two-dimensional slices. However, it is much more difficult to rigorously satisfy static equilibrium in a three dimensional analysis and simplistic and restrictive

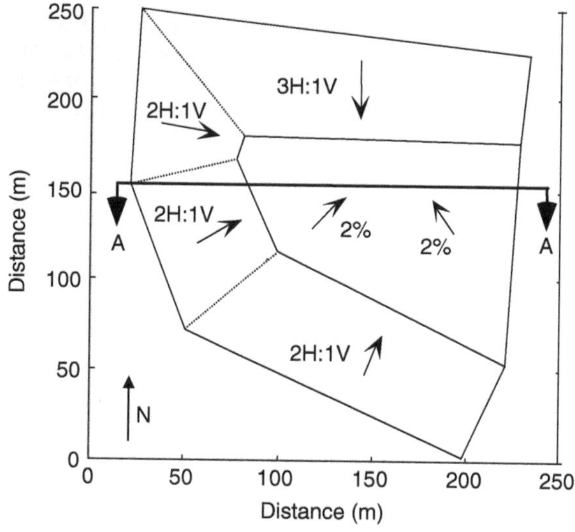


(a) Plan view showing grades for base liner system

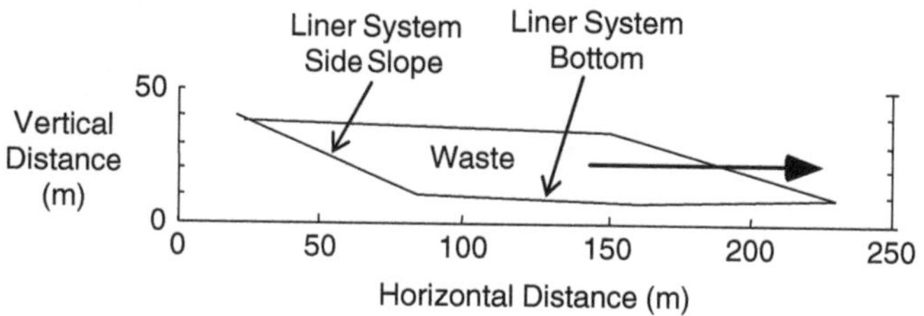


(b) Representative cross-section A-A for stability analysis

Figure 4. Illustration of selecting a representative two-dimensional cross-section for a stability analysis (from Gilbert et al. 1998).

assumptions are therefore necessary. Loehr et al. (1998) and Stark et al. (1998) provide a more detailed discussion of these issues.

### 2.4 *Effective normal stresses*

The shear strengths for materials and interfaces in a liner or cover system are generally a function of the effective normal stresses acting on the shear surfaces. Fluid pressures, including liquids and gases, can reduce the effective stresses and are an important consideration. In addition, the relationship between shear strength and effective normal stress is not necessarily linear. For computer analyses, non-linear relationships between shear strength and effective normal stress can and should be included. For simplified analyses, which all assume a linear relationship between shear strength and effective normal stress, the parameters describing the relationship should be carefully selected to correspond to the range of effective normal stresses in the field.

### 2.5 *Displacement and strain compatibility*

A key assumption in a limit-equilibrium analysis is that all of the resisting forces can be mobilized to their maximum limit, regardless of the deformation required to do so. There are two aspects of slopes in liner and cover systems that require particular attention in selecting maximum limits: strain-softening materials and tensile reinforcement.

Most of the interfaces and materials in liners and covers, including GCLs, exhibit strain softening in shear (Fig. 8). Deformations in a slope can therefore cause the available or limiting shear strength to drop from the peak to the residual value. Deformations will occur as the slope is loaded and waste and/or soil materials compress and geosynthetic materials go into tension. A method of analysis that explicitly accounts for compatibility between displacements, strains and forces is needed to capture this behavior (e.g., Wilson-Fahmy and Koerner 1993; Long et al. 1994; Gilbert and Byrne 1996; Gilbert et al. 1996; and Liu and Gilbert 2003). Based on limited results from deformation analyses, the following practical guidance is provided to implicitly consider displacement and strain compatibility in a limit equilibrium analysis:

- For a veneer slope (Fig. 1a) that is stable without a buttress at the toe or tensile reinforcement, the limiting shear strength will be the peak shear strength;
- For a veneer slope that relies upon compression developing at the toe to buttress it or tension developing in geosynthetic layers to support it, the limiting shear strength will be the residual shear strength;
- For a waste slope (Fig. 1b), the limiting shear strength will be the peak shear strength on the bottom and the residual shear strength on the side slope.

When tensile reinforcement is included in the slope, such as a geogrid layer, the limiting tension will not necessarily be the ultimate tensile capacity because the strain needed to mobilize this tension my cause severe distress to the slope. Therefore, the limiting tension generally corresponds to the tension developed at an acceptably small strain level in the reinforcement, such as 5 to 10%.

### 2.6 *Seismic loads*

A limit-equilibrium analysis is well-suited for static loads. A common approach to simplistically consider seismic or dynamic loads in a limit-equilibrium analysis is to represent the dynamic forces with a pseudo-static force. The dynamic forces, specifically the peak ground acceleration, are estimated by considering the characteristics of possible earthquakes, such as magnitude and distance from the site, and the attenuation and amplification that can occur in soil and waste layers at the project site (e.g., EPA 1995 and 2006; Rathje 1998; and Matasovic et al. 1998).

It is excessively conservative to use the peak ground acceleration for an earthquake event directly to calculate the pseudo-static force because this peak force is applied for a very short time during an earthquake, meaning that very little displacement will accumulate on the shear surface during this time. Therefore, the peak ground acceleration is reduced in practice to implicitly account for the short duration over which it acts and the amount of accumulated displacement that a slope can tolerate over the duration of an earthquake. The acceleration used in a pseudo-static analysis to realize tolerable displacements in the slope, on the order of tenths of a meter, ranges from 60% of the peak ground acceleration for a base liner down to 20 to 10% of the peak ground acceleration for a cover. Guidance on this approach for cover and linter systems is provided in EPA (1995), Kavazanjian (1998) and EPA (2006).

The failure surface under earthquake loading may not be the same as that under static loading since the horizontal seismic force will change the effective normal stress acting on inclined shear surfaces. Therefore, as with static loading, a variety of different possible failure surfaces need to be investigated in a seismic analysis.

Since accumulated displacements along shear surfaces of tenths of meters are accommodated in seismic analysis, it is generally appropriate to use residual shear strengths. In addition, the tensile reinforcement can be strained to higher levels and a limiting tension approaching the ultimate tension may be possible.

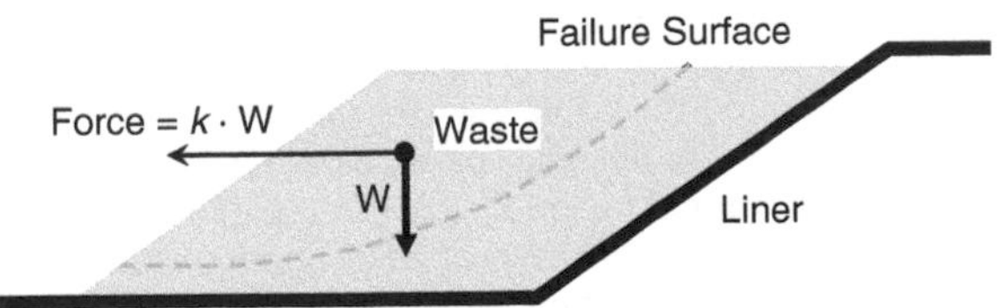


Figure 5. Schematic showing pseudo-static method for seismic forces, where W is the weight of the sliding mass and k is the seismic coefficient (the seismic acceleration divided by the acceleration due to gravity).

Table 1. Possible design conditions to consider in evaluating factor of safety.

| | Possible design conditions to be checked |
|---|---|
| Strengths | Likely Values for Shear Strengths Considering Pore Water or Pore Gas Pressures and Displacements<br>Conservative Estimates for Shear Strengths Considering Pore Water or Pore Gas Pressures<br>Conservative Estimates for Shear Strengths Considering Displacements |
| Loads | Permanent Configurations of Waste Containment System<br>Interim Configurations during Construction and Waste Placement<br>Temporary Loading from Construction Equipment<br>Earthquakes<br>Likely Values for Tensile Strengths Considering Displacements<br>Conservative Estimates for Tensile Strengths Considering Damage and Creep |

The result of a pseudo-static analysis, even when the short duration of the loading is implicitly accounted for, is conservative. If this approach indicates that there is a potential problem in the event of an earthquake, then a more rigorous deformation analysis may be warranted (EPA 1995; Bray and Rathje 1998; EPA 2006).

## 3 DESIGN CRITERIA

Published guidance on design criteria for slope stability analyses is intentionally broad so that slopes can be designed effectively on a project-specific basis. In the United States for liner and cover systems, there are no federal regulatory requirements as to what is an acceptable design but there can be local (state) requirements. Therefore, one consideration in a stability analysis is to identify all applicable regulatory requirements.

However, a slope stability analysis is not a cook-book procedure and is not simply a check to satisfy regulatory requirements. The consequences of a slope failure can range from minor, such as periodically re-grading a cover slope, to major, such as loss of an operating permit, significant environmental or economic damage, or even loss of life. The costs of improving stability can also range from minor to major. Therefore, the criteria for design need to be considered carefully for each individual project in terms of the consequences and the costs. In addition, the criteria need to be considered by all the stakeholders, including the owner, operator and regulator as well as the designer.

The most commonly-used indicator of stability is the factor of safety, which is defined generally as follows:

$$\text{FS} = \frac{\text{Available Shear Strength}}{\text{Shear Stress Required for Stable Equilibrium}} \tag{1}$$

A value of 1.0 for the FS indicates that the slope is marginally stable; a value less than 1.0 indicates that it is not stable and a value greater than 1.0 indicates that it is stable. The denominator in Eq. (1), the shear stress needed achieve equilibrium, is obtained from the stability analysis.

Note that the factor of safety here, which is the recommended convention in practice (Wright 1991; Duncan and Wright 2005), is defined in terms of the available shear strengths for materials and interfaces in the slope. Resisting forces provided by tensile reinforcement are included in determining the shear stress required for stable equilibrium (the denominator in Eq. 1). Since the tensile forces used for reinforcement in the analysis are assigned by considering allowable strains, they implicitly incorporate an additional factor of safety compared to the ultimate tensile capacity.

The factor of safety should be checked for a variety of different conditions related to the strengths and the loads (Table 1). For each combination, the appropriate factor of safety depends on how probable that combination is to occur in the design life of the slope and on the consequences of a slope failure. Table 2 provides a framework to organize these considerations.

Table 2. Considerations for evaluating factor of safety.

| Consideration | Factors | Categories |
|---|---|---|
| What are the consequences of a slope failure? | Human Health<br>Environmental Impact<br>Cost<br>Regulatory Involvement | *Major*: Human health and the environment will likely be impacted; costs of a failure will exceed the costs of construction; repairs will require corrective action with regulatory oversight.<br>*Moderate*: Human health and the environment will not likely be impacted; costs of a failure are comparable to costs of construction; repairs may require corrective action with regulatory oversight.<br>*Minor*: Human health and the environment will not be impacted; costs of a failure are significantly less than the cost of construction; repairs will be considered as operation and maintenance or construction activities. |
| How probable is it that the loads will be greater than those for the loading condition being checked? | Temporary versus Permanent Loads<br>Static versus Dynamic Loads<br>Uncertainty in Slope Geometry during Construction and Operation<br>Uncertainty in Weights of Materials<br>Uncertainty in Occurrence, Magnitude, and Site-Specific Ground Motion for Earthquake Loads | *Likely*: Greater than 50-percent chance in design life.<br>*Unlikely*: About 10-percent chance in design life.<br>*Rare*: Less than 1-percent chance in design life. |
| How probable is it that the strengths will be less than those for the condition being checked? | Uncertainty in Strengths Estimated from Laboratory Tests<br>Changes in Strength with Time due to Changes in Pore Water or Gas Pressures<br>Compatibility of Strains and Displacements between Materials<br>Rate of Loading | *Likely*: Greater than 50-percent chance in design life.<br>*Unlikely*: About 10-percent chance in design life.<br>*Rare*: Less than 1-percent chance in design life. |

The appropriate factor of safety could range from near 1.0 for a situation where the consequences of a slope failure are minor and the combination of loads and strengths is a rare possibility to greater than 2.0 for a situation where the consequences are major or the combination of loads and strengths is likely. No single value for an allowable factor of safety is appropriate in all situations. Both incurring a larger risk than desired (that is, designing a slope with too small of a factor of safety) as well as incurring a larger cost than necessary (that is, designing a slope with too high of a factor of safety) are possibilities that need to be considered and balanced.

The selection of an appropriate set of design conditions and associated factor of safety values for a given project requires careful judgment by the design engineer in consultation with the owner, operator and regulator. Additional guidance on selecting factor of safety values is provided in Duncan (1992), Koerner and Soong (1998), Liu et al. (1997), Duncan and Wright (2005) and USEPA (2006). A project-specific risk analysis where the consequences and probability of failure are assessed may be warranted, particularly in cases where the consequences of a slope failure are major and/or the costs of construction are large.

## 4 DESIGN SHEAR STRENGTH FOR GCLS

The following conditions can all affect the shear strength for a GCL in a containment system: saturation of the bentonite, total normal stress, water or gas pressures, shear rate, magnitude of shear displacement, the quality of internal reinforcement, the properties of soil or geosynthetic materials forming interfaces on either side of the GCL, and the possibility bentonite extruding into adjacent interfaces. These issues are discussed here in the context of selecting design shear strengths for GCLs.

### 4.1 *Unconsolidated versus consolidated bentonite*

The bentonite in as-produced GCLs has a water content ranging from 10 to 30% and is not saturated with water. Hence, the unconsolidated condition for a GCL corresponds to unsaturated bentonite, which has been commonly referred to as "dry" or "unhydrated" bentonite. To put this water content in a physical context, bentonite will only be saturated at this water content when it is consolidated under more than 10,000 kPa of normal stress (or nearly 1,000 m of soil). In this condition, the bentonite behaves much like sand; it is relatively stiff in compression and it has a relatively high shear strength.

A reasonable design assumption is that the bentonite will become fully saturated with water in the field. At a water content of 10 to 30%, the bentonite has a matric suction that is extremely high relative to most natural soils and the bentonite will tend to draw water from adjacent soil layers. For example, Daniel et al. (1993) demonstrate that a GCL will generally hydrate to moisture contents in excess of 100% even when it is in contact with very dry soils having moisture contents less than 10%. GCLs in liners can also become hydrated with the waste liquids they are intended to contain, while GCLs in covers can become hydrated from precipitation and moisture in landfill gases.

A possible exception to this typical field condition is a GCL that is encased between two geomembranes and installed in an arid environment. If the geomembranes are intact, then it is possible in theory that the bentonite will not become saturated for a very long time, if ever (e.g., GSE 2001). However, caution is warranted because it is difficult in practice to keep the bentonite from having any access to water. For example, full-scale test slopes were constructed in the field with bentonite encased between two geomembranes to investigate the practicality of this approach (Daniel et al. 1998). The test slopes were located in Cincinnati, Ohio, a humid environment with about 600 mm of rainfall annually, and were intended to simulate a landfill cover. In one of the three slopes, the bentonite became saturated, probably due to surface water entering penetrations in the overlying geomembrane that were made for instrumentation. One of the notable features of this test slope is that while the penetrations consisted of several small holes in the geomembrane, the bentonite became saturated over the entire 9-m wide by 20-m long slope within about 500 days.

The total normal stress under which the bentonite in a GCL is consolidated before it is sheared will be project specific. This normal stress could be less than 10 kPa for a GCL in a cover and more than 1,000 kPa for a GCL in a base liner. The shear strength can vary significantly and non-linearly with changes in normal stress. It is consequently important to select design shear strengths that are representative of the normal stress or range of normal stresses to be encountered in the field.

Table 3. Estimated representative consolidation times for GCLs.

| GCL Drainage Conditions | Time Required for 95-Percent Consolidation* Final normal stress at end of loading: 10 kPa | Final normal stress at end of loading: 100 kPa | Final normal stress at end of loading: 1,000 kPa |
|---|---|---|---|
| Drainage on both sides of GCL | 1.5 days | 4 days | 5 days |
| Drainage on one side of GCL | 6 days | 16 days | 20 days |

*Normal stress increased linearly from zero to the final value over the time period; Consolidation rate data based on Shan (1993) and Gilbert et al. (2004)

It is reasonable in design to assume that the bentonite is fully consolidated in typical field conditions. The time required for consolidation under a normal stress is relatively short compared to the time it takes to apply a normal stress in the field. For example, Table 3 presents the estimated time required for 95-percent consolidation of a saturated GCL for "ramp" loading where the normal stress is increased linearly from zero to the final normal stress. The times depend both on the drainage conditions at the boundary of the GCL and the normal stress applied to the GCL (both the coefficient of consolidation and the thickness of the GCL depend on the normal stress). The times presented in Table 3 indicate that a GCL will come to equilibrium very quickly relative to typical load application rates in the field. For example, a normal stress of 200 kPa corresponds to approximately 20 m of waste; it would be difficult to place 20 m of waste in less than 16 days.

Caution is warranted in interpreting laboratory data because full hydration and a consolidation can take several weeks in the laboratory. While several weeks is practically instantaneous in the field (Table 3), it is a long time in laboratory testing and there is a tendency to conduct tests more quickly; 24-hour and 48-hour long consolidation times are commonly used. These "quick" tests will generally not produce results that are consistent with field conditions. For example, a typical test conducted with a 48-hour consolidation time would only achieve about 50% of the full consolidation for a final normal stress of 1,000 kPa. The strength measured in this test would consequently correspond to an effective normal stress of 500 kPa and could be as low as 50% of the actual strength that would be available at a normal stress of 1,000 kPa in the field. Furthermore, the test data will be difficult to interpret because the bentonite will continue to consolidate as it is sheared.

Another difficulty in interpreting laboratory test data is related to the sequence of hydration and application of the normal stress during consolidation. A GCL could hydrate under a small normal stress after installation and then consolidate under larger normal stresses as overlying fill or waste is placed. Conversely, a GCL could hydrate under the final or permanent normal stress, e.g., a GCL in a landfill cover may become hydrated over time under the weight of the cover. For the bentonite alone, Shan (1993) presents data showing that the final void ratio of the bentonite is not sensitive to the order of hydration and normal stress application. Bentonite during hydration and consolidation can extrude through carrier geotextiles into interfaces with adjacent materials and affect the shear strength for those interfaces. For woven geotextiles, this extrusion occurs regardless of the consolidation sequence. For example, Figure 6 shows bentonite that has extruded through the woven geotextile of a needle-punched GCL when it was hydrated under the final normal stress. However, there is an extreme loading case that can cause extrusion of bentonite even through thick non-woven, geotextiles: hydration under a low normal stress (less than 10 kPa) and then subsequent sudden loading under a much higher normal stress. While it is possible to implement this loading sequence in the laboratory, it is not a realistic representation of field conditions and should therefore be avoided in laboratory testing. Therefore, the sequence of hydration and application of normal stress should not affect the available shear strength for a GCL in the field, but it can affect laboratory results if unrealistic conditions are imposed.

### 4.2 *Undrained versus drained shear*

Shear stresses will generally be applied slowly enough in the field to allow for dissipation of shear-induced pore water pressures and correspond to drained conditions because the bentonite layer

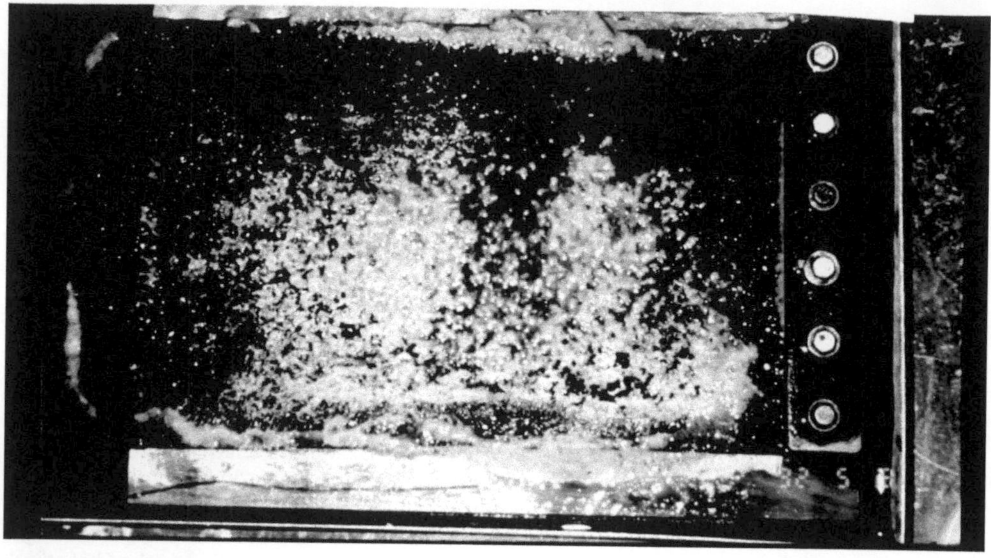

Figure 6. Photograph showing bentonite that has extruded through the woven geotextile of a needle-punched GCL; specimen was hydrated with distilled water under 14 kPa of normal stress (photograph taken from Gilbert et al. 1997).

in a GCL is so thin. As with the normal stresses, shear stresses are typically applied to GCLs by building fill or waste on a slope. It is practically impossible to place the fill or waste quickly enough to preclude dissipation of shear-induced pore water pressures. Based on laboratory test data (Shan 1993, Gilbert et al. 1997 and Gilbert et al. 2004), shear-induced pore water pressures will dissipate within several days at low normal stresses around 10 to 20 kPa (such as a GCL in a landfill cover) and within several months at high normal stresses around 1,000 kPa (such as a GCL in a base liner). In practice, it will easily take longer than several days to build a landfill cover and longer than several months to construct a 50 to 100-m high waste fill.

While drained conditions are generally most applicable for design shear strengths, many laboratory shear tests on GCLs do not produce drained conditions. Laboratories commonly use shear rates between 0.01 to 1.0 mm/min, while shear rates of 0.0001 to 0.001 mm/min are required to produce drained conditions (Gilbert et al. 1997). To illustrate the potential for misusing laboratory data in design, Figure 7 shows how the measured shear strength varies with shear rate for an unreinforced GCL at two different normal stresses. The shear strength in the field under drained conditions could be significantly less than a shear strength measured in the laboratory under partially drained or undrained conditions, particularly at low normal stresses. Therefore, caution is warranted in using existing laboratory test data to estimate design shear strengths. In addition, designers should be careful to specify that drained conditions are achieved in laboratory testing for new projects, in accordance with all of the relevant testing standards: ASTM D3080 (Direct Shear Testing for Soils), D5321 (Direct Shear Testing for Geosynthetics and Soils), and D6243 (Direct Shear Testing for GCLs).

One possible exception to drained conditions in the field is shear stresses imposed by an earthquake. The high rate of load application in an earthquake will likely correspond to undrained conditions within the bentonite. The load will also be applied cyclically. Therefore, the available shear strength in theory will correspond to an undrained, cyclic shear strength. However, in practice, laboratory testing to measure this shear strength for clay soils is rare, whether for natural clays, compacted clay liners or GCLs.

In addition to the difficulty and expense in conducting undrained, cyclic tests, there are several explanations for why this type of testing is not necessary in practice for GCLs. First, the relatively high rate of loading in an earthquake, which is at least an order of magnitude faster than that used in a conventional undrained test, tends to offset the strength degradation associated with

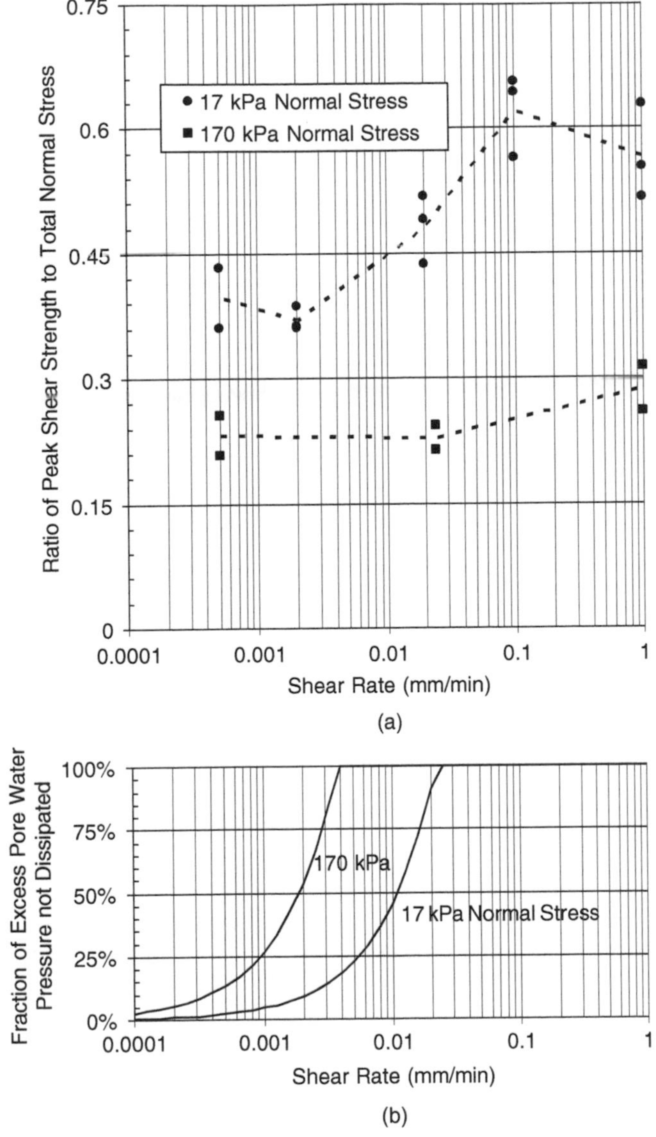


Figure 7. Effect of shear rate on (a) shear strength and (b) dissipation of shear-induced water pressures (data taken from Gilbert et al. 1997).

cyclical loading for normally consolidated clays. Hence, the cyclic, undrained shear strength for earthquake loading is not significantly different than the static, undrained shear strength measured in a conventional test. For example, Lai et al. (1998) conducted consolidated (hydrated), undrained, cyclic tests on an unreinforced GCL using a direct simple shear device. Based on an interpretation of their data, the estimated undrained, cyclic shear strength for earthquake loading at a period of 0.5 seconds is essentially equal to the static, undrained shear strength measured using a typical time to failure of 5 minutes. Second, the drained, residual shear strength for bentonite provides a lower-bound estimate of its shear strength under a variety of static loading conditions (Gilbert et al. 2004). Hence, a consolidated-drained direct shear test provides information that can be used as a conservative estimate of the static, undrained shear strength for the bentonite. Third, the governing shear strength for a GCL will generally occur at interfaces where drainage is rapid and drained

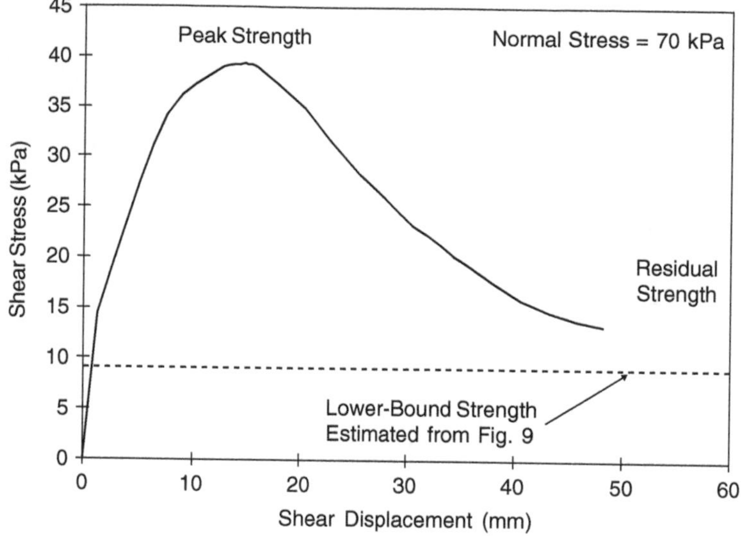


Figure 8. Example of strain softening for the interface between a GCL and a textured geomembrane (data from Gilbert et al. 1996).

conditions may exist even in an earthquake. If a realistic (versus conservative) estimate for the shear strength of a GCL during an earthquake is needed in design, then laboratory testing to measure the cyclic shear strength using a rate of loading representative of an earthquake is recommended.

### 4.3 *Peak versus residual shear strength*

When sheared under consolidated, drained conditions, GCLs, internally and at interfaces with other materials, can exhibit a residual strength substantially lower than the peak strength (Fig. 8). These reductions in shear strength with increasing shear displacement, loosely termed strain softening, can be due to the alignment of clay particles, the failure of internal reinforcement, and the polishing or failure of geosynthetic and soil materials at interfaces with the GCL. The challenge in design is that the shear strength that will be available in the field is not known. The use of a peak shear strength in design could be un-conservative, while the use of a residual shear strength in design could be overly conservative.

The following approach is recommended for considering strain softening in selecting shear strengths for design:

1. Estimate both peak and residual shear strengths for all possible shear surfaces in a liner or cover system.
2. Establish the peak and residual shear strengths for the system of components in a liner or cover. The peak shear strength for the system is the smallest peak strength among all possible shear surfaces in the system. The residual shear strength for the system is the residual shear strength for the shear surface with the lowest peak shear strength.
3. Check stability in design using a range of possible shear strengths, including the peak and residual shear strengths for the system.

A check of the stability using the residual shear strength for the liner or cover system is an important consideration in design. There are numerous case histories in both natural slopes (Skempton 1964 and Bjerrum 1967) and landfill liners (Byrne et al. 1992 and Stark et al. 2001) where a slope would not have failed if the peak shear strength were mobilized. Furthermore, the consequences of a slope failure are greater if there is a sudden drop in shear strength due to strain-softening once the slope fails. For example, the waste in the Kettleman Hills failure moved more than 10 m along the liner system, destroying the liner system and causing tens of millions of dollars of damage (Mitchell et al. 1990 and Seed et al. 1990). Koerner and Soong (2000) provide additional examples of high-consequence slope failures in landfills due to large deformations. Therefore, both the

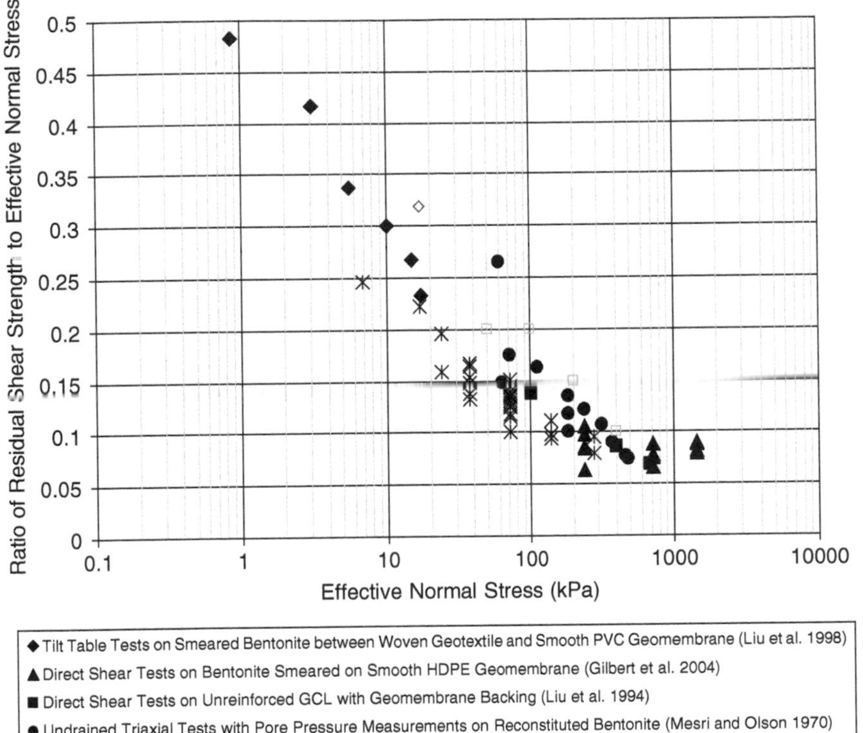


Figure 9. Summary of data to provide a lower-bound estimate for the drained, residual shear strength at interfaces and internally for all types of GCLs.

potential for and the consequence of a slope failure in a liner or cover system can be significant if residual shear strengths are not considered in design.

The residual shear strength for all types of GCLs, both internally and at interfaces with other materials, can generally be represented, as a lower bound, by the drained, residual shear strength of sodium bentonite. Published data that establish this lower-bound shear strength are summarized on Figure 9. These data represent the consolidated-drained shear strength for either internal shearing of sodium bentonite or sodium bentonite shearing against a smooth surface, such as a smooth geomembrane. They also represent shearing to relatively large deformations, and are therefore indicative of the residual shear strength. The lower-bound shear strength estimated from Figure 9 is shown on Figure 8 for illustration.

Two field case histories with unreinforced GCLs provide benchmarks for validating the lower-bound shear strength shown on Figure 9: a field test of a cover slope in Cincinnati, Ohio (Daniel et al. 1998) and the failure of a base liner in a Municipal Solid Waste landfill in Youngstown, Ohio (Stark et al. 1998). Both slopes had a factor of safety close to 1.0, so the applied shear stresses to the bentonite layer in the field are very close to the available shear strength. These field data, which are included on Figure 9, are very consistent with the laboratory data and demonstrate the value in design of establishing a lower-bound shear strength data for GCLs.

The residual shear strength can only be mobilized along a potential failure surface involving a GCL if the peak shear strength along that surface is exceeded. Hence, the residual shear strength for a system of surfaces in a liner or a cover system is the residual strength of the surface with the lowest peak strength, providing that any compressive or tensile forces in the individual components are negligible. An example of this concept is shown on Figure 10. This double liner system has a primary composite liner, consisting of a textured geomembrane over a reinforced geosynthetic

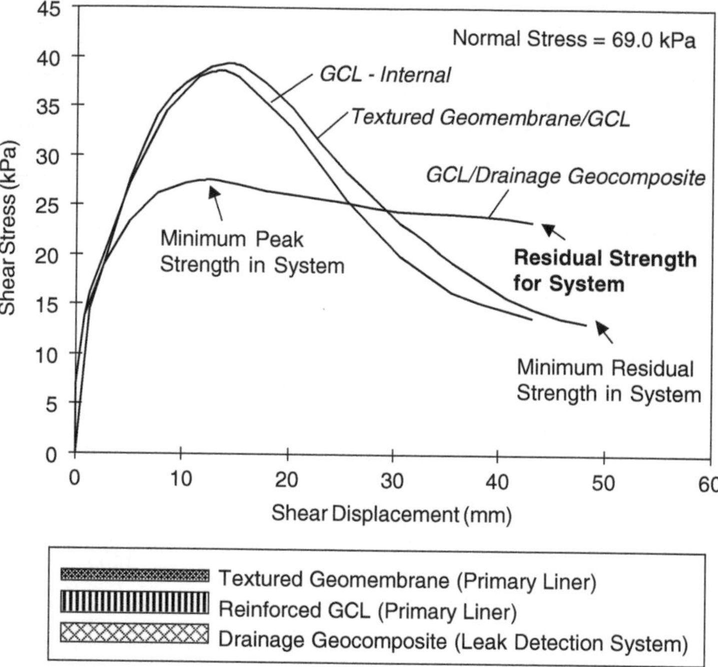


Figure 10. Example of how to establish the residual shear strength for a system of failure surfaces involving a GCL (figure taken from Gilbert 2001).

clay liner (GCL), underlain by a leak detection system with a drainage geocomposite (nonwoven geotextiles bonded to a geonet). Direct shear tests were run on each of the possible slip surfaces: between the textured geomembrane and the GCL; within the GCL; and between the GCL and the geocomposite. The most significant strain softening occurs internally within the GCL due to failure of the reinforcing fibers at large displacements. The textured geomembrane interface with the GCL has a greater peak strength than the peak internal strength of the GCL; the failure surface moves into the GCL and the same strain softening occurs. However, the peak strength for the interface between the GCL and the drainage geocomposite is less than that for the GCL; the reinforcing fibers in the GCL are not stressed to failure and there is little strain softening. Therefore, the residual strength for this system is that for the interface between the GCL and the drainage geocomposite and not the minimum residual strength for slippage occurring internally within the GCL. In a sense, the interface between the GCL and the geocomposite protects or isolates the GCL from undergoing strain softening.

The residual shear strength for a system of surfaces will generally be greater than the minimum residual shear strength for any surface. In the example on Figure 10, the residual shear strength for the system is no only larger than the smallest residual shear strength, but it is actually not much smaller than the peak shear strength for the system (the minimum peak shear strength) even though some of the components exhibit significant strain softening. Therefore, the stability of a slope may not be much different if it is checked with both the minimum peak shear strength for any possible failure surface and the residual shear strength for the system of failure surfaces.

The concept of the residual shear strength for the system is useful in design because it can be used to prevent significant strain softening from occurring within the containment system. Interfaces and materials can be intentionally included in a liner system to isolate possible materials, such as a reinforced GCL, that are particularly susceptible to strain softening.

### 4.4 *Internal versus interface shear strength*

Failure surfaces in the field are generally free to form anywhere in a liner or cover system. Within a GCL, the governing failure surface, defined as the shear surface with the smallest shear strength, is usually located at the boundary between the bentonite and one of the carrier materials. For example,

Figure 11. Photograph showing bentonite that is smeared through the woven geotextile into the interface with a smooth HDPE geomembrane; specimen was hydrated with distilled water under 14 kPa of normal stress.

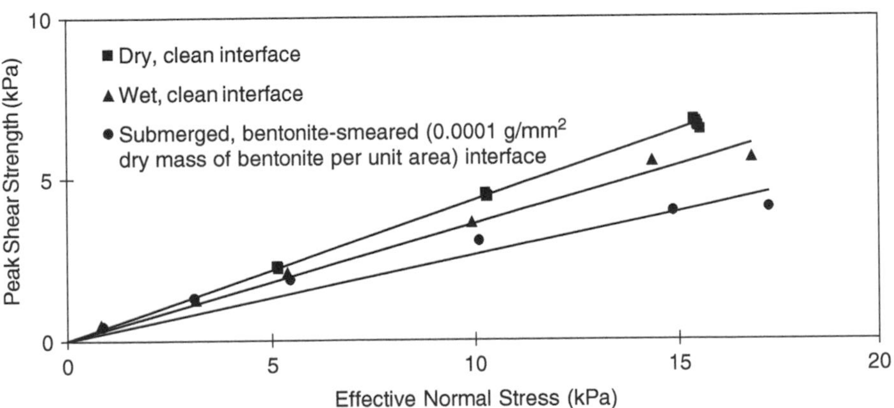


Figure 12. Effect of bentonite on the consolidated-drained shear strength for the interface between the woven geotextile of a reinforced GCL and a smooth PVC geomembrane (data from Liu et al. 1998).

the laboratory data on Figure 10 for the internal shear strength of the reinforced GCL correspond to a shear surface that formed at the interface between the bentonite and the woven geotextile encasing the bentonite. When interfaces between the GCL and other materials are considered, the governing failure surface can either be at the interface or within the GCL. For example, the laboratory data on Figure 10 for the interface between the textured geomembrane and the reinforced GCL correspond to a shear surface that formed internally within the GCL because the shear strength at the interface was greater than that within the GCL.

The governing failure surface may depend on the normal stress. For example, consider the laboratory data on Figure 10 for the interface between the textured geomembrane and the reinforced GCL. At this normal stress of 69 kPa, the governing failure surface formed internally within the GCL. However, at normal stresses less than about 15 kPa, the governing failure surface formed at the interface between the textured geomembrane and the woven geotextile of the GCL (Gilbert et al. 1996). Daniel et al. (1998) provide field evidence from large-scale test slopes, intended to model cover systems with a GCL, where failure surfaces formed at the interface between a textured geomembrane and a reinforced GCL at low normal stresses.

The bentonite in the GCL can affect the shear strength of an interface with the GCL if the bentonite can extrude into the interface (Figure 6). Figure 11 shows how bentonite from a reinforced GCL has extruded into an interface with a geomembrane; the shear strength for this reinforced GCL is governed by the shear strength of unreinforced bentonite in this case. Figure 12 shows how the presence of bentonite in an interface can affect the shear strength of the interface. The consolidated-drained shear strength for this interface between a woven geotextile and a smooth geomembrane, which is a relatively weak interface without the presence of bentonite, is reduced about 40% when a thin layer of bentonite is smeared on the interface. Also, note that the drained shear strength of the interface between two geosynthetics is affected by moisture even when there is no bentonite present; this reduction in shear strength is possibly due to swelling and softening of the geotextile fibers when submerged in water. The data on Figure 12 highlight the importance of running laboratory tests for design purposes using conditions that best represent the range of conditions that may be encountered in the field.

In summary, all possible failure surfaces over all possible normal stresses need to be considered in establishing design shear strengths for liners and covers with GCLs. Likewise, laboratory testing should be conducted in a way so that the location of the failure surface is not inadvertently constrained by the test device or test method and so that the conditions at all possible shear surfaces represent the possible conditions that will be encountered in the field.

### 4.5 *Summary of design shear strength*

The bentonite in a GCL in the field will typically be saturated, consolidated and sheared under drained conditions. Caution is warranted in interpreting laboratory data for GCLs because complete consolidation and drainage during shear have commonly not been achieved in the laboratory in order to expedite the testing time. Designers should be careful to specify clearly that consolidated-drained conditions are to be achieved in all testing programs. Possible variations on the consolidated-drained shear strength, such as an encapsulated GCL that may not hydrate fully if at all or a GCL subjected to undrained, cyclic shear loading in an earthquake, should be considered after the consolidated-drained shear strength is established. The consolidated-drained, residual shear strength provides a conservative estimate of the shear strength that will be mobilized in most, if not all, practical situations of concern for slope stability.

Both peak and residual shear strengths need to be estimated for all possible shear surfaces. With this information, the residual shear strength for the cover or liner system, which is the residual shear strength for the surface with the lowest peak shear strength, should be established. A range of design shear strengths, including both the peak and residual shear strengths for the system, should be considered in design. The residual shear strength for the system is important because both the potential for and the consequence of a slope failure are related to this strength.

All possible failure surfaces over all possible normal stresses need to be considered in establishing design shear strengths for liners and covers with GCLs. Laboratory testing should be conducted in a way so that the location of the failure surface is not inadvertently constrained by the test device or test method and so that the conditions at all possible shear surfaces represent the possible conditions that will be encountered in the field.

## 5 APPROACHES TO MITIGATE STABILITY PROBLEMS

When there is a concern about slope stability, the following approaches can be used to mitigate these concerns:

- flatten the grade of a slope
- provide a buttress at the toe of a slope
- reinforce a slope with tensile reinforcement;
- reduce fluid pressures in a slope;
- replace weak shear surfaces in a slope; and
- reduce consequences of a slope failure.

The optimal approach depends on its effectiveness and cost, recognizing that approaches that reduce the volume for waste in the landfill or restrict operations in the landfill carry indirect as well as direct costs.

### 5.1 *Flattening slope*

Reducing the grade of a slope is generally an effective means to improve stability when the loads are due to the weight of soil or waste. Flattening a slope can both reduce the shear stress, or the driving force, and increase the normal stress and consequently the shear strength, or the resisting force. The impact that flattening a slope has on the factor of safety depends on the contribution of the slope angle to the driving and resisting forces. For example, the impact of flattening the slope on a cover system (Fig. 1a) is generally greater than the impact of flattening the side slope in a base liner system (Fig. 1b).

### 5.2 *Buttressing slope*

Buttressing a slope at the toe is effective if the buttress force is large relative to the total resisting force; the wider the buttress relative to the height of the slope, the more effective it will be. One consideration with a buttress is that the buttress will compress when it is loaded, meaning that there will be displacement along shear surfaces and the possibility of mobilizing residual shear strengths.

A buttress can either be permanent, such as the toe of a cover, or temporary, such as the waste configuration on a base liner during filling. If the stability of the slope depends on operations, such as waste placement, versus construction, then the design engineer has less control in assuring that the buttress is constructed properly. Therefore, more conservatism is generally warranted in these situations.

### 5.3 *Reinforcing slope*

Reinforcing a slope with tensile reinforcement will be effective if the reinforcing force is large relative to the total resisting force. Generally, geogrids or high-strength woven geotextiles are required to have much of an impact on the stability. In addition, the reinforcement needs to be anchored on the stable side of a possible failure surface in order to develop tension. The development of tension in the reinforcement will result in displacements along shear surfaces and deformations in the slope that need to be considered.

### 5.4 *Reducing fluid pressures*

Fluid pressures can arise from groundwater, leachate, or landfill gas. Small changes in groundwater or gas pressures can have a large impact on the stability of a cover slope where the total normal stress is small. Means for reducing fluid pressures include passive drainage and active pumping. An important consideration is that the effectiveness of a drain is only as good as its outlet. Several slope failures have occurred when outlets for drainage layers did not function properly. Another consideration is that passive and active drainage systems will have a limited life and require continual maintenance and monitoring if they are to be relied upon for decades or longer.

### 5.5 *Replacing weak shear surfaces*

GCLs can provide some of the weakest shear surfaces in a liner or cover system. One viable alternative to improving stability is to replace a GCL with a compacted clay liner. Other alternatives include using a reinforced GCL versus an unreinforced GLC, using a reinforced GCL encapsulated with non-woven geotextiles to prevent the migration of bentonite to adjacent interfaces, or encapsulating an unreinforced GCL with geomembranes to minimize hydration of the bentonite.

### 5.6 *Reducing consequences of failure*

In some cases, particularly when the costs of improving stability are excessive, the best approach to mitigate concerns about stability is to reduce the consequences of a slope failure if it were to occur instead of trying to prevent the possibility of a failure. Regardless of how likely it is that a slope failure will occur, it is always helpful in design to consider the potential consequences and ways to mitigate those consequences.

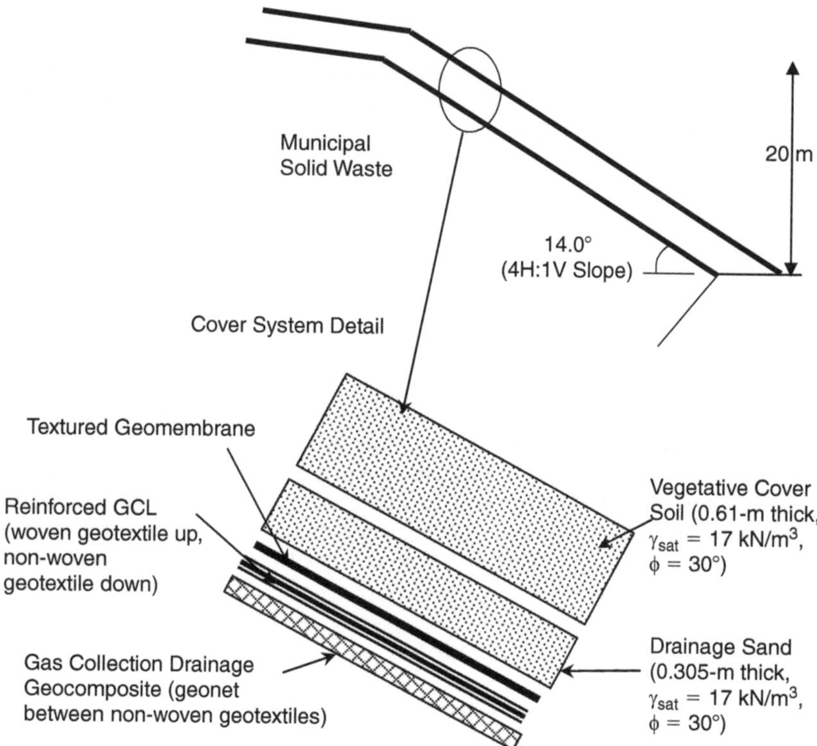


Figure 13. Configuration for example cover slope.

## 6 DESIGN EXAMPLES

Three examples are provided below to illustrate and highlight the important points to be considered in designing slopes with GCLs. These examples are all based on actual cases from design practice.

### 6.1 *Example cover slope*

The geometry and properties for a slope in a cover system are shown on Figures 13 and 14. This municipal solid waste landfill is located in the Midwestern United States. The average annual rainfall at the site is approximately 900 mm. The cover slope is permanent and will need to function during the entire, potentially unlimited, post-closure care period for this landfill. The consequences of failure for the cover slope are moderate in that it would be costly to repair but would not pose a significant threat to humans or the environment.

The design cases to be considered are summarized in Table 4. These cases are specifically focused on shear surfaces that are in or at interfaces with the GCL; additional cases including water pressures in the overlying drainage layer would be considered in a comprehensive stability analysis. The peak shear strength for the system of failure surfaces involving the GCL corresponds to the peak shear strength for the interface between the GCL and the drainage geocomposite (Fig. 14). Likewise, the residual shear strength for the system corresponds to the residual shear strength for this same interface. The gas collection system, which is passively vented, is expected to operate with a maximum of 4 kPa of gas pressure in the gas collection layer below the GCL (Fig. 13).

#### 6.1.1 *Design check for base case*

The effect of the slope angle on the factor of safety is shown on Figure 15 for the expected or base case design conditions, neglecting the effect of a toe buttress or geosynthetic tension. These calculations are based on a simplified, infinite slope analysis. In addition, it is assumed

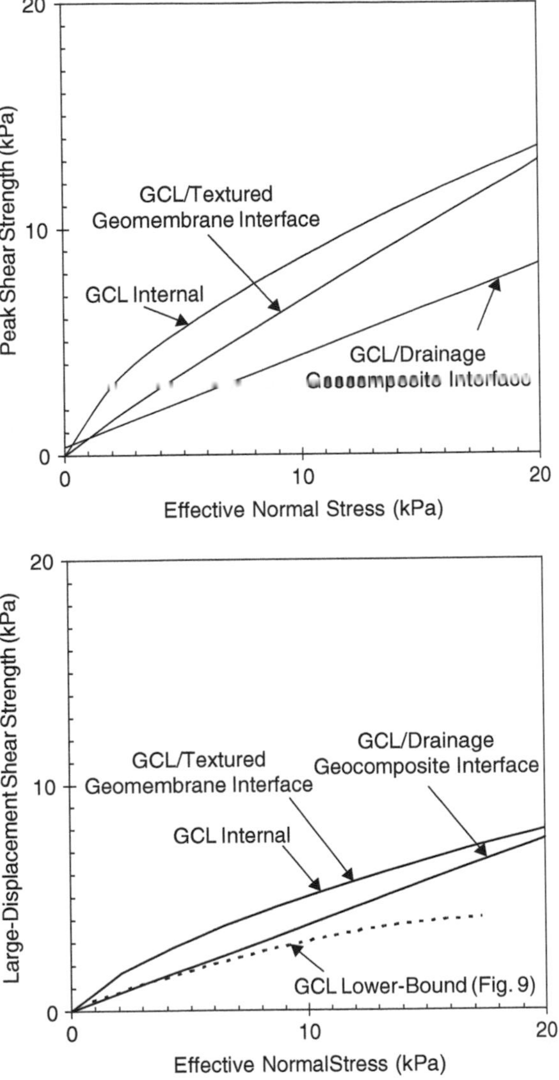


Figure 14. Shear strength envelopes for GCL in example cover slope.

that the drainage geocomposite in the gas collection system is saturated with gas, so that the gas pressure is treated the same as an equivalent water pressure in establishing the effective normal stress at the interface between the GCL and the drainage geocomposite. For the design grade of 4 horizontal to 1 vertical (4H:1V), the factor of safety is less than the target value of 1.5 (Fig. 15). The slope would need to be flattened to about 5H:1V to achieve a factor of safety greater than 1.5. The downside to this approach is the loss of airspace.

The effect of counting on a buttress at the toe of the slope is shown on Figure 16. These calculations are based on a simplified wedge analysis where the failure surface is assumed to daylight horizontally at the toe of the slope and the inter-slice force is parallel to the slope. While the buttress increases the factor of safety for short slopes, its effect is essentially negligible for the 20-m high slope because the driving load increases but the size of the buttress does not as the height of the slope increases. Furthermore, if the buttress is engaged, then there will be displacement in the slope due to compression at the toe. With the residual shear strength for the system to account for this displacement, the factor of safety is actually lower if the toe buttress is mobilized with the 20-m high slope. One possible alternative to improve stability would be to break the 20-m high slope

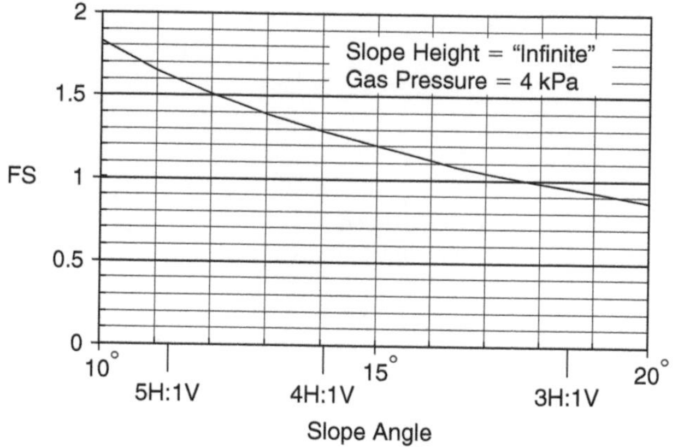


Figure 15. Factor of safety versus slope angle for example cover slope; no toe buttress.

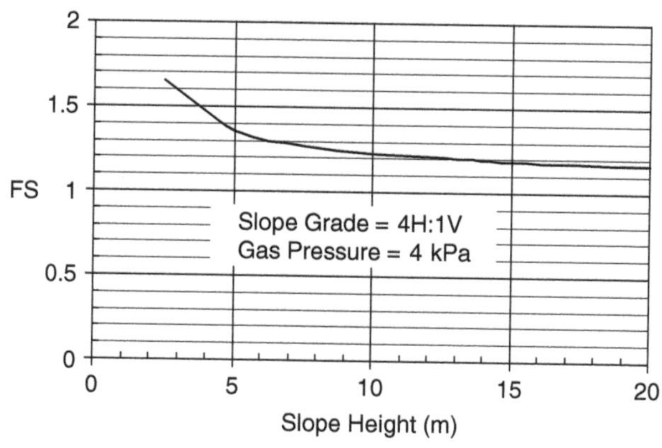


Figure 16. Factor of safety versus slope height for example cover slope.

into shorter slopes, say 5-m high, with benches. However, the net effect would be about the same loss of air space as flattening the slope to 5H:1V due to the width of the benches.

Tension developing in the geomembrane overlying the critical shear surface between the GCL and the drainage geocomposite could increase the stability. However, while they have tensile strength, geomembranes are not manufactured specifically to carry permanent tensile loads. Therefore, slopes should generally be designed so that a geomembrane does not develop significant tension. Furthermore, the effect of tension in the geomembrane is very small. An allowable tension for a geomembrane would be on the order of 5 to 10 kN/m, which would have very little impact on the factor of safety (Fig. 17)

Tensile reinforcement in the form of a woven geotextile or a geogrid could be included in the slope above the geomembrane as an alternative to increase the factor of safety. This layer could be put directly over the textured geomembrane or between the drainage sand and the vegetative cover soil (Fig. 13); if it is on the geomembrane, then the effect of the geogrid on the shear strength at the interface between the sand and the textured geomembrane would need to be considered. In the stability analysis, the design tension for the reinforcement layer is selected on the basis of minimizing deformation in the slope and accounting for factors such as creep and installation damage. Therefore, this design tension already includes a factor of safety and it is not reduced further by the factor of safety applied to the shear strengths in the stability analysis. A geogrid with an allowable tension of greater than about 70 kN/m would be sufficient to increase the factor of safety to 1.5 (Fig. 17).

Table 4. Design cases for example cover slope.

| Design case | Load | Resistance | | | Target FS |
|---|---|---|---|---|---|
| | | Shear strength | Added support | Fluid pressure | |
| Base Case – Expected | Weight of Cover | Peak for System<br>Residual for System | None<br>Toe Buttress & Allowable Geosynthetic Tension | 4 kPa Landfill Gas | $\geq 1.5$ |
| Gas Kick – Conservative | Weight of Cover | Peak for System<br>Residual for System | None<br>Toe Buttress & Allowable Geosynthetic Tension | 6 kPa Landfill Gas | $\geq 1.3$ |
| Gas Collection Failure – Extreme | Weight of Cover | Peak for System<br>Residual for System | None<br>Toe Buttress & Allowable Geosynthetic Tension | 10 kPa Landfill Gas | $\geq 1.0$ |
| Residual Strength – Extreme | Weight of Cover | Residual for System | None | 4 kPa Landfill Gas | $\geq 1.0$ |
| Unreinforced Bentonite – Extreme | Weight of Cover | Lower-Bound Shear Strength for GCL | None | None | $\geq 1.0$ |
| Earthquake – Extreme | Weight of Cover & Earthquake | Residual for System | Toe Buttress & Ultimate Geosynthetic Tension | 4 kPa Landfill Gas | $\geq 1.0$ |
| Construction – Expected | Weight of Sand Drainage Layer | Peak for System<br>Residual for System | None<br>Toe Buttress & Allowable Geosynthetic Tension | None | $\geq 1.3$ |
| Construction – Conservative | Weight of Sand Drainage Layer | Peak for System<br>Residual for System | None<br>Toe Buttress & Allowable Geosynthetic Tension | 4 kPa Landfill Gas | $\geq 1.0$ |

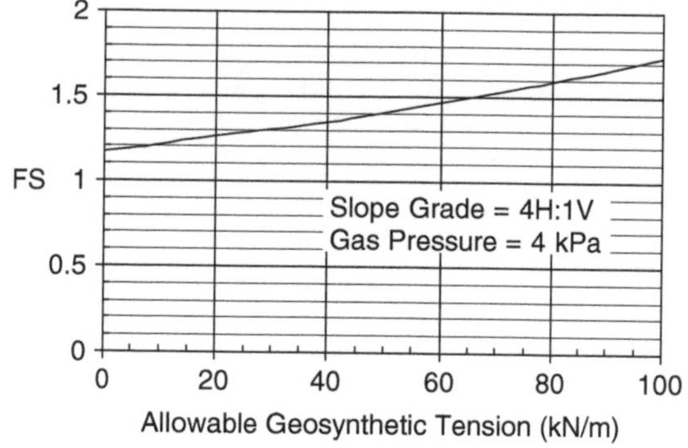


Figure 17. Factor of safety versus tension in geosynthetic reinforcement for example cover slope.

Another alternative to increase the factor of safety for the expected design conditions would be to replace the geocomposite in the gas collection system with a sand layer. Due to the relatively high shear strength for the interface between a non-woven geotextile and sand, the governing shear surface would move up to the interface between the textured geomembrane and the GCL (Fig. 14). The factor of safety for the 4H:1V slope increases to 2.0 for an infinite slope with the peak shear strength; the factor of safety is again lower in this case if the toe buttress is mobilized and the shear strength drops to the residual shear strength. The disadvantage of this alternative is the potential loss in air space by making the cover system thicker.

In summary for the base case, the 4H:1V slope does not have a factor of safety that is above the target minimum of 1.5. Three alternatives will increase the factor of safety to greater than 1.5: (1) flattening the slope to 5H:1V; (2) reinforcing the slope with a geogrid placed above the geomembrane; and (3) replacing the geocomposite drainage layer in the gas collection system with a sand drainage layer. These alternatives will be examined further for the other design cases in Table 4.

6.1.2 *Design check for conservative and extreme cases*

The conservative and extreme design cases will be analyzed for each of the three design alternatives. The effect of gas pressure on the factor of safety is shown on Figure 18. Small increases in gas pressure can have a large impact on the factor of safety. The greatest sensitivity is for the alternative of replacing the geocomposite drainage in the gas collection layer. This sensitivity is caused by the steepness of the shear strength envelope; the interface between the textured geomembrane and the GCL is stronger but also more sensitive to the effective normal stress than the interface between the GCL and the geocomposite (Fig. 14). In all alternatives, the factor of safety is greater than 1.3 for the conservative estimate of the maximum gas pressure, 6 kPa (Fig. 18). However, none of the alternatives provide a factor of safety greater than 1.0 for an extreme gas pressure of 10 kPa. One solution for this extreme design case would be to accept and plan for the risk of slope failure in the rare event that the gas pressure approaches 10 kPa. Another solution would be to put an active gas collection system in to better control the gas pressure. A final solution would be flatten the slope further or to add stronger geosynthetic reinforcement.

There are two extreme design cases for a shear strength that is weaker than expected. One extreme case is mobilization of the residual shear strength for the system, say due to deformations during construction. In this case, the large-displacement shear strength for the interface between the GCL and the geocomposite is taken as the residual shear strength for the system. In addition, the effect of any added support from the toe or tensile reinforcement is removed as an extreme, say due to poor quality construction. A second case is mobilization of the consolidated-drained strength of hydrated bentonite, say due to creep degradation or failure of the needle-punched reinforcing fibers over long-term loading. If the consolidated-drained strength of the bentonite is mobilized, it is assumed that its shear strength will not be affected by gas pressures in the gas collection layer. The factors of safety for these cases are summarized in Table 5. In all cases, the factors of safety are greater than the target minimum value of 1.0. While these design checks are extreme, they

Table 5. Factors of safety for smaller than expected shear strengths.

| Design Case | Factor of safety | | |
|---|---|---|---|
| | Flatten slope to 5H:1V | Add geogrid | Replace geocomposite with Sand |
| Residual shear strength for system | 1.4 | 1.1 | 1.4 |
| Lower-bound GCL shear strength | 1.3 | 1.1 | 1.1 |

Table 6. Seismic design information.

| Design Alternative | Yield acceleration | FS for design earthquake |
|---|---|---|
| Flatten slope to 5H:1V | 0.08 g | 1.1 |
| Add geogrid | 0.15 g | 1.2 |
| Replace geocomposite with sand | 0.11 g | 1.1 |

are warranted and reassuring due to the essentially unlimited lifetime for this slope. The results indicate that even if all of the man-made materials perform poorly in the long-term, the slope will still be marginally stable.

The design load associated with an earthquake at this site is a peak horizontal acceleration of 0.18 g in bedrock. This design acceleration corresponds to a value that has a 10% probability of being exceeded in 250 years based on United Sates Geological Survey maps (http://geohazards.cr.usgs.gov/eq/), and it is due primarily at this site to the possibility of a distant, large magnitude earthquake. A simplified and conservative response analysis (e.g., Matasovic et al. 1998 and Bray and Rathje 1998), based on a relationship developed by Harder (1991) for earth dams, gives a peak acceleration of 0.45 g for the landfill cover when the bedrock acceleration is 0.18 g. Therefore, the peak design acceleration for the landfill cover to account for the possibility of an earthquake is 0.45 g.

The seismic yield accelerations, i.e., the horizontal acceleration that causes the slope to slip, are summarized in Table 6 for the three alternative cover designs. Note that the design gas pressure is assumed to be present during the seismic event since this pressure will be continuously present over a long period of time after closure of the landfill. Even though all three alternatives provide a similar factor of safety for static loading where the residual shear strength is mobilized (Table 5), they provide a wide range of resistances to a seismic event. The differences in the yield accelerations are due to differences in how the horizontal force from the earthquake and the subsequent reduction in the normal stress on the shear surface affects the resistance to sliding. For the alternative with the geogrid, the contribution of resistance from the tensile reinforcement is not affected by a reduction in the normal stress on the shear surface; this alternative provides for the greatest seismic resistance (Table 6). The first and third alternatives are differentiated by different shear strength envelopes. The envelope for the large-displacement or residual shear strength is steeper for the interface between the GCL and the geocomposite, the governing shear surface if the slope is flattened, than it is for the interface between the textured geomembrane and the GCL, the governing shear surface if the gecomposite is replaced with sand (Fig. 14). Hence, replacing the geocomposite with sand provides greater seismic resistance (Table 6).

With all three design alternatives, the landfill cover will yield or slip during the design earthquake event since the peak design acceleration of 0.45 g is well above the yield accelerations (Table 6). However, the consequence of yielding will only be significant if the displacement is large. Kavazanjian (1998) provides an approach to account for displacement in a simplified seismic analysis; the peak cover acceleration is reduced by a factor of 0.1 to 0.2, which gives a representative acceleration that will produce about 0.3 m of displacement. With a reduction factor of 0.15, the design acceleration is 0.07 g and the factors of safety are all greater than the target value of 1.0 (Table 6). Therefore, all of the design alternatives provide adequate seismic resistance.

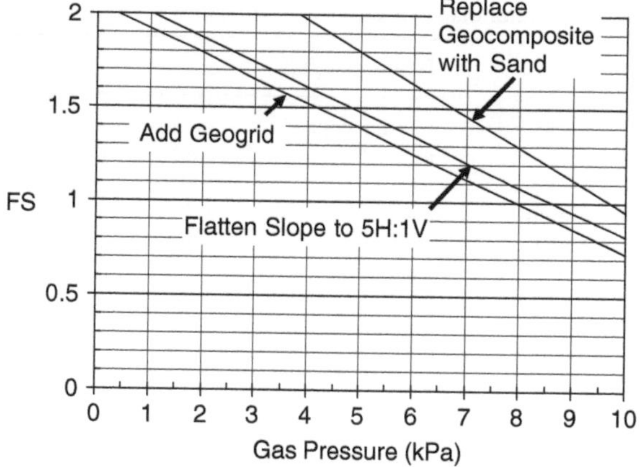


Figure 18. Factor of safety versus gas pressure for alternative designs in example cover slope.

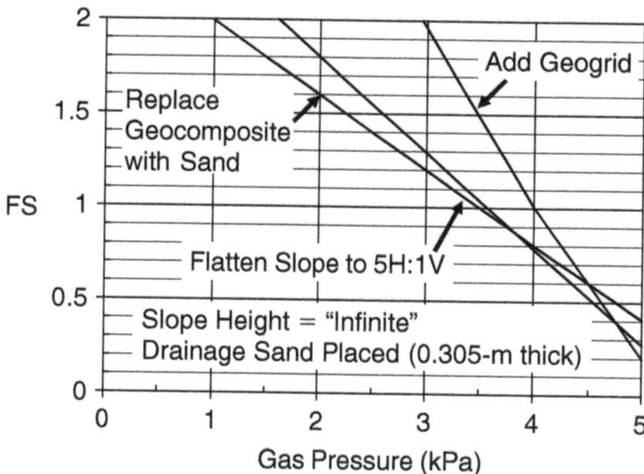


Figure 19. Factor of safety versus gas pressure for alternative designs in example cover slope during construction.

Construction of the slope is the final design consideration. The most critical time during construction will be when the sand layer is placed over the textured geomembrane before the overlying cover soil is placed (Fig. 13). At this time, the total stress on the potential shear surface is small, meaning that a small gas pressure can potentially cause instability. The effect of gas pressure on the stability of the slope during construction is shown on Figure 19. For all three alternatives, the factor of safety is greater than the target value of 1.3 if there is no gas pressure below the GCL, but less than or equal to the target value of 1.0 if there is 4 kPa of gas pressure. The analysis with the geogrid assumes that the geogrid will be placed directly on top of the textured geomembrane. On possible solution to provide for stability during construction would be to monitor and control gas pressure carefully during construction so that it stays below 3 kPa. Another solution would be to place the drainage sand up the slope in stages, bringing the vegetative cover soil up the slope sequentially with the drainage sand. A final solution would be to try placing the sand and be prepared for the possibility of a slope failure during construction.

## 6.2 *Example base liner slope*

The geometry and properties for a slope in a base liner system are shown on Figures 20 and 21. This municipal solid waste landfill is located in the Southwestern United States in an area with an average annual rainfall of approximately 600 mm. The landfill is being constructed in a valley with an open end (to the right in Fig. 20). The base liner on the side slope of the valley wall is graded at 2H:1V, while the base liner on the bottom of the landfill is graded at 2% toward the open end

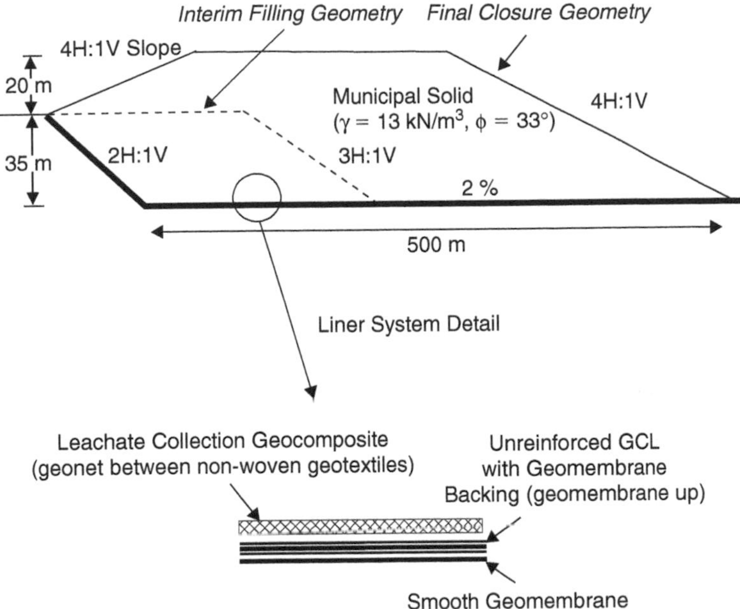


Figure 20. Configuration for example base liner slope.

of the valley where leachate is collected. Waste will be placed 20 m above the lip of the valley and will have above-grade slopes of 4H:1V in the final configuration.

The base liner consists of an unreinforced GCL that is encapsulated between two geomembranes (Fig. 20). The geomembrane-backed GCL is underlain by a smooth geomembrane and overlain by a geocomposite drainage layer, both on the side slope and the bottom of the landfill. The bentonite in the GCL is intended to stay dry between the two geomembranes, except in localized areas where there is a leak through either the upper or the lower geomembrane. The overlying leachate collection system is designed to maintain a head of less than 0.3 m on the liner.

The waste will first be placed in a 35-m thick fill below the lip of the valley. Once the valley is filled to a depth of 35 m, the waste will then be placed above grade (Fig. 20). The design cases for stability include the final closure geometry as well as an interim filling geometry when the waste is being placed below the lip of the valley (Fig. 20). The consequences of a slope failure, both during filling and after closure, are major due to the potential environmental impacts and the high costs of corrective action.

The specific design cases to be considered are summarized in Table 7. The peak and residual shear strengths for the system of failure surfaces involving the GCL depend on whether the bentonite is hydrated. If the bentonite is dry (as-delivered), then the governing shear surface is at the interface between the geocomposite and the smooth geomembrane (Fig. 21). If the bentonite becomes hydrated, then the lower-bound shear strength from Fig. 9 is assumed to apply (Fig. 21). If the bentonite remains dry, then the residual shear strength is assumed to apply for the base liner on the valley side slope since the waste here is buttressed by the waste on the valley bottom.

### 6.2.1 *Design check for final closure geometry*

The critical failure surface for the final closure geometry is calculated using Spencer's method (Duncan and Wright 2005), and the results for the expected conditions are shown on Figure 22. The critical failure surface cuts down through the waste and an angle of about 35 degrees to the horizontal and then exits along the base liner. The factor of safety is 1.6 and greater than the target minimum of 1.5.

The sensitivity of the factor of safety to the available shear strength in the base liner on the bottom of the landfill is shown on Figure 23. The factor of safety is greater than the target minimum value of 1.3 if the residual shear strength for the system is mobilized. Furthermore, even if the GCL

Table 7. Design cases for example base liner slope.

| Design Case | Load | Liner resistance | | Target FS |
|---|---|---|---|---|
| | | Shear strength | Fluid pressure | |
| Final Closure – Expected | Weight of Waste – Final Geometry | Bottom – Peak for System with Dry GCL<br>Side Slope – Residual for System with Dry GCL | 0.3-m Leachate Head on Bottom | ≥1.5 |
| Final Closure – Conservative | Weight of Waste – Final Geometry | Bottom – Residual for System with Dry GCL<br>Side Slope – Residual for System with Dry GCL | 0.3-m Leachate Head on Bottom | ≥1.3 |
| Final Closure – Conservative | Weight of Waste – Final Geometry | Bottom – Peak for System with Dry GCL<br>Side Slope – Residual for System with Dry GCL | 10-m Leachate Head on Bottom | ≥1.3 |
| Final Closure – Extreme | Weight of Waste – Final Geometry | Lower-Bound Shear Strength for GCL | 0.3-m Leachate Head on Bottom | ≥1.0 |
| Interim Waste Filling – Expected | Weight of Waste – Interim Geometry | Bottom – Peak for System with Dry GCL<br>Side Slope – Residual for System with Dry GCL | 0.3-m Leachate Head on Bottom | ≥1.3 |
| Interim Waste Filling – Extreme | Weight of Waste – Interim Geometry | Lower-Bound Shear Strength for GCL | 0.3-m Leachate Head on Bottom | ≥1.0 |
| Interim Waste Filling – Extreme | Weight of Waste – Interim Geometry | Bottom – Peak for System with Dry GCL<br>Side Slope – Residual for System with Dry GCL | 3-m Leachate Head on Bottom | ≥1.0 |

becomes fully hydrated and its shear strength drops to the smallest possible value (Fig. 9), the factor of safety is at about 1.0 and the slope is marginally stable.

The effect of an elevated leachate head is shown on Figures 24 and 25. As the leachate head acting on the base liner on the bottom increases, the failure surface dives more steeply through waste; e.g., about 40 degrees to the horizontal on Fig. 24 versus 35 degrees to the horizontal on Fig. 22. The steeper dip for the failure surface puts more of the surface along the base liner where the higher leachate head has reduced the effective normal stress and, hence, the shear strength. The factor of safety for the final closure geometry is not very sensitive to the leachate head, and it is equal to the target minimum value of 1.3 if the leachate head is as high as 10 m. The factor of safety is still greater than 1.0 for leachate heads as high as 20 m above the base liner (Fig. 25). Providing that post-closure care includes monitoring for and controlling large leachate mounds (greater than 20-m high), the slope should be stable even if the leachate collection system fails in the long-term.

### 6.2.2 *Design check for interim waste filling geometry*

The critical failure surface for the landfill during filling follows the interface between the drainage geocomposite and the GCL down the side slope and out over the bottom of the landfill (Fig. 26). If the waste is to be placed up to the crest of the side slope (elevation 0 m on Figs. 24 and 26), then the waste buttress on the bottom needs to be placed out a distance of 132 m from the toe of the side slope in order to achieve a factor of safety of 1.3 (Fig. 26).

Since the waste cannot be placed instantaneously into the configuration shown on Fig. 26, the length of the buttress required is shown on Figure 27 as a function of the height of the waste above the toe of the side slope. This curve provides useful guidance for the operator in filling the landfill. However, there is not the same assurance that this guidance will be followed as there is for

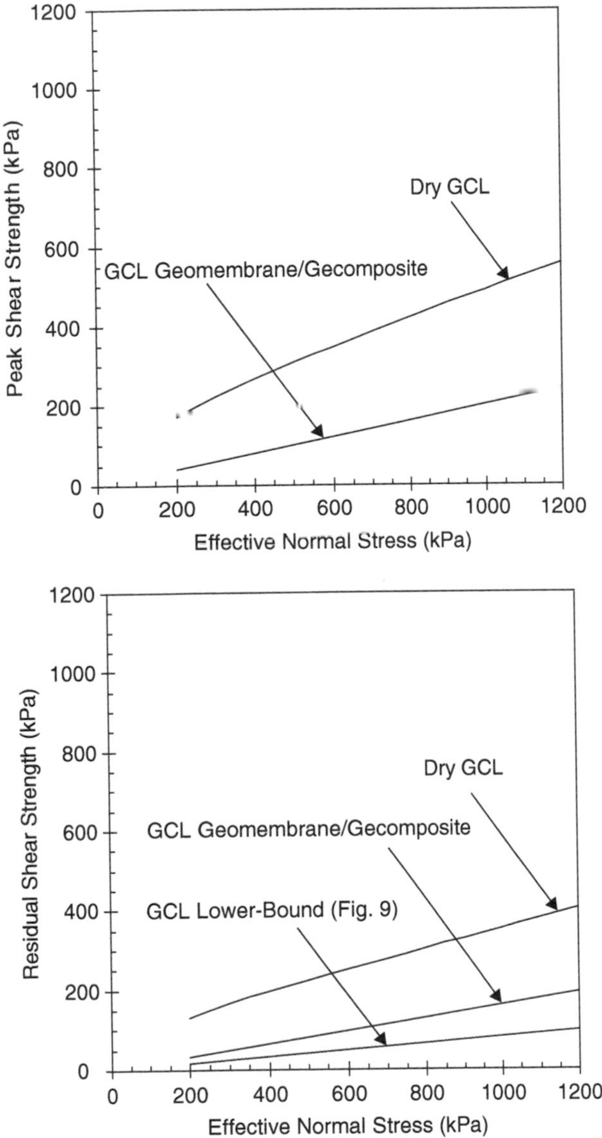


Figure 21. Shear strength envelopes for GCL in example base liner slope.

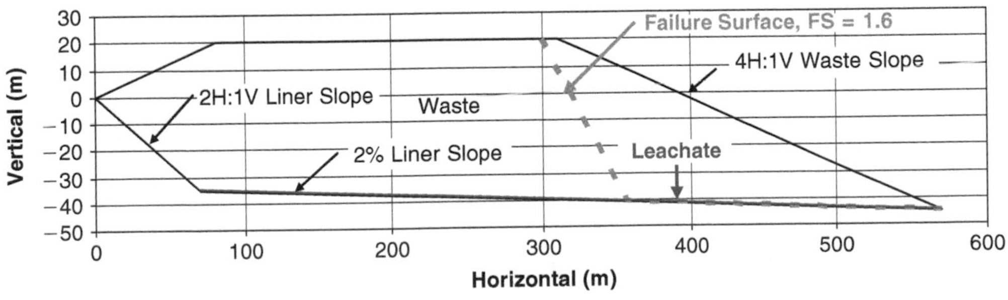


Figure 22. Critical failure surface for final closure geometry with expected conditions: peak shear strength for liner system on bottom of landfill assuming the GCL is dry and a 0.3-m head of leachate on the base liner on bottom.

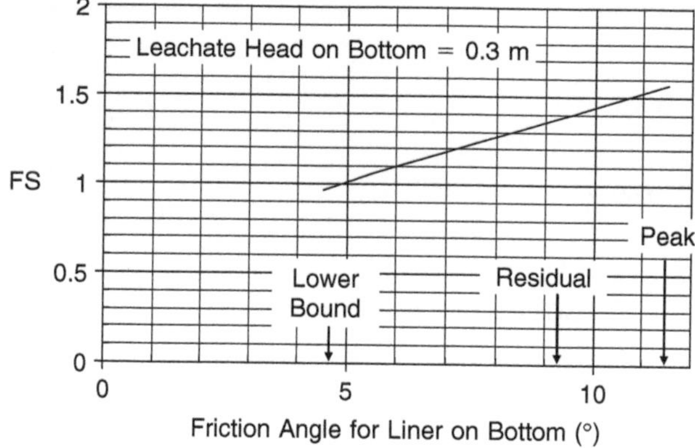


Figure 23. Factor of safety for final closure geometry versus shear strength for base liner on bottom of landfill.

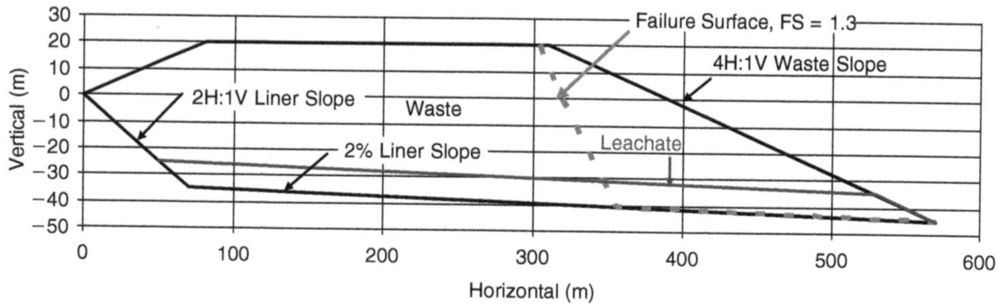


Figure 24. Critical failure surface for final closure geometry with elevated leachate head acting on base liner on bottom; peak shear strength for liner system on bottom of landfill assuming the GCL is dry.

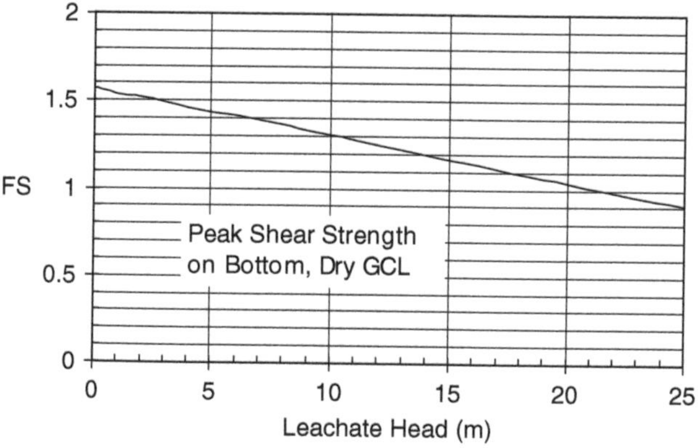


Figure 25. Factor of safety for final closure geometry versus leachate head acting on base liner on bottom of landfill.

a permanent slope, such as the final closure geometry. Therefore, evaluating the sensitivity of the stability to a variety of factors is important.

The sensitivity of the factor of safety to the length of the buttress is shown on Figure 28. If the waste is placed up to the crest of the side slope and the buttress is less than about 100 m, then there is a high potential for a slope failure. There are two notable features about the geometry of the slope in this case, which is shown on Figure 29. First, the geometry does not look perilous or marginally stable; one of the consequences of using materials with relatively low interface

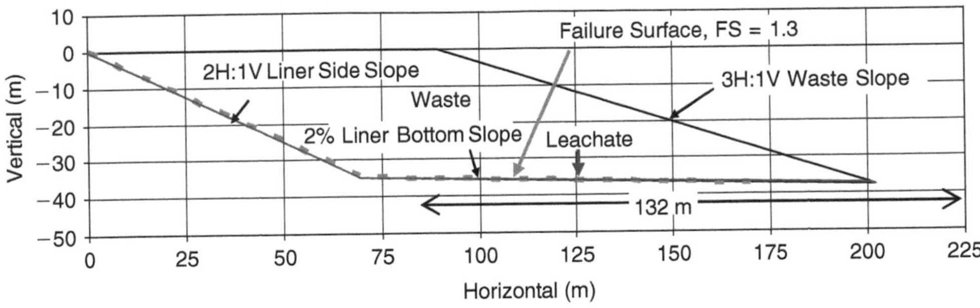


Figure 26. Critical failure surface for interim filling geometry with expected conditions: peak shear strength for liner system on bottom of landfill and residual shear strength on side slope assuming the GCL is dry and a 0.3-m head of leachate on the base liner on bottom.

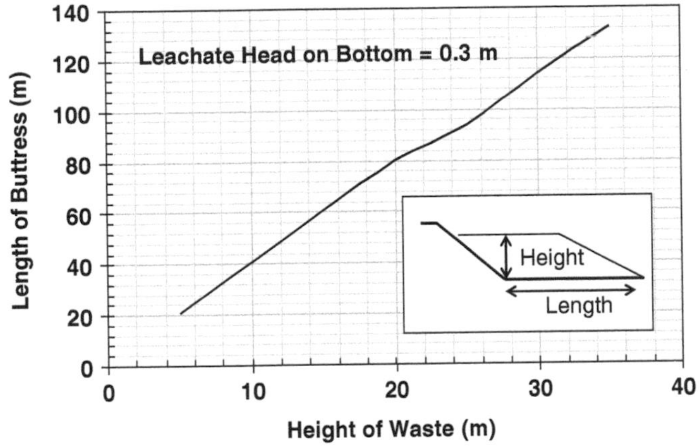


Figure 27. Required length of buttress to achieve a factor of safety of 1.3 for interim filling geometry with expected conditions: peak shear strength for liner system on bottom of landfill and residual shear strength on side slope assuming the GCL is dry and a 0.3-m leachate head on bottom base liner.

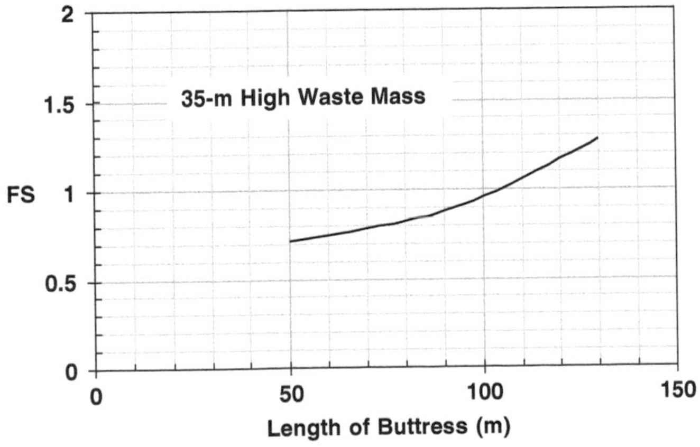


Figure 28. Factor of safety versus length of waste buttress from toe of liner side slope for interim filling geometry with expected conditions: peak shear strength for liner system on bottom of landfill and residual shear strength on side slope assuming the GCL is dry and a 0.3-m leachate head on bottom base liner.

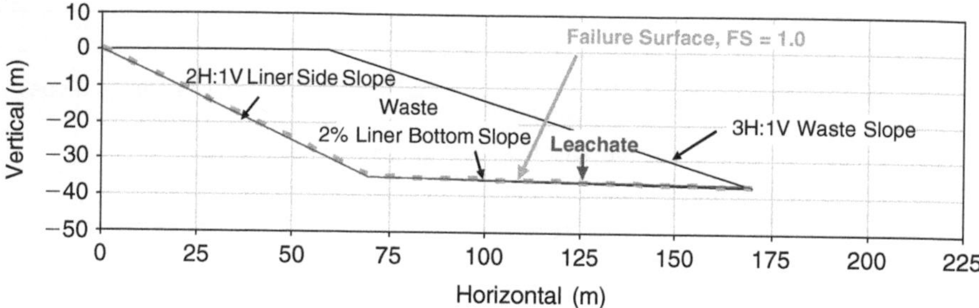


Figure 29. Critical failure surface for interim filling geometry with expected conditions: peak shear strength for liner system on bottom of landfill and residual shear strength on side slope assuming the GCL is dry and a 0.3-m head of leachate on the base liner on bottom.

Table 8. Sensitivity results for interim waste filling geometry (Fig. 23).

| Design Case | FS |
|---|---|
| Base Case – Expected Conditions | 1.3 |
| Residual versus Peak Shear Strength on Liner Bottom | 1.1 |
| Hydrated versus Dry GCL on Liner Side Slope | 1.2 |
| Hydrated versus Dry GCL on Liner Bottom | 0.75 |
| Hydrated versus Dry GCL on Liner Bottom and Side Slopes | 0.60 |
| 3-m versus 0.3-m of Leachate Head Acting on Liner Bottom | 1.2 |
| 10-m versus 0.3-m of Leachate Head Acting on Liner Bottom | 0.99 |
| Interim Waste Filling – Extreme | ≥1.0 |
| Interim Waste Filling – Extreme | ≥1.0 |

shear strengths in liners is that human intuition does not provide a useful guide to workers at the project site since their intuition is calibrated with natural and engineered slopes that are generally in much stronger materials. Second, the volume of waste that would be involved in a slope failure, approximately 400,000 m$^3$ for every 100-m width, is large and the consequences of a failure would be significant. One positive conclusion from Figure 28 is that the factor of safety becomes less sensitive to the length of the buttress as the length decreases, which provides some margin for variations in operations.

If the landfill is filled following the guidance on Fig. 28, there are events that could possibly cause a slope failure during filling. In order to consider these possibilities, a sensitivity analysis is summarized in Table 8 for the 35-m high waste slope. The most significant concern is the potential for the encapsulated GCL to become hydrated in the base liner on the bottom of the landfill (Table 8). The factor of safety is most sensitive to the shear strength along the bottom of the landfill, which provides the buttress, since most of the resistance to sliding is provided along this shear surface (it is longer and the effective normal stress is higher compared to the shear surface on the side slope). One alternative to mitigate the risk from this possibility would be to increase the buttress length (that is, develop Fig. 28 to achieve a factor of safety of at least 1.0 using the hydrated shear strength for the GCL). Another alternative would be to monitor the moisture content of the bentonite and adjust the filling plan accordingly if the moisture content increases.

## 6.3 *Example liner construction*

The geometry and properties for a slope that will be constructed to form a base liner are shown on Figure 30. The governing shear surface is the bentonite layer in the unreinforced GCL. Figure 9 provides a means to estimate its shear strength under the effective stress due to the overlying gravel. Once the waste is completely filled in the landfill, there will be minimal shear stress in the liner system on the side slope (like water in a bathtub). However, if the layer of gravel forming the leachate collection system were placed up the 20-m high slope before any waste were placed, then

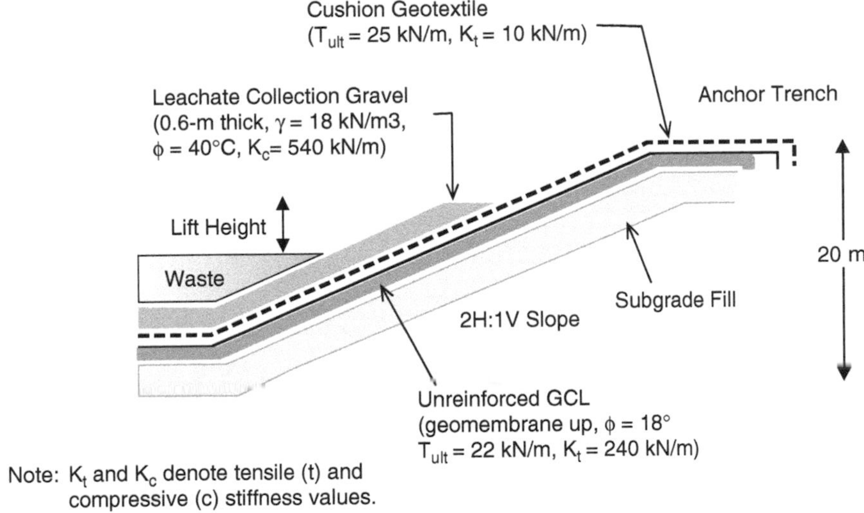


Figure 30. Configuration for example liner construction slope.

tension will develop in the geosynthetic layers overlying the bentonite (the cushion geotextile and the geomembrane backing on the GCL) since the slope angle (26.6°) is greater than the secant friction angle for the bentonite ($\phi = 18°$). A limit equilibrium analysis indicates that a total tensile force equal to 47 kN/m, in addition to the force supplied by the toe buttress, would be required for a factor of safety of 1.0. This tensile force is equal to the sum of the ultimate tensile capacities for these geosynthetic layers (Fig. 30) and therefore too large.

An approach to reduce the tensile forces developing in the geosynthetic layers is to sequentially place the gravel layer up the slope in lifts as the waste is filled into the landfill (Fig. 30). As the gravel is placed, compression will develop at the toe of each unsupported gravel segment and tension will accumulate in the geosynthetic layers. These support forces will require displacement of the gravel layer. For a given displacement down the slope, the magnitudes of compression and the tension that develop will be related to the axial stiffness for each material. Hence, the total support force will be distributed between compression and tension according to the relative axial stiffness values for the materials (Fig. 30).

A simplified, elastic column analysis (e.g., Liu and Gilbert 2003) was conducted on this slope in order to develop a plan for placing the gravel layer. Since the cushion geotextile has such a small stiffness in comparison to the geomembrane, it carries very little of the tensile force. The tensile force that accumulates in the geomembrane is shown on Figure 31 versus the total vertical height of the gravel layer as it is placed up the slope in lifts (sequentially bringing the waste up after each lift is placed). The smaller the lift height, the less tension that accumulates in the geomembrane (Fig. 31). If the allowable tension is 50% of the ultimate tension, then a lift height of 4 m would be required (that is, the gravel could be placed about 9 m up the dip of the slope in front of the waste). If the allowable tension is 33% of the ultimate tension, than the lift height would need to be 2 m or less (Fig. 31).

## 7 CONCLUSIONS

While the potentially low shear strength for GCLs poses a challenge in designing stable slopes for cover and liner systems, it is possible and practical to achieve sound designs.

The most important step in design is to use shear strengths for GCLs that are representative of field conditions. The bentonite in a GCL in the field will typically be saturated, consolidated and sheared under drained conditions. The consolidated-drained, residual shear strength for unreinforced bentonite (Fig. 9) provides a conservative or lower-bound estimate of the GCL shear strength that will be mobilized in most, if not all, practical situations of concern for slope stability. Both

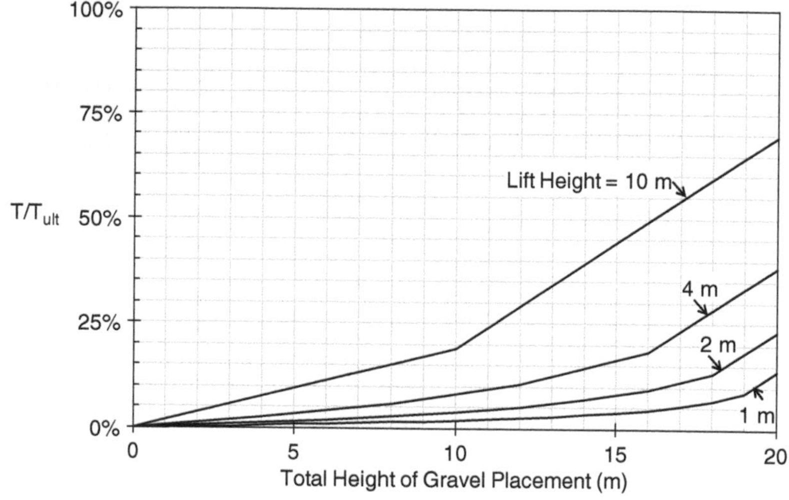


Figure 31. GCL Geomembrane tension versus height of waste placement for different lift heights in example liner construction slope.

peak and residual shear strengths need to be estimated for all possible shear surfaces. A range of design shear strengths, including both the peak and residual shear strengths for the system, should be considered in design. The residual shear strength for a system of possible shear surfaces is important because both the potential for and the consequence of a slope failure are related to this strength. Finally, all possible failure surfaces over all possible effective normal stresses need to be considered in establishing design shear strengths.

Stability analyses should consider a wide-range of possible geometries, loading conditions, fluid pressures, and shear strengths. Sensitivity analyses are very important in practice to manage uncertainties in field conditions. Stability analyses should be evaluated within the context of how probable a combination of shear strengths and loading conditions is to occur in the lifetime of the slope and how significant the consequences would be if the slope were to fail. Design criteria should not be treated as prescriptive and should reflect the judgment and input of the engineers, owners, operators and regulators on a project-specific basis.

Finally, if there is a concern about slope stability, then there is a variety of approaches that can be used to mitigate these concerns: flattening the grade of a slope, providing a buttress at the toe of a slope, reinforcing a slope with tensile reinforcement, reducing fluid pressures in a slope, replacing weak shear surfaces (such as the GCL) in a slope; and reducing the consequences of a slope failure.

## REFERENCES

Achilleous, E. 1988. *PC STABL5M, User Manual*," Informational Report, School of Civil Engineering, Purdue University, West Lafayette, IN, 132 p.

ASTM D 3080. Standard Test Method for Direct Shear Test of Soils Under Consolidated Drained Conditions. American Society of Testing and Materials, West Conshohocken, Pennsylvania.

ASTM D 5321. Standard Test Method for Determining the Coefficient of Soil and Geosynthetic and Geosynthetic and Geosynthetic Friction by the Direct Shear Method. American Society of Testing and Materials, West Conshohocken, Pennsylvania.

ASTM D 6243. Standard Test Method for Determining the Internal and Interface Shear Resistance of Geosynthetic Clay Liner by Direct Shear Method. American Society of Testing and Materials, West Conshohocken, Pennsylvania.

Bjerrum, L. 1967. Progressive Failure in Slopes of Overconsolidated Plastic Clay and Clay Shales. *Journal of the Soil Mechanics and Foundations Division,* ASCE, Vol. 93, SM5, 3–49.

Bray, J.D. and Rathje, E.M. 1998. Earthquake-Induced Displacements of Solid-Waste Landfills. *Journal of Geotechnical and Geoenvironmental Engineering*, ASCE, 124(3), pp. 242–253.

Byrne, R. J., Kendall, J., and Brown, S. 1992. Cause and Mechanism of Failure, Kettleman Hills Landfill B-19, Unit IA. *Proc., ASCE Spec. Conf. on Performance and Stability of Slopes and Embankments - II*, Vol. 2, 1188–1215.

Corps of Engineers 2003. Engineering and Design – Slope Stability. *Engineer Manual EM 1110-2-1902*, Department of Army, Corps of Engineers, Office of the Chief of Engineers.

Daniel, D. E., Shan, H. Y., and Anderson, J. D. 1993. Effects of Partial Wetting on the Performance of the Bentonite Component of a Geosynthetic Clay Liner. *Geosyn. '97*, St. Paul, MN.

Daniel, D. E., Carson, D. A., Bonaparte, R., Koerner, R. M., and Scranton, H. B. 1998. Slope Stability of Geosynthetic Clay Liner Test Plots. *Journal of Geot.and Geoenv. Eng.*, ASCE, 124(7), 628–637.

Duncan, J. Michael and Wright, Stephen G. 2005. *Soil Strength and Slope Stability*, John Wiley and Sons, New York.

EPA 1989. *Final Covers on Hazardous Waste Landfills and Surface Impoundments*. Technical Guidance Document, EPA/530/SW-89/047, U.S. Environmental Protection Agency, Office of Solid Waste and Emergency Response, Washington, D.C., 39 p.

EPA 1995. *RCRA Subtitle D (258) Seismic Design Guidance for Municipal Solid Waste Landfill Facilities*. EPA/600/R-95/051, U.S. Environmental Protection Agency, Office of Research and Development, Washington, D.C., 143 p.

EPA 2006. Technical Guidance for RCRA/CERCLA Final Covers. EPA/540-R-04-007, U.S. Environmental Protection Agency, Office of Solid Waste and Emergency Response, Washington, D.C.

Fox, P. J., Rowland, M. G., and Scheithe, J. R. 1998. Internal Shear Strength ofThree Geosynthetic Clay Liners. *Journal of Geot. and Geoenv. Eng.*, ASCE, 124(10), 933–944.

Geo-Slope International Ltd. 2004. SLOPE/W, Slope Stability Analysis Software.

Gilbert, R. B. and Byrne, R. J. 1996. Strain-Softening Behavior of Waste Containment System Interfaces. *Geosynthetics International*, 3(2), 181–203.

Gilbert, R. B., Fernandez, F. F., and Horsfield, D. 1996. Shear Strength of a Reinforced Clay Liner. *Journal of Geot. and Geoenv. Engineering*, ASCE, 122(4), 259–266.

Gilbert, R. B., Long, J. H. and Moses, B. E. 1996. Analytical Model of Progressive Failure in Waste Containment Systems. *International Journal for Numerical and Analytical Methods in Geomechanics,* 20(1), 35–56.

Gilbert, R. B., Scranton, H. B., and Daniel, D. E. 1997. Shear Strength Testing for Geosynthetic Clay Liners. *Testing and Acceptance Criteria for Geosynthetic Clay Liners*, L. Well, ed., American Society for Testing and Materials, Philadelphia, 121–138.

Gilbert, R. B., Wright, S. G. and Liedtke, E. 1998. Uncertainty in Back Analysis of Slopes. *Journal of Geotechnical Engineering*, ASCE, 124(12), 1167–1176.

Gilbert, R. B., Wright, S. G., Shields, K. M. and Obermeyer, J. E. 2004. Lower-Bound Shear Strength for Geosynthetic Clay Liners in Base Liners. *Geosynthetics International*, Thomas Telford, 11(3), 200–211.

GSE 2001. The GSE Gundseal Design Manual. GSE Lining Technology, Inc., Houston, Texas.

Giroud, J. P. and Beech, J. F. 1989. Stability of Soil Layers on Geosynthetic Lining Systems. *Proceedings*, Geosynthetics '89, San Diego, pp. 35–46.

Giroud, J. P., Williams, N. D., Pelte, T., and Beech, J. F. 1995. Stability of Geosynthetic-Soil Layered Systems on Slopes. *Geosynthetics International*, Vol. 2, No. 6, pp. 1115–1148.

Harder, Jr., L.S. 1991. Performance of Earth Dams During the Loma Prieta Earthquake. Proceedings of the 2nd International Conference on Recent Advances in Geotechnical Earthquake Engineering and Soil Dynamics, St. Louis, Missouri, pp. 1613–1629.

Kavazanjian, E., Jr. 1998. Current Issues in Design of Geosynthetic Cover Systems. Proc. Sixth International Conference on Geosynthetics, Atlanta, Georgia, Vol. I, pp. 219–226.

Koerner, R. M. and Hwu, B. L. 1991. Stability and Tension Considerations Regarding Cover Soils on Geomembrane Lined Slopes. *Geotextiles and Geomembranes,* Vol. 10, pp. 335–355.

Lam, L. and Fredlund, D. G. 1993. A General Limit Equilibrium Model for Three-Dimensional Slope Stability Analysis. *Canadian Geotechnical Journal*, 30(6), 905–919.

Lai, J., Daniel, D. E. and Wright, S. G. 1998. Effects of Cyclic Loading on Internal Shear Strength of Unreinforced Geosynthetic Clay Liner. *Journal of Geotechnical and Geoenvironmental Engineering*, ASCE, Vol. 124, No. 1, 45–52.

Liu, C-N. and Gilbert, R. B. 2003. A Simplified Method for Estimating Geosynthetic Loads During the Placement of Cover Soil. *Geosynthetics International*, IGS, Vol. 10, No. 1.

Liu, C.-N., Gilbert, R. B., Thiel, R. S. and Wright, S. G. 1997. What Is an Appropriate Factor of Safety for a Cover Slope? *Proceedings,* Geosynthetics 97, IFAI, San Diego, California, 481–498.

Loehr, J.E., McCoy, B.F., and Wright, S.G. 2004. Quasi 3-D Method for Slope Stability Analysis of General Sliding Bodies. *Journal of Geotechnical and Geoenvironmental Engineering*, ASCE, Vol. 130, No. 6, 551–560.

Long, J. H., Gilbert, R. B. and Daly, J. J. 1994. Geosynthetic Loads in Landfill Slopes – Displacement Compatibility. *Journal of Geotechnical Engineering* , ASCE, Vol. 120(11), 2009–2025.

Matasovic, N. Kavazanjian, E., Jr., and Anderson, R.L. 1998. Performance of Solid Waste Landfills in Earthquakes. *Earthquake Spectra*, EERI, Vol. 14, No. 2, pp. 319–334.

Mesri, G. and Olson, R. E. 1970. Shear Strength of Montmorillonite. *Geotechnique*, 20(3), 261–270.

Mitchell, J. K., Seed, R. B. and Seed, H. B. 1990. Kettleman Hills Waste Landfill Slope Failure. I: Liner System Properties. *J. of Geotech. Engrg.,* ASCE, 116(4), 647–668.

Morgenstren, N. R. and Price, V. E. 1965. "The Analysis of the Stability of General Slip Surfaces," *Geotechnique*, Vol. 17, No. 1, 11–26.

Seed, R. B., Mitchell, J. K. and Seed, H. B. 1990. Kettleman Hills Waste Landfill Slope Failure. II: Stability Analyses," *J. of Geotech. Engrg.,* ASCE, 116 (4), 669–690.

Shan, H.-Y. 1993. Stability of Final Covers Placed on Slopes Containing Geosynthetic Clay Liners," Ph.D. Dissertation, University of Texas, Austin, Texas, USA.

Sharma, S. 1994. *XSTABL, Reference Manual, Version 5*. Interactive Software Designs, Inc., Moscow, Idaho.

Skempton, A. W. 1964. Long-Term Stability of Clay Slopes. *Geotechnique,* Vol. 14, No. 2, 77–102.

Spencer, E. 1967. A Method of Analysis of the Stability of Embankments Assuming Parallel Inter-Slice Forces. *Geotechnique*, Vol. 17, No. 1, 11–26.

Stark, T. D., Arellano, D., Evans, W. D., Wilson, V. L., and Gonda, J. M. 1998. Unreinforced Geosynthetic Clay Liner Case History. *Geosynthetics International*, IFAI, 5(5), 521–544.

Stark, T. D., and Eid, H. T. 1998. Performance of Three-Dimensional Slope Stability Methods in Practice. *Journal of Geot. and Geoenv. Engrg.,* 124(11), 1049–1060.

Stark, T. D., Eid, H. T., Evans, W. D., and Sherry, P. E. 2001. Municipal Solid Waste Slope Failure. II: Stability Analyses. *J. of Geotech. Engrg.,* ASCE, 126 (5), 408–419.

Wilson-Fahmy, R. F. and Koerner, R. M. 1993. Finite Element Analysis of Stability of Cover Soil on Geomembrane Lined Slopes. *Proc.* Geosynthetics '93, Vancouver, IFAI, 3, 1425–1438.

Wright, S.G. 1991. *UTEXAS3 – A Computer Program for Slope Stability Calculations*. Department of Civil Engineering, University of Texas, Austin, Texas.

# CHAPTER 10

# Hydrologic performance of final covers containing GCLs

C.H. Benson
*University of Washington, Seattle, WA, USA*

J. Scalia
*University of Wisconsin, Madison, WI, USA*

## 1 INTRODUCTION

Geosynthetic clay liners (GCLs) are used frequently as a hydraulic barrier layer in landfill final covers due to their thinness, ease of installation, and perceived resistance to environmental distress relative to barriers constructed with compacted clay. A variety of field studies have been conducted over the last decade in Europe and the USA to assess percolation rates characteristic of covers employing GCLs. Studies that have been published in the open literature are reviewed in this chapter. Recommendations regarding the use of GCLs in final covers are also provided. Characteristics of the field sites are summarized in Table 1.

## 2 COVERS RELYING SOLELY ON A GCL

### 2.1 *Georgswerder study in Germany*

Melchior (2002) constructed two triangular lysimeters near Hamburg, Germany to evaluate percolation rates for covers employing a GCL as the barrier layer. Hamburg has a humid and seasonal climate with 600–750 mm of precipitation annually. Each lysimeter had an area of 100 $m^2$ and 8% slope. One lysimeter was used to test a needle-punched GCL (NP) containing natural sodium (Na) bentonite, and the other tested a stitch-bonded (SB) GCL containing Na-activated bentonite.

Table 1. Characteristics of sites where field performance of covers containing GCLs have been evaluated. NP = needle punched, SB = stitch bonded, CA = Ca bentonite, LNP = laminated needle punched.

| Study | Location | Climate | Barrier Layer | Annual Percolation (mm/yr) | GCL Type | Long Term Precipitation Rate (mm/yr) |
|---|---|---|---|---|---|---|
| Melchior (2002) | Hamburg, Germany | Oceanic | Conventional GCL | 620–750 | NP<br>SB | 188<br>222 |
| Wagner et al. (2002) | Esch-Belval, Luxembourg | Oceanic | Conventional GCL | 713–1037 | | 6.2 |
| Henken-Mellis (2002) | Aurach, Germany | Continental | Conventional GCL | 750 | CA | 14 |
| Benson et al. (2007) | Wisconsin USA | Continental Humid | Conventional GCL | 892 | NP | 299–450 |
| | | | Laminated GCL | | LNP | <5 |
| Albright et al. (2004) | California, USA | Arid | GCL-geom. composite | 112 | NP | <0.1 |
| | Boardman, USA | Semi-Arid | GCL-geom. composite | 215 | NP | 0 |

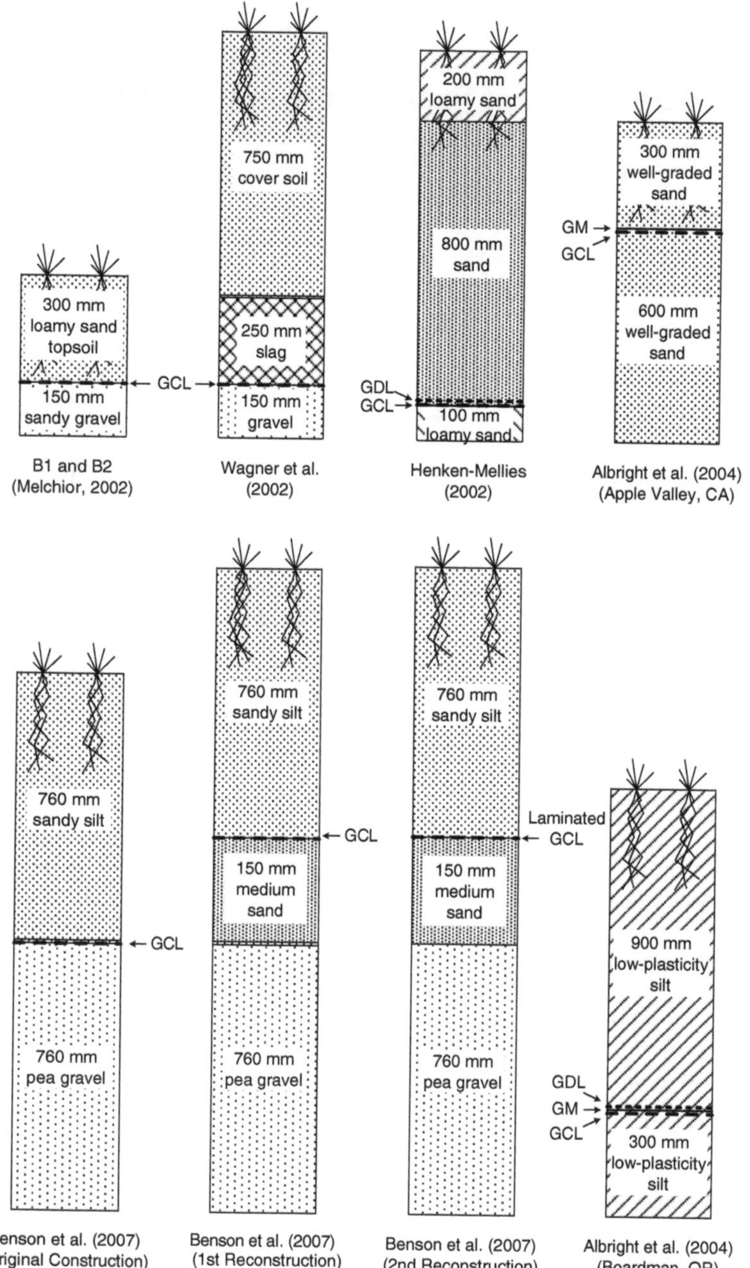


Figure 1. Profiles of final covers with GCLs evaluated in Europe and the USA. GCL = geosynthetic clay liner, GDL = geosynthetic drainage layer, GM = geomembrane.

The cover profile consisted of (top to bottom) 300 mm of loamy sand topsoil, the GCL, and 150 mm of sandy gravel, as depicted in Figure 1.

Annual percolation rates reported by Melchior (2002) are shown as a function of time in Figure 2. During the first year of operation, the percolation rate was less than 6 mm/yr for both lysimeters (<1% of precipitation). After the first year, however, the percolation rate increased substantially, and ultimately reached 220 mm/yr for the NP GCL and 190 mm/yr for the SB GCL. The peak daily percolation rate was 15 mm/d for both lysimeters.

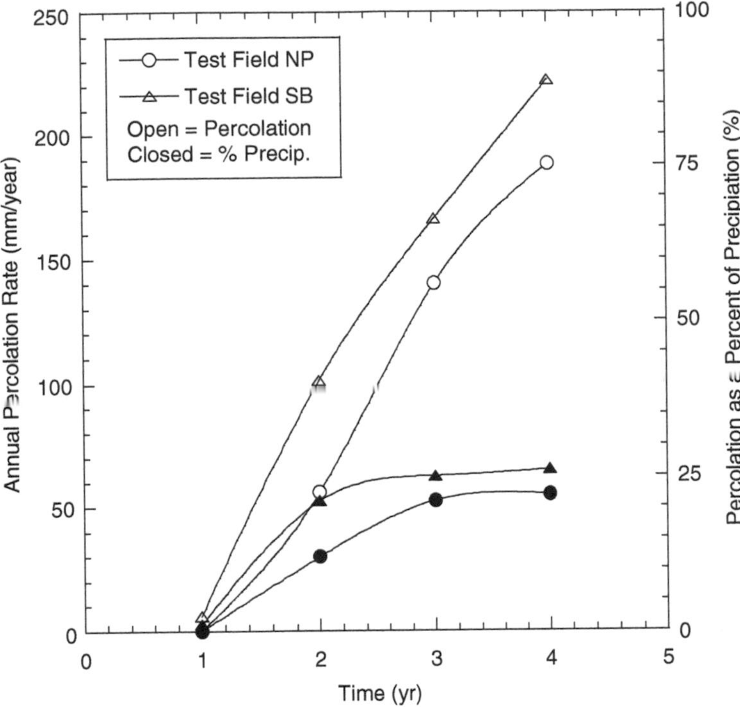


Figure 2. Annual percolation rate and percolation as a percentage of precipitation from final covers with GCLs evaluated by Melchior (2002) in Hamburg, Germany.

For a GCL with a typical hydraulic conductivity of $2 \times 10^{-9}$ cm/s, a percolation rate <1 mm/yr is expected under saturated steady-state unit-gradient conditions. In contrast, a percolation rate of 15 mm/d corresponds to $1.7 \times 10^{-5}$ cm/s under saturated steady-state unit-gradient conditions. Thus, the GCLs in both lysimeters had higher hydraulic conductivity than originally anticipated.

Exhumations were conducted periodically to determine if changes in the GCL were occurring. Within the first year, root penetrations and cracks in the bentonite were visible in both GCLs. By the end of the monitoring period, nearly all of the Na originally in the exchange complex of the bentonite was replaced by calcium (Ca) and magnesium (Mg), and the free swell of the bentonite was comparable to that of Ca bentonite rather than Na bentonite. Cation exchange was attributed to cations in water percolating downward through the cover soils and into the GCL. Hydraulic conductivity of GCLs exhumed at the end of the study ranged from $1.1 \times 10^{-5}$ to $3.0 \times 10^{-4}$ cm/s. This range is consistent with the hydraulic conductivity corresponding to the peak percolation rate measured in the field.

## 2.2 *Esch-Belval study in Luxembourg*

Wagner et al. (2002) evaluated a needle-punched GCL containing Na bentonite as the barrier layer in a final cover profile at a blast furnace dust dump in Luxembourg, where the annual precipitation is 700–1100 mm/yr. The test section was located on a 5% slope and consisted of (top to bottom) a 750-mm-thick layer of cover soil, the GCL, and 250 mm of electric furnace slag (Fig. 1). A 3 m × 5 m pan lysimeter was used to collect percolation from the cover. Data were collected for 2 yr.

The annual percolation rate was 1.4 mm/yr in the first year and 6.2 mm/yr in the second year. These annual percolation rates correspond to hydraulic conductivities of $4 \times 10^{-9}$ cm/s and $2 \times 10^{-8}$ cm/s under steady-state unit-gradient conditions. Thus, the hydraulic conductivity of the GCL apparently increased during the first two years of study. However, samples of the GCL were not exhumed and thus the hydraulic conductivity of the GCL cannot be confirmed.

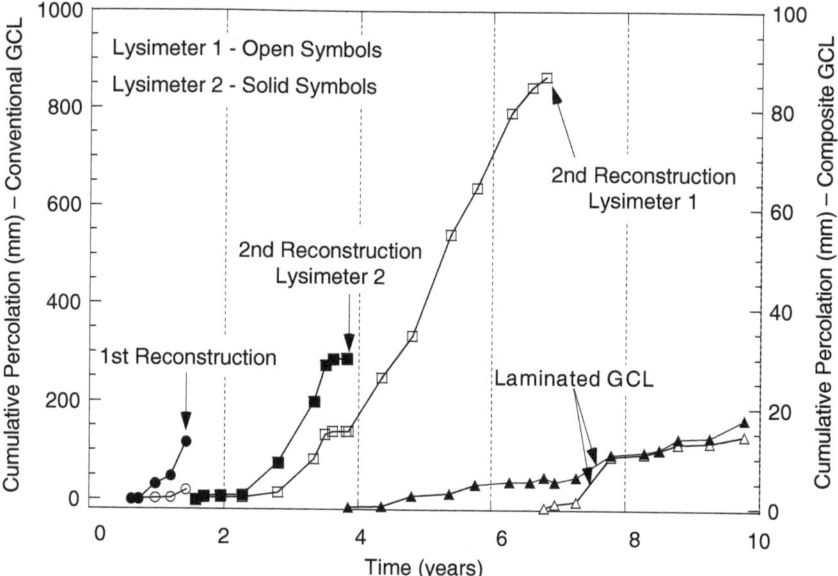


Figure 3. Percolation rates recorded in lysimeters beneath a final cover employing a GCL in Wisconsin, USA. Left vertical ordinate is cumulative percolation through a conventional GCL system; right vertical ordinate is cumulative percolation through a laminated GCL system (adapted from Benson et al. 2007).

### 2.3 *Aurach study in Germany*

Henken-Mellies et al. (2002) evaluated percolation rates from a test cover installed at a landfill in Aurach, Germany, where the average annual precipitation rate is 750 mm/yr. The cover employed a GCL containing Ca bentonite as the barrier layer and was composed of (top to bottom) 200 mm of loamy sand topsoil, 800 mm of sand, a geocomposite drainage layer, the GCL, and 300 mm of loamy sand (Fig. 1). The GCL had higher dry mass per unit area (9.5 kg/m$^2$) than is typical of GCLs containing Na bentonite (3.5–5.5 kg/m$^2$). Percolation from the test section was monitored for 3 yr, during which 14 mm of percolation (0.5% of precipitation) was recorded. No temporal pattern in percolation rate was observed, and the highest percolation rates occurred within the first 6 months after installation.

### 2.4 *Wisconsin study in USA*

Benson et al. (2007) report percolation rates recorded in two pan lysimeters (4.3 × 4.9 m) installed beneath a cover containing a GCL used for closure of a coal ash landfill in Wisconsin, USA. The cover profile consisted of (top to bottom) 760 mm of sandy silt, the GCL, and 760 mm of pea gravel (Fig. 1). A non-reinforced GCL containing granular natural Na bentonite was employed. Percolation rates recorded in both lysimeters are shown in Figure 3.

Percolation rates in both lysimeters remained low (<13 mm/year) for the first few months, but increased to as much as 299 mm/yr within the next 4–7 months. Consequently, the GCLs over both lysimeters were replaced. A 150-mm-thick layer of quartz sand was also installed as a buffer between the GCL and the underlying pea gravel to eliminate stress concentrations that might affect the GCL (Fig. 1). Percolation rates remained low for 9–15 months after reconstruction, but then increased again to rates similar to those measured prior to replacing the GCL (Fig. 3).

GCL samples were collected from the field site during reconstruction activities to evaluate how the GCL had changed since installation. Hydraulic conductivity tests conducted on the exhumed GCLs confirmed that the hydraulic conductivity of the GCL had increased. The exhumed GCLs had hydraulic conductivities ranging between $1.4 \times 10^{-6}$ and $9.1 \times 10^{-5}$ cm/s, whereas tests on samples of new GCL had hydraulic conductivities ranging between $2.7 \times 10^{-9}$ and $7.8 \times 10^{-9}$ cm/s. The new GCL that was tested was the same type used in the original installation and the reconstruction.

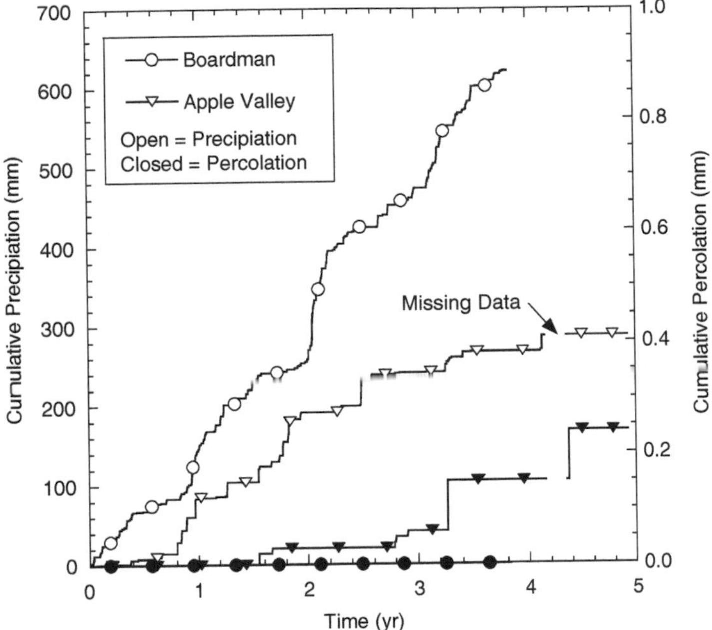


Figure 4. Cumulative precipitation and percolation for final cover test sections with GCLS in Boardman, OR and Apple Valley, CA, USA. Test sections at both sites had a composite GCL-geomembrane barrier layer.

Examination of the bentonite showed that nearly all of the Na originally in the exchange complex was replaced by Ca and Mg, and that the swell index of the bentonite was comparable to that of Ca bentonite. Benson et al. (2007) attributed the high percolation rates observed in the field (and the high hydraulic conductivity of the GCL) to the loss of swelling capacity incurred by replacement of Na by Ca and Mg coupled with dehydration of the bentonite. Shrinkage cracks that formed when the bentonite dried did not swell shut when the bentonite rehydrated, resulting in the high hydraulic conductivities. Cation exchange was attributed to cations in water that percolated downward through the cover soils and into the GCL.

## 3 COVERS WITH A GCL-COMPOSITE BARRIER

### 3.1 *Wisconsin study in USA*

The GCLs in the field tests described by Benson et al. (2007) were replaced a second time after the percolation rates were found to be much higher than expected. The second reconstruction employed a GCL containing natural Na bentonite laminated with a 0.1-mm-thick polyethylene geofilm. The reconstruction was conducted in two phases. In the first phase, the laminated GCL was installed with the geofilm downward. The geofilm was oriented upward in the second phase. Otherwise the cover profiles were identical to those employed with the original GCL (Fig. 1).

Percolation rates for both orientations of the GCL are shown in Figure 4. Percolation rates remained low in both lysimeters after the laminated GCL was installed. The peak percolation rate was less than 18 mm/yr, and the average percolation rates were 2.6 and 4.1 mm/yr for the two lysimeters.

### 3.2 *California study in USA*

A cover with a composite barrier layer consisting of a GCL overlain by a geomembrane was evaluated at a field site in Apple Valley, California, USA as part of the Alternative Cover Assessment Program (Albright et al. 2004). Apple Valley, California is an arid location, and receives 112 mm of precipitation annually (on average). Percolation from the cover was collected using a 10 m × 20 m pan lysimeter (Benson et al. 2001).

The cover profile consisted of (top to bottom) 300 mm of well-graded sand, a textured high-density polyethylene (HDPE) geomembrane, a needle-punched GCL containing Na bentonite, and 600 mm of well-graded sand (Fig. 1). An 11-mm-diameter hole was placed in the geomembrane near the center of the test area to simulate a construction defect. No percolation was transmitted by the cover during the first 1.5 yr of operation. After 5 yr, only 0.2 mm of cumulative percolation had been transmitted (Fig. 4).

The GCL was uncovered and samples were collected 5 yr after installation. No roots or desiccation cracks were present, indicating that the geomembrane was protecting the GCL. The geomembrane was covered with 300 mm of cover soil, which provided sufficient surcharge to maintain good contact between the geomembrane and the GCL. Hydraulic conductivity of the GCL samples ranged between $1.0 \times 10^{-9}$ and $3.0 \times 10^{-9}$ cm/s, which is typical of a new GCL (i.e., no change in hydraulic conductivity was observed). However, only 23% of the Na originally in exchange complex remained when the GCLs were sampled, having been replaced by Ca and Mg. The replacement of Na by Ca and Mg was evident in the swell index of the bentonite (13.0–16.5 ml/2 g), which fell between swell indices typical of Ca (6–10 mL/2 g) and Na bentonites (34–36 mL/2 g). Thus, the GCL was not immune to cation exchange, even though the GCL was overlain by a geomembrane. Melchior (2002) and Meer and Benson (2007) have reported similar observations for sites where a GCL was covered by a geomembrane. They suggest that upward migration of Ca and Mg from the underlying subgrade is responsible for cation exchange.

### 3.3 *Oregon study in USA*

A cover with a composite barrier layer consisting of a needle-punched GCL containing Na bentonite overlain by a textured HDPE geomembrane was evaluated at a field site in Boardman, Oregon, USA as part of the Alternative Cover Assessment Program (Albright et al. 2004). Boardman, Oregon is semi-arid, with 215 mm of precipitation annually (on average).

The cover profile consisted of (top to bottom) a 900-mm-thick layer of low-plasticity silt, a geocomposite drainage layer, a textured HDPE geomembrane, the GCL, and 300 mm of low-plasticity silt (Fig. 1). An 11-mm-diameter hole was placed in the geomembrane near the center of the test area to simulate a construction defect. No percolation was transmitted from the cover over a 4-yr monitoring period (Fig. 4).

The GCL was uncovered and samples were collected 6.7 yr after installation. No roots were present in the GCL and hydration of the bentonite was modest. The bentonite still appeared granular and had relatively low water content (16–20%). The subgrade was also very dry at the time of the exhumation, which probably precluded significant hydration of the bentonite. Despite the lack of hydration and a dry subgrade, approximately % of Na originally in the exchange complex of the bentonite had been replaced by Ca and Mg, and the swell index of the bentonite (16.0–17.0 mL/2 g) was between swell indices typical of Ca and Na bentonites. As with the site in Apple Valley, CA, cation exchange at the Boardman site apparently was due to upward migration of divalent cations from the underlying subgrade.

Hydraulic conductivity of the exhumed GCLs was measured using two different permeant liquids: 0.01 M $CaCl_2$ solution (i.e., so called "standard water" that is stipulated in ASTM D 5084) and deionized (DI) water. When the Ca solution was used as the permeant liquid, the hydraulic conductivity of the exhumed GCLs ranged from $8.5 \times 10^{-8}$ to $1.9 \times 10^{-6}$ cm/s. In contrast, when DI water was used, the hydraulic conductivity was between $1.3 \times 10^{-9}$ to $2.3 \times 10^{-9}$ cm/s (i.e., in the range for a new GCL). This contrast in hydraulic conductivities suggests that a transition was occurring in the bentonite in response to cation exchange, with significantly higher hydraulic conductivities being realized when additional exchange occurred during permeation with the Ca solution. Similar behavior might have occurred in the field if additional Ca-for-Na exchange from the subgrade had continued.

## 4 PRACTICAL IMPLICATIONS

### 4.1 *Cation exchange and hydraulic conductivity*

The evidence presented herein and by others has shown that cation exchange is common in GCLs used in covers regardless of whether they are covered with a geomembrane. The source of cations involved in exchange most likely is different for GCLs used without and with a geomembrane (i.e., cations eluted from overlying cover soils or cations migrating upward from the subgrade). However, the end result is the same: Na naturally in the bentonite is replaced by Ca (primarily) and Mg. This Ca-for-Na exchange can (but does not always) result in a large increase in hydraulic conductivity. GCLs that hydrate and then undergo exchange, but do not dehydrate, appear to maintain low hydraulic conductivity (e.g., Apple Valley, CA site). In contrast, GCLs that hydrate and undergo exchange and then dehydrate (e.g., Georgswerder Germany study, Wisconsin USA study) or GCLs that undergo exchange without significant hydration (Oregon USA study) can become very permeable. This suggests that GCLs should be installed under conditions that promote rapid hydration and prevent dehydration. Conditions that preclude or minimize cation exchange should also be selected if practical.

## 4.2 *Promoting hydration and preventing dehydration*

The most practical means to promote hydration is to place the GCL on a moist compacted subgrade. In general, GCLs placed on a subgrade prepared at optimum water content (or wetter) will be fully hydrated within 60–90 d (USEPA 1996, Bradshaw 2008), with greater hydration occurring when the water content of the subgrade is wet of optimum water content (EPA 1996). Hydration can also be achieved by direct wetting of the GCL after placement, but this practice can result in construction difficulties and displacement of bentonite within the GCL when overlying geosynthetics and cover soils are placed. GCLs that are prehydrated in the factory are also available (e.g., Kolstad et al. 2004a), but they tend to be more costly than conventional GCLs.

The only effective means currently available to prevent dehydration of GCLs is to cover the GCL with a geomembrane or to use a laminated GCL with the geofilm oriented upward. For GCLs covered with a geomembrane, placing cover soil over the geomembrane is particularly important. Surcharge provided by the cover soil ensures good contact between the geomembrane and GCL. Without good contact, dehydration of the bentonite can occur as moisture migrates upward and evaporates into the space between the geomembrane and GCL. For this reason, cover soil should be placed on the geomembrane as soon as practical.

This strategy to prevent dehydration effectively implies that GCLs used in final covers should be part of a composite barrier (geomembrane over GCL or GCL laminated with a geofilm). As illustrated earlier, the field performance of final covers with GCL composite barriers is excellent. Percolation rates less than 1 mm/yr are typical in arid regions, and percolation rates less than 5 mm/yr have been recorded for laminated GCLs in humid climates. Even lower percolation rates are anticipated in humid regions for GCLs overlain by geomembranes, but field data for such covers are not available.

## 4.3 *Conditions precluding cation exchange and increases in hydraulic conductivity*

Cation exchange can be precluded or limited to acceptable levels if the cover soils are sufficiently sodic. For example, Mansour (2001) describes a case history at a semi-arid site in California where a GCL was deployed in a final cover with sodic cover soils. After 5 yr of service, the GCL was exhumed and evaluated. The average hydraulic conductivity of the exhumed GCL was $1.9 \times 10^{-9}$ cm/s and the swell index was characteristic of Na bentonite.

Benson and Meer (2008) describe a series of experiments that were conducted to determine how the relative abundance of monovalent and divalent cations in water contacting a GCL affects cation exchange, bentonite swell, and the hydraulic conductivity. The intent was to identify pore water conditions conducive to maintaining low hydraulic conductivity in the presence of wetting and drying. GCL specimens were subjected to cyclic wetting and drying, with wetting accomplished via permeation with salt solutions prepared using various proportions of NaCl and $CaCl_2$. Relative abundance of monovalent and divalent cations in water permeating the GCL was quantified using RMD (Kolstad et al. 2004b):

$$\mathrm{RMD} = \frac{\mathrm{M_m}}{\sqrt{\mathrm{M_d}}} \tag{1}$$

where $M_m$ is the total molarity of monovalent cations in the pore water and $M_d$ is the total molarity of divalent cations in the pore water.

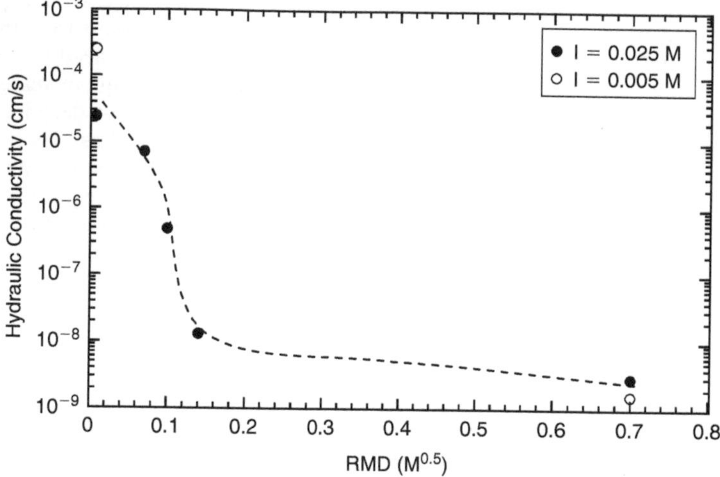


Figure 5. Equilibrium hydraulic conductivity of GCLs subjected to cyclic wetting and drying using permeant solutions with different RMD and two ionic strengths (I) (adapted from Benson and Meer 2008).

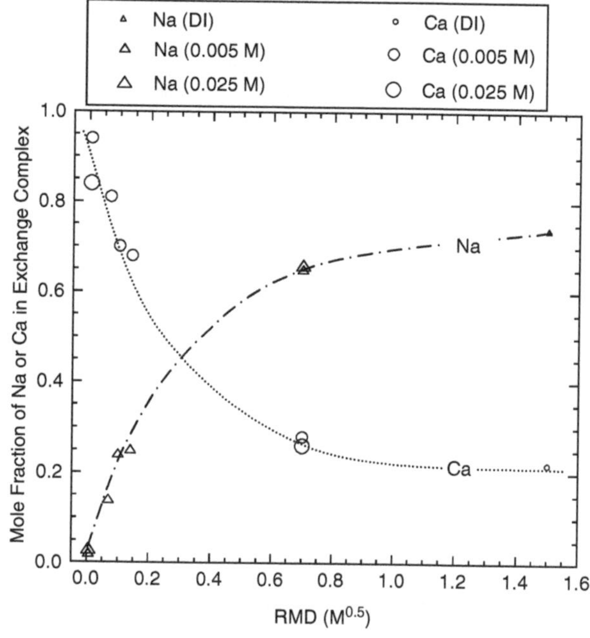


Figure 6. Mole fractions of Na and Ca in the exchange complex of GCLs subjected to cyclic wetting and drying using permeant solutions with different RMD and two ionic strengths (I). Control tests conducted with DI water. Graph adapted from Benson and Meer (2008).

Results of hydraulic conductivity tests by Meer and Benson (2007) are shown in Figure 5. The hydraulic conductivity after wet-dry cycling was strongly related to the RMD of the permeant solution, with lower hydraulic conductivity obtained as the sodicity of the solution increased (i.e., higher RMD). Moreover, when RMD of the permeant solution was at least 0.15 $M^{0.5}$, only a small increase in hydraulic conductivity was observed. Low hydraulic conductivity was maintained at high RMD because less Na was replaced by Ca (Fig. 6). As a result, the swelling capacity of the bentonite was retained, which permitted desiccation cracks in the bentonite to swell shut during rehydration.

### 4.4 *Recommended usage of GCLs in final covers*

Based on the observations that have been reported, the following recommendations are made regarding the use of GCLs in final covers:

- GCLs should not be used in final covers without an overlying geomembrane or geofilm unless the RMD of the pore water in the cover soil is at least 0.15 $M^{0.5}$. RMD of the pore water can be assessed using ASTM D 6141.
- When covering a GCL with a geomembrane, the overlying cover soil should be placed as soon as possible after deployment of the geomembrane. The cover soil provides a surcharge that ensures good contact between the geomembrane and GCL, thereby limiting the potential for dehydration of the bentonite.
- GCLs should be placed on a moist subgrade to promote rapid hydration of the bentonite. More rapid hydration occurs when the water content of the subgrade is at optimum water content or higher.

## ACKNOWLEDGEMENT

Financial support for this study was provided by a consortium of US government agencies (US National Science Foundation, US Environmental Protection Agency, US Nuclear Regulatory Commission, and US Department of Energy), the Environmental Research and Education Foundation, Colloid Environmental Technologies Corporation, Waste Connections, Inc., and Veolia Environmental Services. This support is gratefully acknowledged. The conclusions of this study are solely those of the authors. Endorsement by the supporting institutions is not implied and should not be assumed.

## REFERENCES

Albright, W. et al. 2004. Field water balance of landfill final covers. J. Environ. Quality, 33(6), 2317–2332.

Benson, C., Abichou, T., Albright, W., Gee, G., and Roesler, A. 2001. Field evaluation of alternative earthen final covers, International J. Phytoremediation, 3(1), 1–21.

Benson, C. and Meer, S. 2008. Relative abundance of monovalent and divalent cations and the impact of desiccation on geosynthetic clay liners, J. Geotech. and Geoenvironmental Eng., in press.

Benson, C., Thorstad, P., Jo, H., Rock, S. 2007). Hydraulic performance of geosynthetic clay liners in a landfill final cover. J. Geotech. Geoenviron. Eng., 133(7), 814–827.

Bradshaw, S. 2008. Effect of cation exchange during subgrade hydration and leachate permeation, MS Thesis, University of Wisconsin, Madison, WI, USA.

Henken-Mellies, W., Zanzinger, H., and Gartung, E. 2002. Long-term field test of a clay geosynthetic barrier in a landfill cover system. Clay Geosynthetic Barriers, H. Zanzinger, R. Koerner, and E. Gartung, eds., 303–309.

Kolstad, D., Benson, C., Edil, T., and Jo, H. 2004a. Hydraulic conductivity of a dense prehydrated GCL permeated with aggressive inorganic solutions, Geosynthetics International, 11(3), 233–240.

Kolstad, D., Benson, C., and Edil, T. 2004b. Hydraulic conductivity and swell of non-prehydrated GCLs permeated with multi-species inorganic solutions, J. Geotech. and Geoenvironmental Eng., 130 (12), 1236–1249.

Mansour, R. 2001. GCL performance in semi-arid climate conditions. Proc. Sardinia 2001, Eighth International Waste Management and Landfill Symposium, T. Christensen, R. Cossu, and R. Stegmann, eds., CISA, Cagliari, Italy, 219–226.

Meer, S. and Benson, C. 2007. Hydraulic conductivity of geosynthetic clay liners exhumed from landfill final covers. J. Geotech. Geoenviron. Eng., 133(5), 550–563.

Melchior, S. 2002. Field studies and excavations of geosynthetic clay barriers in landfill covers. Clay geosynthetic barriers, H. Zanzinger, R. Koerner, and E. Gartung, eds., Swets and Zeitlinger, Lesse, 321–330.

USEPA 1996. Hydration of GCLs adjacent to soil layers. Report of 1995 Workshop on Geosynthetic Clay Liners. US Environmental Protection Agency, Office of Research and Development, Washington, D.C.

Wagner, J. and Schnatmeyer, C. 2002. Test field study of different cover sealing systems for industrial dumps and polluted sites. Appl. Clay Sci., 21, 99–116.

# CHAPTER 11

# Oxygen diffusion through geosynthetic clay liners

A. Bouazza
*Monash University, Melbourne, Victoria, Australia*

## 1 INTRODUCTION

Engineered cover systems commonly are used in reactive mining waste (sulphide bearing waste) storage facilities to separate buried waste from the surface environment. The purpose of these cover systems is to prevent possible contamination of the environment following the closure of the mine storage facility which might be caused by the oxidation of reactive sulphide bearing mine residues in presence of oxygen and /or water. In this respect, the cover system will have two main objectives: 1) to reduce or control water infiltration; and 2) to limit oxygen ingress into sulphide bearing mine waste. The design of such cover systems is, in most cases, site specific and depends usually on the climatic conditions prevailing at a given mine site. In areas where humid climates tend to prevail, the conventional approach to cover systems (usually multi-layered systems) is to construct a "resistive barrier" that utilises a liner with a low saturated hydraulic conductivity or composite liners, in combination with a number of other soil layers, to reduce or control the water infiltrations and oxygen ingress into the waste. Capillary barrier cover systems can also be used to provide a proper capping of the mine waste (Bussiere et al., 2001, 2003, Yanful et al., 2003). These barriers consist usually of a fine grained soil layer sandwiched between two coarse grained soil layers. To work efficiently, the fine grained layer needs to remain saturated under drainage conditions. This can be achieved by the presence of the coarse grained layer underneath the fine grained layer. The role of the upper coarse grained layer is to drain water away and minimise loss of water by evaporation.

In recent years, geosynthetic clay liners have been increasingly included in the construction of mine cover systems as replacement to soil resistive barriers, at least in humid or wet climates (Aubertin et al., 2000, Kim and Benson, 2004, Renken et al., 2005). In this respect, knowledge of oxygen diffusion through GCLs plays an important role in the assessment of the effectiveness of the cover system in minimizing oxygen migration. Oxygen diffusion plays also a role in the estimation of fugitive emissions of biogas through municipal solid waste landfill cover systems and in the optimization process of passive methane oxidation.

## 2 BACKGROUND

The movement of gases in a porous media such as GCL occurs by two major transport mechanisms: advective flow and diffusive flow. In advective flow, the gas moves in response to a gradient in gas pressure (see Chapter 7) and is controlled mainly by the gas permeability of the resistive barrier. Recent studies have shown that the advective gas flux in geotextile supported GCLs vary largely with the change of degree of saturation and tends to approach a so-called zero advective flow condition when the degree of saturation is very high (>85%) (Didier et al., 2000; Bouazza and Vangpaisal, 2003; Vangpaisal and Bouazza, 2004). Work reported by Aubertin et al. (2000), Rahman et al. (2002) and Bouazza and Rahman (2004) has indicated that substantial low gas diffusivity can be achieved if a needle punched GCL has a high degree of saturation (S > 85% to 90%).

Gas movement by diffusion dominates advection when the pressure difference is small. The diffusion of gas is a molecular process. When a gas is more concentrated in one region of a mixture more than another, it is likely that this gas diffuses into the less concentrated region. Thus the molecules move in response to a partial pressure gradient or concentration gradient of the gas. This movement will tend to occur through the air filled pores if the porous medium is partially saturated whereas in highly saturated medium it will occur partly in the gaseous phase and partly in the liquid phase. Both mechanisms of transport are reviewed in details in Aubertin et al. (2000).

Gas diffusion can be modelled using Fick's laws in a similar fashion to diffusion transport of dissolved contaminants through porous media. For one dimensional diffusion process, the mass diffusive flux is given by Fick's first law (equation 1):

$$J_g = -n_e D_e \frac{\partial c}{\partial z} \tag{1}$$

where, $J_g$ is the mass diffusive flux of a gas [M/L$^2$T], $n_e$ is the porosity available for diffusion (*i.e.* effective porosity), $c$ is the gas concentration difference in the gaseous phase [M/L$^3$], $D_e$ is the effective diffusion coefficient of gas [L$^2$/T], $z$ is a distance (thickness, or height, etc.) [L], $\partial c/\partial z$ is the concentration gradient [M/L$^4$]. The minus sign in Equation 1 means that mass transfer over time occurs in the direction of decreasing concentration. Under one-dimensional transient conditions, the principle of conservation of mass requires that the change in mass flux of a diffusing solute across an infinitesimal GCL element ($\partial J/\partial z$) must be equal to the time rate of change of concentration within the element, i.e.:

$$\frac{\partial J}{\partial z} = n_e \frac{\partial c}{\partial t} \tag{2}$$

Equation 2 assumes that there is no change in porosity of the GCL element with respect to time (i.e. $\partial n/\partial x = 0$). Equating 1 and 2 for the mass flux and eliminating $n_e$ from both sides; gives Fick's second law for diffusion through an inert porous medium (i.e in a material where oxygen consumption or production is non existent) under transient condition:

$$\frac{\partial c}{\partial t} = D_e \frac{\partial^2 c}{\partial^2 z} \tag{3}$$

Equations 1 and 3 can be used for the determination of effective diffusion coefficient $D_e$, which is dependent on the pores and fluid characteristics as total porosity, tortuosity, degree of saturation, molecular weight, etc.

It is often practical to normalize $D_e$ by the gas diffusion in air $D_a$ measured at the same temperature and pressure. The ratio $D_e/D_a$ is lower than 1 owing to the reduction of the cross sectional area available for gas diffusion and the tortuosities of the air filled pores.

## 3 OXYGEN DIFFUSION MEASUREMENT

Measurement of oxygen diffusion can be performed in a diffusion cell which consists of a stainless steel cylinder comprising 3 chambers (Figure 1). The upper chamber is used as a source reservoir where atmospheric air is stored at the beginning of the test and used as a source for diffusing oxygen. The middle chamber is designated for the porous media (GCL), through which diffusion occurs and the bottom chamber is used as a collector reservoir. The cell is made airtight by clamping the top and bottom capping plates with three clamping rods. At the start of the test, the bottom chamber is purged with humidified nitrogen by flowing nitrogen through a water bottle and a water trap, the valve in the upper chamber is opened to fill it with atmospheric oxygen ($O_2$ content = 20.9%). As diffusion process occurs in response to a concentration gradient, the concentration of oxygen in the upper chamber is expected to decrease with time whereas the bottom concentration is anticipated to increase by the same amount if the condition of conservation of mass is to be fulfilled and no consumption of oxygen occurs in the GCL. Two Gas Analysers are used to monitor the oxygen concentration in the upper and lower chambers at regular time intervals. The gas analysers work in a closed circuit by pushing the same amount of gas that was sucked in. Further information on the testing procedures can be found in Aubertin et al. (2000) and Bouazza and Rahman (2007).

It is important to note that when the top air valve is closed a finite mass of gaseous oxygen becomes enclosed in the upper chamber. This gas is then allowed to diffuse through the GCL. If the initial concentration of oxygen in the upper chamber is considered as $c_o$, then after time $T$ the concentration $c_T$ can be expressed as:

$$c_T = c_o - \frac{1}{h_g}\int_0^t J_o(c,T)\,dT \tag{4}$$

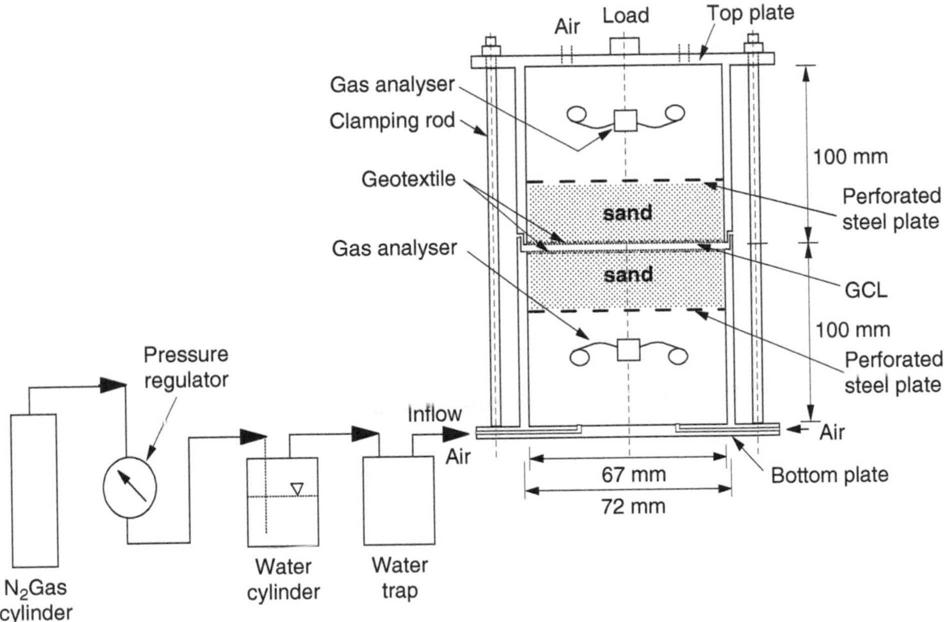


Figure 1. Schematic diagram of the diffusion cell.

where $J_0(c, T)$ is the surface flux at $z = 0$; $h_g$ is height of the gas space above the GCL sample. The height is calculated by adding the free space above the sand and the void space within the sand and geotextile, which can be estimated by multiplying the porosity of the sand and geotextile by their respective thicknesses. Since the oxygen accumulating at the base was not flushed out during the test, the base concentration can be simulated with a zero outflow velocity using the expression derived by Rowe and Booker (1990). The concentration in the lower chamber could be simulated with a zero outflow velocity using the following expression:

$$c_b = \int_0^t \frac{J(x)}{n_b h_b} \, dT \tag{5}$$

where $c_b$ is oxygen concentration at the base; $h_b$ is the height of the space below GCL sample including open space and the void space in the sand layer and geotextile; $n_b$ is the porosity of the gas filled space below the GCL.

## 4 OXYGEN DIFFUSION COEFFICIENT

Gaseous oxygen concentration profiles observed in the diffusion tests can be simulated using different modelling methods including numerical, analytical or mathematical methods. A common approach to simulate diffusion test results is to use Pollute®, a finite layer contaminant transport program (Rowe and Booker, 1999). The two boundary conditions discussed above are allowed in Pollute® and can be used to analyse the laboratory test results. While purging with nitrogen, some of the residual oxygen is usually retained in the bottom cell. With the boundary conditions used, Pollute® considers the bottom oxygen concentration as zero at the start. Therefore, adjustments can be made with the data series to obtain the actual concentration gradient for Pollute® using the superposition principle applicable to linear diffusion, as in equation 2. The same adjustments can also be applied to the simulation results before any comparison is made with the laboratory values. The effective diffusion coefficient $D_e$ is estimated by fitting the laboratory data with the theoretical solution provided by Pollute® for each diffusion test. A typical variation of oxygen concentration with time is presented in Figure 2. The test results indicate a decrease in source concentration of

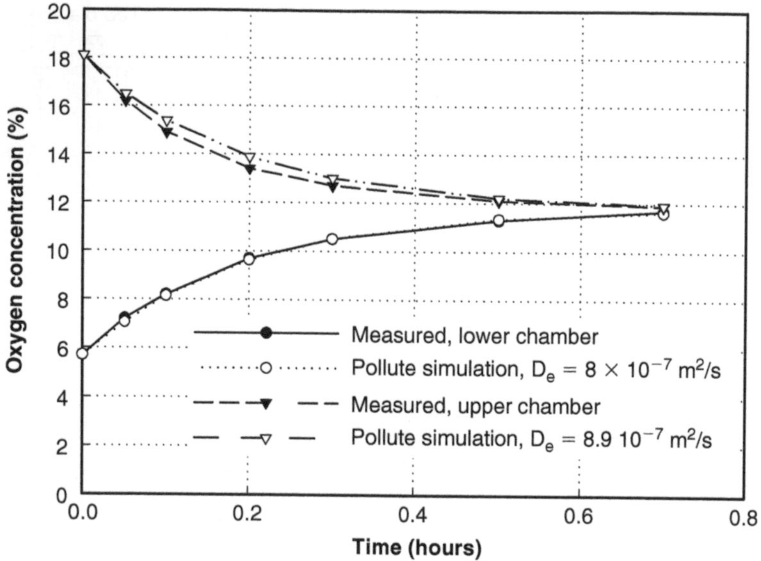


Figure 2. Typical oxygen concentration in upper and lower chambers as function of time at degree of saturation = 40% (from Bouazza and Rahman, 2007).

oxygen with time in the upper chamber whereas the concentration in the lower chamber increases by the same amounts. The experiments are continued until a steady state condition is attained so that oxygen concentration is the same at all depths and no further diffusion could occur. Figure 2 also shows how closely the Pollute® solution fits the experimental results.

The modelling of the time concentration data allows the computation of the effective oxygen diffusion coefficients ($D_e$) at various degree of saturation. The measured values of $D_e$ can then be plotted as a function of degree of saturation as shown in Figure 3. Bouazza and Rahman (2007) reported on the variation of oxygen diffusion against the degree of saturation of three GCLs referred to as GCL1, GCL2, and GCL3, respectively (Figure 3). These GCLs consisted of essentially dry bentonite (powder in GCL1 or granular in GCL2) sandwiched between polypropylene geotextiles. The geotextiles are held together as a composite material by needle-punching for GCL1 and GCL2, whereas GCL3 is stitch bonded. Diffusion data from Soltani (1998) and Aubertin et al. (2000) are also included in Figure 3. Most of their data has been obtained at high degree of saturation (S > 88%). Soltani (1998) investigated needle punched and stitch bonded GCLs over a range of water saturation varying from 88% to 97%. Aubertin et al (2000) conducted their tests on needle punched GCLs, mostly at 100% saturation, only one test was conducted at lower saturation (S = 71%). The test results shown in Figure 3 indicate that the decrease of gas diffusion is associated with an increase in degree of saturation. A decrease of around 4 orders of magnitude in the diffusion coefficient is observed as the degree of saturation increased from 20% to 97%, this is in line with the fact that the diffusion coefficient of oxygen in water ($D_{aw} = 2.2 \times 10^{-9}$ $m^2/s$) is much lower than in air ($D_a = 1.8 \times 10^{-5}$ $m^2/s$). It also appears from Figure 3 that GCL3 achieved a higher diffusion coefficient than GCL1 and GCL2. This might be caused by the way that GCLs are held together as a composite material. Stitch bonding is used in GCL3, whereas needle punching is used in GCL1 and GCL2. As GCL3 hydrates, the bentonite will tend to become partly confined along the stitch lines, and swell freely between them. This results in zones (along the stitches) with less bentonite available to mitigate gas diffusion. In contrast, the bentonite will tend to swell uniformly in the needle punched GCLs. Notwithstanding the variations observed between the two types of GCLs, the diffusion coefficient was found to vary by two orders of magnitude for water saturation varying from 20% to approximately 80%. Whereas for S >80% a change of only 15% in saturation was needed to achieve an additional two orders of magnitude variation in the diffusion coefficient. This indicates that GCLs should have a high degree saturation to be effective as a barrier against oxygen. It is interesting to note that the result reported by Soltani (1998) for the stitch bonded GCL gives comparatively a high diffusion value. For needle punched GCLs, the data reported by

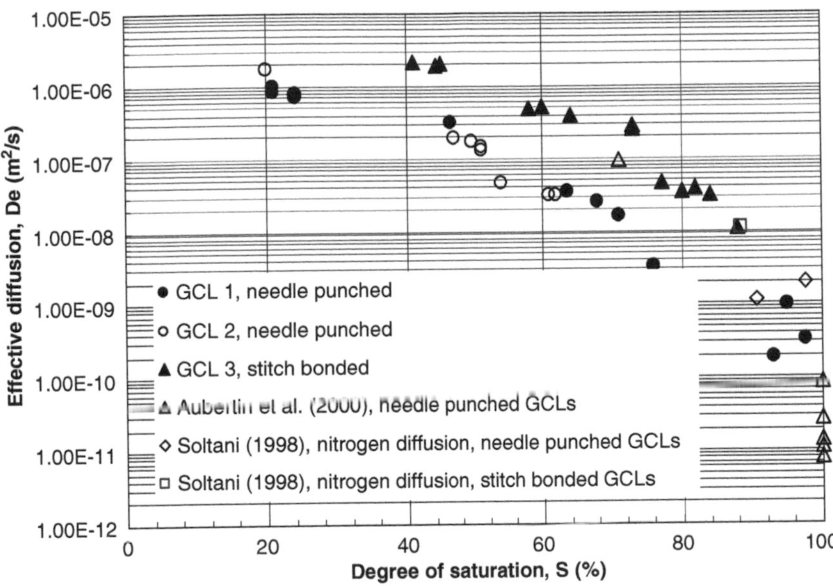


Figure 3. Effective diffusion coefficient of oxygen versus degree of saturation in different GCLs (from Bouazza and Rahman, 2007).

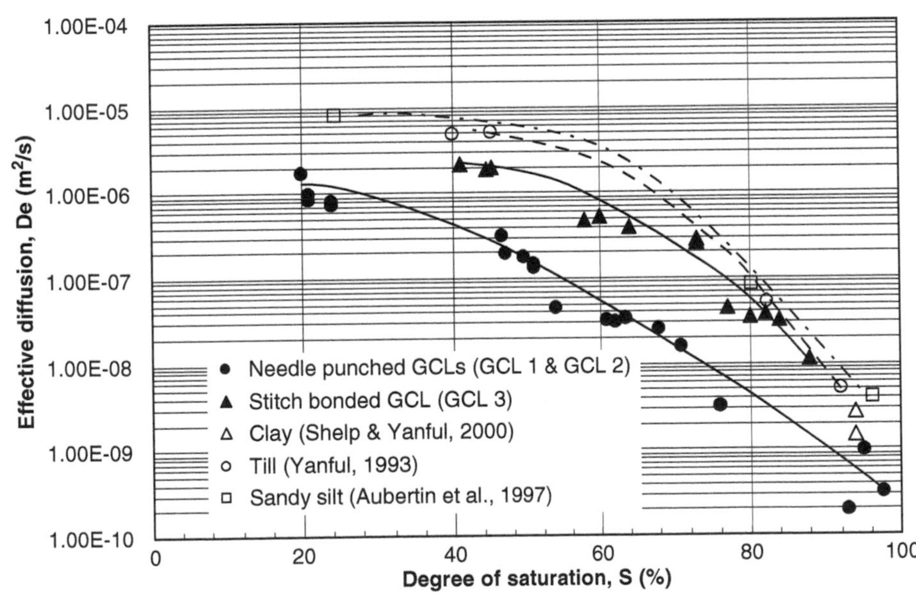


Figure 4. Effective diffusion coefficient of oxygen versus degree of saturation: comparison between present study and soils (from Bouazza and Rahman, 2007).

Soltani (1998) and Aubertin et al. (2000) seem to corroborate the findings on the higher range of saturation reported by Bouazza and Rahman (2007).

Soils traditionally used in mine covers are different in textures and are heavily dependent on the way they are placed and compacted. Soils oxygen diffusion data collected from several studies reported in the literature seems to plot well above the needle punched GCLs investigated in the present study (Figure 4). A larger difference in terms of effective diffusion is noticed at the lower

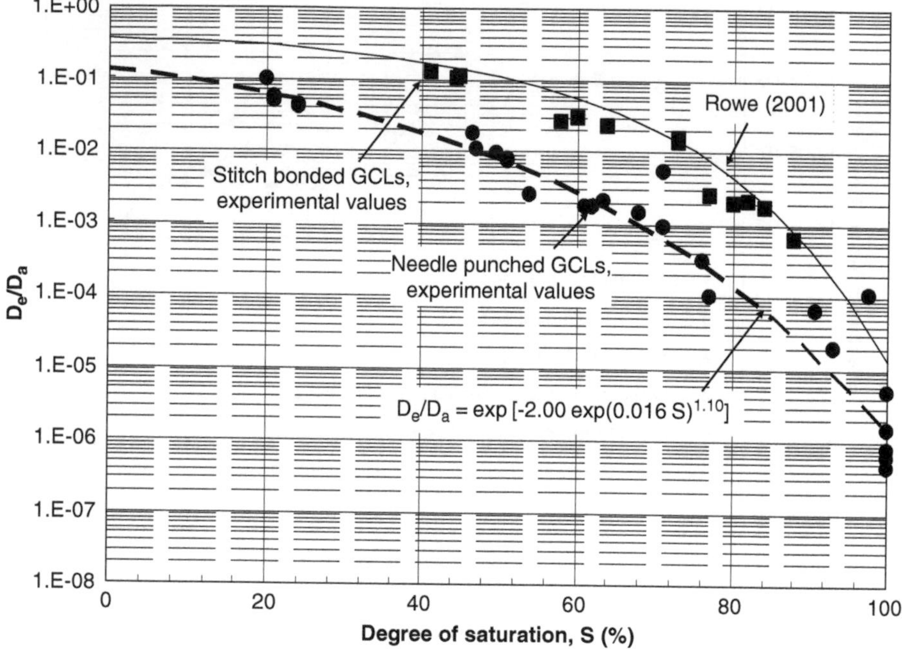


Figure 5. Diffusion coefficient as a function of degree of saturation (from Bouazza and Rahman, 2007).

range of saturation where soils by nature contain larger air filled pores. Interestingly, at very high saturation ($S \geq 90\%$) it seems that this difference is largely reduced and the variation between the two materials becomes minimal. In the case of the stitch bonded GCLs, their effective diffusion is close to that of soils reported in Figure 4 due to the fact that the stitch bonding tends to provide a preferential flow path to gas.

Numerous studies have been conducted on gas diffusion through soils, Rowe (2001) summarised the results of these studies in the form of a relationship relating the ratio of diffusion coefficients $D_e/D_a$ (effective diffusion coefficient/diffusion coefficient in air at the relevant temperature) to the degree of saturation ($S$) in soils. This relationship is given by:

$$D_e/D_a = \exp\,[-1.03\exp\,(0.017S)^{1.64}] \tag{6}$$

Equation 6 is plotted in Figure 5 together with the results obtained by Bouazza and Rahman (2007) (normalised with respect to $D_a = 1.8 \times 10^{-5}\,m^2/s$ at 20°C) including the data from Soltani (1998) and Aubertin et al. (2000). Equation 6 seems to capture the variations observed in stitch bonded GCLs. This was expected since their diffusion behaviour was close to the one encountered in conventional soils as shown in Figure 4. In the case of the needle punched GCLs, their experimental diffusion values are significantly lower than the values for soils, at the same degree of saturation, suggesting that they might provide a more robust barrier to oxygen. The reasons for such disparity could be due to the fact that equation 6 was developed specifically for conventional soils. Needle punched GCLs have a completely different structure where the randomly distributed fibres and bentonite are heavily intermingled to form a more homogeneous material with probably less pore space available for gas migration than for soils. Based on the data presented in Figure 5, a unique relationship between the effective diffusion of needle punched GCLs and degree of saturation appears to exist and can be described by the following equation (Bouazza and Rahman, 2007):

$$D_e/D_a = \exp\,[-2.00\exp\,(0.016S)^{1.10}] \tag{7}$$

Notwithstanding the above, the general trend observed in GCLs is similar to that observed in soils, i.e. $D_e/D_a$ was found to decrease significantly as the degree of saturation increased. Both empirical relationships (equations 6 and 7) between the effective diffusion and the degree of saturation can

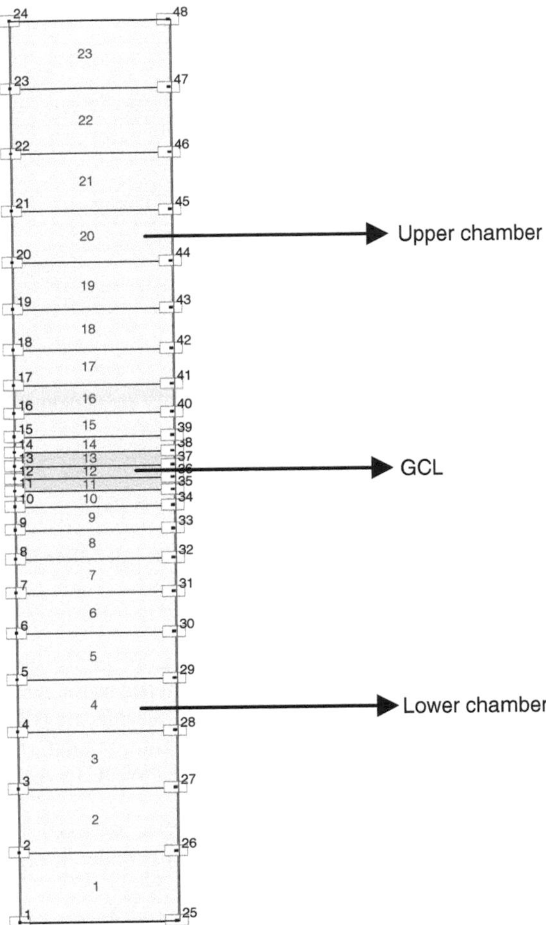


Figure 6. Finite element mesh used in SEEPW®/CTRANW®.

be used to obtain an estimate of the diffusion coefficient value for use in simulations, predictions or calculations of diffusive flux through cover systems containing stitch bonded or needle punched GCLs.

Estimation of $D_e$ can also be made using finite element programs such SEEPW®/CTRANW® (2007) and an empirical model proposed by Collin (1987). SEEPW® is used to simulate the movement of water and pore pressure distribution within a porous media whereas CTRANW® is used to simulate the movement of contaminants through porous materials. CTRANW® is integrated with SEEPW® by using the water velocities from SEEPW® to compute the movement of dissolved constituents in the porewater. Although SEEPW® and CTRANW® were developed specifically for the purpose of simulating transport of solutes, they can be used to simulate gas transport (Kelln, 2001). The theoretical solution provided by SEEPW® and CTRANW® can be fitted to the concentration time data to determine the effective coefficient of diffusion. The effective coefficient of diffusion that results in the best fit to the observed data is selected as the experimental value. A single column of elements can be considered in the simulation (Figure 6).

Width of the elements is selected to make the aspect ratio reasonable. The GCL is considered in three layers and similar spacing is used for adjacent elements both in the upper and lower part. An expansion factor is then used, to the full depth of the cell, to reduce the number of elements. A constant low value of hydraulic conductivity is used in SEEPW® to ensure that the flow velocity approaches zero to make the system purely diffusive. The longitudinal and transverse dispersivity

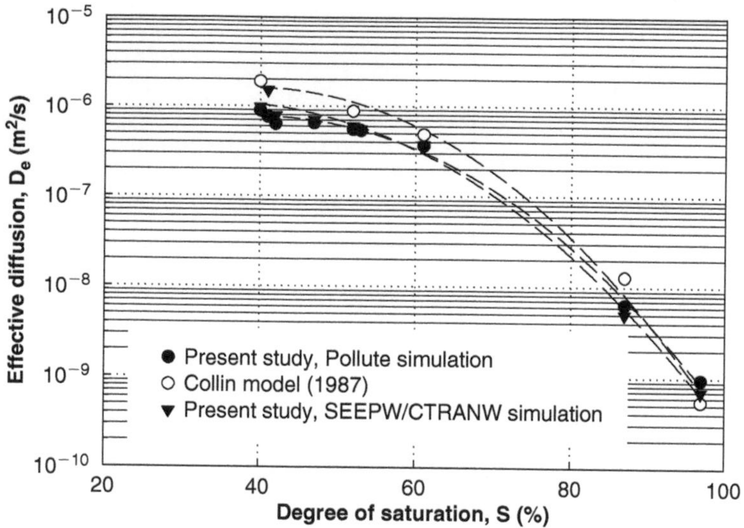


Figure 7. Comparison of measured effective diffusion coefficients with different models (GCL1).

of the reservoir and GCLs are also set to very low values to make the hydrodynamic dispersion coefficient equivalent to the gas diffusion coefficient. Volumetric water content is considered equal to 1 for both reservoirs to represent the condition where the reservoirs are filled with air, for the GCL effective air filled porosity is used in place of volumetric water content. Similar to the approach used with the Pollute® method, solutions obtained from SEEPW®/CTRANW® are compared with the test results before the final results are selected.

Collin (1987) and Collin and Rasmusson (1988) proposed a model for soils that takes into account gas diffusion in both the air and the liquid phase. The model is expressed by:

$$D_e = D_a^o(1 - S)^2[n_a]^{2x_a} + HD_wS^2\theta^{2x_w} \tag{8}$$

where $D_a^o$ is the diffusion coefficient of oxygen in air ($\approx 1.8 \times 10^{-5}$ m$^2$/s at 20°C), $n_a$ is the air porosity, $H$ is a solubility constant of gas in water ($\approx 0.028$ at 20°C), $D_w$ is the diffusion coefficient of oxygen in water ($\approx 2.5 \times 10^{-9}$ m$^2$/s at 20°C), $\theta$ is the volumetric water content ($=nS$). The first term in the left hand side of the above equation is related to the diffusion of oxygen in air filled pores, whereas the second term is related to diffusion in water. Aubertin et al. (1997) pointed out that when $S \leq 85$ to 90% gas diffusion mostly occurs within the air filled pores. In this case, only the first term in the left hand side of the above equation will be used to compute $D_e$. $x_a$ and $x_w$ are empirical parameters which can be obtained from the following equations Collin and Rasmusson (1988):

$$(n_a)^{2x_a} + (1 - n_a)^{x_a} = 1 \tag{9}$$

$$\theta^{2x_w} + (1 - \theta)^{x_w} = 1 \tag{10}$$

Typical values of $x_a$ and $x_w$ are within the range of 0.6 to 0.75 (Aubertin et al., 1997).

Figure 7 shows that the diffusion coefficients obtained from Pollute® and SEEPW®/CTRANW® are nearly identical. The Collin empirical model is plotted along the curves generated by the finite element method and the semi-analytic solution (finite layer). It correlates quite well with the experimental data over the range of saturation investigated. However, further investigations are needed before it can be used to describe the effective diffusion coefficients in geosynthetic clay liners regardless of their structure or form.

## 5 CONCLUDING REMARKS

The oxygen diffusion coefficient follows the same trend for both the needle punched and stitch bonded GCLs (i.e. oxygen diffusion coefficient decreases with the increase in the degree of saturation). Variations of up to 4 orders of magnitude were recorded (approximately from $2 \times 10^{-6}$ m$^2$/s to $2 \times 10^{-10}$ m$^2$/s) over a range of saturation varying from 20% to 100%. Obviously, low oxygen diffusivity can be achieved if GCLs have a high degree of saturation. Therefore for a GCL to be an effective barrier against oxygen, the cover system will need to be designed to maintain the GCL close to saturation.

The way that GCLs are held together as a composite material has a significant effect on the variation of gas diffusivity. For comparable conditions, the effective oxygen diffusion of needle punched GCLs is about one order of magnitude lower than that of a stitch bonded GCL. Unique empirical relationships between the effective diffusion of needle punched and stitch bonded GCLs and degree of saturation appear to exist and can be used to obtain an estimate of the diffusion coefficient value for use in simulations, predictions or calculations of diffusive flux through cover systems containing stitch bonded or needle punched GCLs.

## REFERENCES


Aubertin, M., Aachib, M., Authier, K. 2000. Evaluation of diffusive gas flux through covers with a GCL. *Geotextiles and Geomembranes*, Vol. 18, No. 2-4, pp. 215–233.

Aubertin, M., Bussiere, B., Barbera, B.M, Chapuis, R.P., Monzon, M. , Aachib, M. 1997. Construction and instrumentation of in-situ test plots to evaluate covers built with clean tailings. *Proceedings 4th International Conference on Acid Rock Drainage*, Vancouver, Canada, Vol.2, pp.717–730.

Bouazza, A., Vangpaisal, T. 2003. An apparatus to measure gas permeability of geosynthetic clay liners. *Geotextiles and Geomembranes*, 21(2):85–101.

Bouazza, A., Rahman, F. 2004. Experimental and numerical study of oxygen diffusion through a partially hydrated needle punched geosynthetic clay liner. *Advances in Geosynthetic Clay Liner Technology, ASTM STP 1456*, American Society for Testing and Materials, West Conshohocken, PA., pp.147–158.

Bouazza, A., Rahman, F. 2007. Oxygen diffusion through partially hydrated geosynthetic clay liners. *Geotechnique,* 57(9):767–772.

Bussiere, B., Aubertin, M., Chapuis, R.P. 2001. Unsaturated flow in layered cover systems: a comparison between numerical and field results. *Proceedings 54th Canadian Geotechnical Conference*, Calgary, Canada, Vol. 3, pp. 1612–1619.

Bussiere, B., Aubertin, M., Chapuis, R.P. 2003. The behaviour of inclined covers used as oxygen barriers. *Canadian Geotechnical Journal*, 40:512–535.

Collin, M., 1987. *Mathematical Modelling of Water and Oxygen Transport in Layered Soil Covers From Deposits of Pyritic Mine Tailings*. Licentiate Treatise. S-100 44, 1987, Department of Chemical Engineering, Royal Institute of Technology, Stockholm, Sweden.

Collin, M. and Rasmusson, A., 1988. Gas Diffusivity Models for Unsaturated Porous Media. *Soil Science Society of America Journal*, 52:1559–1565.

Didier, G., Bouazza, A., Cazaux, D. 2000. Gas permeability of geosynthetic clay liners. *Geotextiles and Geomembranes*, 18 (2-4):235–250.

Geo-slope, *SEEPW/CTRANW*, 2007 Computer Software for Geotechnical Engineering. Calgary, Alberta, Canada.

Kelln, C.G. 2001. *A Gas Phase Tracer for Measuring Porosity and Effective Coefficient of Diffusion in Unsaturated Porous Media*. Masters Thesis, University of Saskatchewan, Canada.

Kim, H., Benson, C.H. 2004. Contributions of advective and diffusive oxygen transport through multilayer composite caps over mine waste. *Journal of Contaminant Hydrology*, 71:193–218.

Rahman, F., Bouazza, A., Kodikara, J. 2002. Oxygen diffusion through a needle punched geosynthetic clay liner. Proceedings 4th International Congress on Environmental Geotechnics, Rio de Janeiro, Brasil, Vol. 1, pp. 539–543.

Renken, K., Yanful, E.K., Mchaina, D.M. 2005. Effective oxygen diffusion coefficient and field oxygen concentrations below a geosynthetic clay liner covering mine tailings. *Waste Containment and Remediation*, Geotechnical Special Publication No142, ASCE, GeoFrontiers 05, Austin, USA (CD Rom).

Rowe, R.K. 2001. Liner systems. Chapter 25, *Geotechnical and Geoenvironmental Engineering Handbook*, Kluwer Academic Publishing, Norwell, U.S.A., pp. 739–788.

Rowe, R. K., Booker, J. R. 1990. *1-D Pollutant migration through a non-homogenous soil*, Report No GEOP 90-1, 1990, Geotechnical Research Centre, University of Western Ontario, London, Ontario, Canada.

Rowe, R.K., Booker, J.R. 1999. POLLUTE v6.4- 1-D pollutant migration, distributed by GAEA, Withby, Ontario, Canada.

Shelp, M.L., Yanful, E.K. 2000. Oxygen diffusion coefficient of soils at high degrees of saturation. *Geotechnical Testing. Journal*, 23 (1):36–44.

Soltani, F. 1998. *Etude de l'ecoulement de gas a travers les geosynthetiques bentonitiques utilises en couverture des centres de stockage de dechets*. These de Doctorat, INSA-Lyon, France.

Vangpaisal, T., Bouazza, A. 2004. Gas permeability of partially hydrated geosynthetic clay liners. *Journal of Geotechnical & Geoenvironmental Engineering*, 129 (1): 93–102.

Yanful, E. K. 1993. Oxygen diffusion through soil covers on sulphidic mine tailings. *Journal of Geotechnical Engineering*, 119 (8):1207–122.

Yanful, E.K., Mousavi, S.M. and Yang, M. 2003. Modeling and measurement of evaporation in moisture retaining soil covers. *Advances in Environmental Research*, Vol. 7, pp. 783–801.

# CHAPTER 12

# Field observations of GCL behaviour

E. Kavazanjian
*Arizona State University, Tempe, AZ, USA*

## 1 INTRODUCTION

Field observations of geosynthetic clay liner (GCL) performance can be divided into two broad categories: physical integrity and containment capability. Physical integrity refers to the continuity of a GCL as a barrier in a waste containment system, including stability (or instability) and other physical phenomenon such as puncture and panel shrinkage and separation that may impact GCL integrity. Containment capability refers to the ability of a physically intact GCL to contain liquids and gases, including phenomenon that impact the in-place saturated hydraulic conductivity of a GCL. Together these two categories of GCL performance describe the overall ability of a GCL to function as a barrier layer in a containment system.

Field performance of GCLs can be assessed indirectly through surveys of the performance of containment systems containing GCLs and directly through individual case histories of GCL field behavior. Typical GCLs deployed in the field are composed of sodium bentonite, have a nominal thickness of 6 mm, and have a bentonite content on the order of 4500 g/m$^2$at a water content of approximately 20%. The GCLs described in the field survey and case histories cited below may be assumed to have these properties unless otherwise specified.

## 2 GCL PERFORMANCE IN COMPOSITE LINER SYSTEMS

Bonaparte et al. (2002) conducted a comprehensive survey of waste containment system performance for the United States Environmental Protection Agency (USEPA). The study reported on operator-identified performance problems at approximately 2000 U.S. landfills and surface impoundments designed and constructed under US federal Resource Conservation and Recovery Act (RCRA) Subtitles C and D regulations. A total of 85 performance problems were identified at 79 landfill and 5 surface impoundment facilities. Of these 85 problems, four involved GCLs, including one case of GCL hydration beneath the bottom geomembrane liner during construction, one case of sliding along a geosynthetic drainage net/GCL interface in the liner system during operation, one case of sliding along a top soil/GCL interface in the final cover system after a rainfall event, and one case in which a the overlapped joints of a GCL beneath a geomembrane in a final cover system separated due to settlement. While operated-reported problems may not provide a complete picture of performance problems at landfills, considering the large number of facilities in the study and the small number of reported problems, this study indicates generally good field performance of GCLs with respect to gross stability at engineered facilities.

Bonaparte et al. (2002) also reported flow rates through primary liners into the leak detection system (LDS) for active landfill cells with double liner systems that employed geomembranes in the upper (primary) liner (either as a single geomembrane or as a composite liner). Comparison of reported flow rates from this study for liner systems with single geomembrane primary liners to liner systems with composite geomembrane/GCL primary liners provides an indication of the field performance of GCLs as the low permeability layer in a composite liner system. Comparison of flow rates for liner systems with composite primary liners that employ GCLs as the low permeability soil layer to composite primary liners that employ compacted clay as the low permeability soil layer provides an assessment of the field performance of GCLs compared to compacted clay as the low permeability soil layer in composite liners.

Table 1 summarizes the mean, maximum, and minimum average monthly LDS flows reported by Bonaparte et al. (2002) for single geomembrane (GM), geomembrane-compacted clay liner

Table 1. Leak detection system flow rates for double liner systems at US landfills (from Bonaparte, et al., 2002).

| Stage (LDS flow rate) | Initial (L/h/day) | Active (L/h/day) | Post-Closure (L/h/day) |
|---|---|---|---|
| *Geomembrane* | | | |
| Mean average monthly flow | 307 | 187 | 127 |
| Minimum average monthly flow | 4 | 0 | 1 |
| Maximum average monthly flow | 2144 | 1603 | 328 |
| *Sand/Geomembrane/Compacted Clay* | | | |
| Mean average monthly flow | 114 | 142 | 64.4 |
| Minimum average monthly flow | 1.2 | 22.7 | 0 |
| Maximum average monthly flow | 1192 | 672 | 274 |
| *Sand/Geomembrane/GCL* | | | |
| Mean average monthly flow | 133 | 22.5 | 0.3 |
| Minimum average monthly flow | 0 | 0 | 0 |
| Maximum average monthly flow | 984 | 284 | 0.9 |

(GM-CCL), and geomembrane-geosynthetic clay liner (GM-GCL) composite systems overlain by sand liquid collection layers for the initial, active, and post-closure periods of landfill operation. The initial period corresponds to the first few months after the start of waste disposal in a cell. During this initial period, there is generally insufficient waste to significantly impede the flow of rainfall into the leachate collection system. The active period is when the cell is being filled with waste. The post-closure period is after the final cover system has been placed.

The GM-GCL data was collected from 19 cells at 3 different landfills while the GM data was collected from 41 cells at 19 landfills and the GM-CCL data was collected from 31 cells at 13 landfills. LDS flows were significantly greater for the GM systems compared to the GM-GCL systems for all three periods of landfill operation, with a difference of between 2 and 3 orders of magnitude during the post-closure period. Furthermore, LDS flows through the GM-GCL systems were generally equal to or less than LDS flows through the GM-CCL systems, typically by an order of magnitude during both the active life and post-closure period, indicating superior performance of GCLs compared to CCLs as the low permeability soil layer in a composite liner system. Similar data were reported for GM-GCL systems compared to GM-CCL systems when a geonet liquid collection layer was used.

## 3 CASE HISTORIES OF FIELD PERFORAMNCE

### 3.1 *Physical integrity*

Case histories of phenomena impacting the physical integrity of GCL barriers include observations of instability in GCLs employed in liner and cover systems in landfills, observations of seam separation after GCL deployment, and observations of GCL puncture and leakage. Potential instability is one of the most common performance concerns in containment systems that employ GCLs due to the potential for low interface and internal (In-plane) GCL shear strength in these systems. The potential for puncture and associated leakage in GCLs used in containment systems is also often identified as a serious performance concern for GCLs in composite liner applications due to thin (typically on the order of 6 mm) cross section of GCLs. Separation of the lapped seams of a GCL after field deployment is a relatively new phenomenon that has only recently been identified as a threat to GCL barrier physical.

Perhaps the best known case history of the field performance of GCLs with respect to their stability is the GCL test plots initially constructed by the United States Environmental Protection Agency in the Cincinnati, Ohio area in 1994. A total of 14 GCL test plots were constructed on a 12 m high slope for the specific purpose of evaluating the stability performance of GCLs in landfill cover system applications. All of the plots were configured to represent a "typical" landfill final cover system with 0.9 m of cover soil and a drainage layer overlying a GCL-inclusive barrier

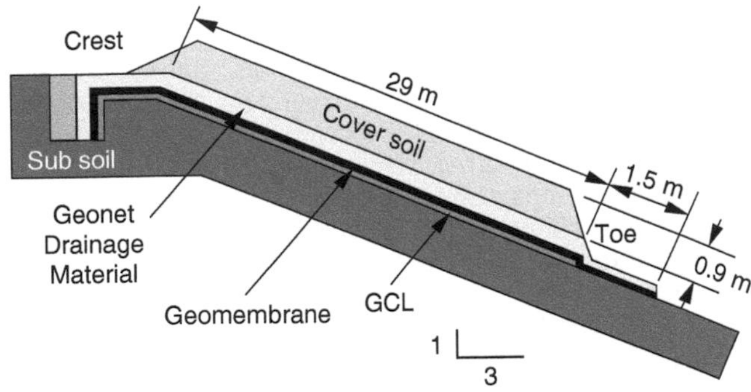


Figure 1. Cross section of Cincinnati GCL test plot on 3H:1V slopes (Daniel et al., 1998).

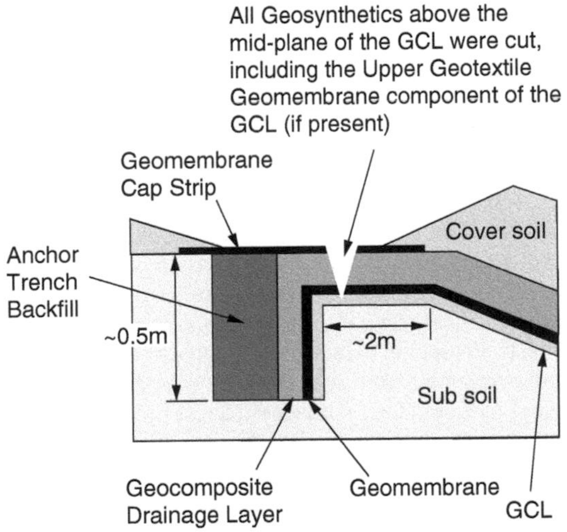

Figure 2. Anchor trench detail for Cincinnati GCL test plots (Daniel et al., 1998).

layer placed on a prepared soil foundation. Five plots had nominal slope inclinations of 3H:1V (3 horizontal to 1 vertical) while the other 10 plots had a nominal inclination of 2H:1V.

The plots were configured so that neither tension in the geosynthetic elements at the head of the plots nor buttressing of soil at the toe of the plots contributed to stability. Figure 1 presents a schematic cross section of the 3H:1V test plots illustrating how the toe buttress effect was eliminated.

Figure 2 shows the detail used at the head of the test plots to eliminate any contribution to slope stability from the tensile strength of the geosynthetic materials. Using this general configuration, the factor of safety (FS) of the test plots could be described by the simple infinite slope equation:

$$FS = \tan\phi/\tan\beta \tag{1}$$

where $\phi$ is the secant friction angle of the interface or soil layer with the lowest peak strength and $\beta$ is the slope angle.

Table 2 summarizes the configurations of the 14 GCL test plots (plus a 15th plot that contained no GCL), including nominal and actual slope angle, the configuration of the soil and geosynthetic layers in the test section, and the type of GCL employed in each test plot. All plots ere two GCL panel-widths wide. Five different types of GCLs, including geotextile encased needle-punched and stitch-bonded reinforced GCLs and a geomembrane-supported GCL, were employed in the test plots. Ten of the plots employed a composite GM/GCL barrier layer while four plots employed only a GCL as the barrier layer. Most of the plots employed a geocomposite drainage layer on top of the barrier layer. However, three of the GCL-only test plots employed a sand drainage layer.

The Cincinnati GCL test plots were maintained for a period of approximately 3 years. At the start of the three-year period, 13 GCL test plots were constructed. The 14th GCL test plot was constructed approximately 7 months into the three year period to replace a test plot that failed. Test plot field performance is summarized in Table 3. All test plots were initially stable and the 3H:1V test plots remained stable for the entire 3-year period. Three of the 2H:1V test plots failed along surfaces either within the GCL or along an interface with the GCL. Two test plots that employed needle-punched reinforced GCLs (Test Plots G and H in Tables 2 and 3) failed along the interface between the GM and the woven geotextile side of the GCL. However, when the same GCLs were used with their non-woven geotextile side in contact with the GM this interface remained stable (Test Plots J and K in Tables 2 and 3). The failures along the GM/woven geotextile interface were attributed to bentonite from the GCL extruding through the geotextile and "lubricating" the interface.

A third 2H:1V test plot, employing a geomembrane-carrier GCL with the bentonite side facing up and overlain by a GM, also failed. Exhumation of this plot (Test Plot F in Tables 2 and 3) indicated that, contrary to the expectation that the combination of an underlying carrier GM and an overlying "encapsulating GM would keep the GCL bentonite dry, a significant portion of the bentonite in the GCL hydrated. Post-failure investigation of this test plot yielded bentonite water contents as high as 188% and suggested that the bentonite hydration was due to a combination of drainage and anchor trench details for the test section, wrinkles (waves) in the deployed GCL, and cuts made in the geomembrane for instrumentation. The 14th GCL test plot (Test Plot P) was constructed in the same manner as the failed test plot with carrier GM GCL (Test Plot F), except that the suspect details of the failed test plot were modified to mitigate the potential for bentonite hydration. This 14th GCL test plot remained stable for the duration of the test program.

Perhaps the most important conclusion to be drawn from the Cincinnati GCL test plots is that laboratory testing appears to be a reliable means for assessing field stability as long as field hydration conditions are taken into account. The field stability performance of all 14 GCL test plots was consistent with laboratory shear test data, including interface shear testing on the geosynthetic materials and laboratory tests conducted to relate the drained shear strength of bentonite to its water content. For the GCLs not protected from hydration by GM encapsulation, when the laboratory interface shear testing using hydrated GCLs indicated a test plot would remain stable it did remain stable and when testing using hydrated GCLs indicated a test plot would fail it did fail. Furthermore, the test data relating bentonite shear strength to water content was consistent with the measured moisture contents following failure of the 2H:1V test plot in which the encapsulated GCL failed as well as with the two test plots with encapsulated GCLs that did not fail if it is assumed the bentonite did not hydrate in these test plots. A second lesson from these test sections is the need to minimize the potential for hydration in an encapsulated GCL system, including minimization of GCL wrinkles and penetrations and attention to drainage and anchor trench details.

Despite the general concern about the stability of landfill liner and cover systems that employ GCL, documented case histories of the failure of landfill liner and cover systems that employ GCLs are rare. To some extent, this may be attributed to legal constraints and the reluctance of engineers to publicize failures. However, it may also be attributable to the considerable attention applied by the engineering profession to the issue of GCL stability. Richardson et al. (1999) report on the field performance of a GM/GCL composite cover system built on the 18 m high 4H:1V face of a municipal solid waste landfill in the state of Michigan in the United States. The cover included a smooth polyvinyl chloride (PVC) GM on top of the woven geotextile side of the GCL. The failure occurred along the GM/GCL interface when a 30 cm drainage layer sand was being placed on top of the geomembrane. While the authors report that a thin film of bentonite was observed on the bottom of the geomembrane, they also report that the interface shear strength measured in the laboratory for this interface could be characterized by a cohesion of 0.67 kpa and a friction angle of 16 degrees over a normal load range of 2.4 to 12 kPa. With this interface shear strength, the factor of safety for the 14 degree inclination of the cover is approximately 1.7. Even without the cohesion, the factor of safety would be greater than 1. The authors suggest that the failure was due to excess pore pressures on the underside of the geomembrane induced by gas generation within the landfill.

GCLs have not been singled out as the cause of a major liner-related landfill failure under static loading, perhaps because the potential for a liner system using a GCL to have a low in-plane strength has been recognized from their initial use in landfill liner systems and is generally accounted for in liner stability assessments during design.

Table 2. Information on Cincinnati GCL test plots (Daniel et al., 1998).

| Test Plot (1) | Type of GCL (2) | Nominal slope inclination (H:V) (3) | Target slope angle (degrees) (4) | Actual slope angle (degrees) (5) | Actual slope length (m) (6) | Actual plot width (m) (7) | Cross section (top to bottom) (8) | GCL side facing upward (9) | GCL side facing downward (10) |
|---|---|---|---|---|---|---|---|---|---|
| A | Gundseal | 3:1 | 18.4 | 16.9 | 28.9 | 10.5 | Soil/GDL/GM/GCL | Bentonite | Geomembrane |
| B | Bentomat ST | 3:1 | 18.4 | 17.8 | 28.9 | 9.0 | Soil/GDL/GM/GCL | Woven GT | Nonwoven GT |
| C | Claymax 500SP | 3:1 | 18.4 | 17.6 | 28.9 | 8.1 | Soil/GDL/GM/GCL | Woven GT | Woven GT |
| D | Bentofix NS | 3:1 | 18.4 | 17.5 | 28.9 | 9.1 | Soil/GDL/GM/GCL | Nonwoven GT | Woven GT |
| E | Gundseal | 3.1 | 18.4 | 17.7 | 28.9 | 10.5 | Soil/GDL/GCL | Geomembrane | Bentonite |
| F | Gundseal | 2:1 | 26.6 | 23.6 | 20.5 | 10.5 | Soil/GDL/GM/GCL | Bentonite | Geomembrane |
| G | Bentomat ST | 2:1 | 26.6 | 23.5 | 20.5 | 9.0 | Soil/GDL/GM/GCL | Woven GT | Nonwoven GT |
| H | Claymax 500SP | 2:1 | 26.6 | 24.7 | 20.5 | 8.1 | Soil/GDL/GM/GCL | Woven GT | Woven GT |
| I | Bentofix NW | 2:1 | 26.6 | 24.8 | 20.5 | 9.1 | Soil/GDL/GM/GCL | Nonwoven GT | Nonwoven GT |
| J | Bentomat ST | 2:1 | 26.6 | 24.8 | 20.5 | 9.0 | Soil/GT/S and/GCL | Woven GT | Nonwoven GT |
| K | Claymax 500SP | 2:1 | 26.6 | 25.5 | 20.5 | 8.1 | Soil/GT/S and/GCL | Woven GT | Woven GT |
| L | Bentofix NW | 2:1 | 26.6 | 24.9 | 20.5 | 9.1 | Soil/GT/S and/GCL | Nonwoven GT | Nonwoven GT |
| M | Erosion Control | 2:1 | 26.6 | 23.5 | 20.5 | 7.6 | Soil | No GCL | No GCL |
| N | Bentofix NS | 2:1 | 26.6 | 22.9 | 20.5 | 9.1 | Soil/GDL/GM/GCL | Nonwoven GT | Woven GT |
| P | Gundseal | 2:1 | 26.6 | 24.7 | 20.5 | 9.0 | Soil/GDL/GM/GCL | Bentonite | Geomembrane |

Note: GDL = geocomposite (geotextile/geonet/geotextile) drainage layer; GM = textured geomembrane: GT = geotextile: GCL = geosynthetic clay liner.

Table 3. Summary of field performance of Cincinnati GCL test plots (Daniel et al., 1998).

| Plot (1) | Slope (2) | Type of GCL (3) | Stability of test plot as of February 1998 (4) | Maximum total post-construction displacement (mm) (5) | Maximum total post-construction differential displacement (mm) (5) |
|---|---|---|---|---|---|
| A | 3H:1V | Gundseal | Stable | <10 | <10 |
| B | 3H:1V | Bentomat ST | Stable | <25 | <25 |
| C | 3H:1V | Claymax 500SP | Stable | <25 | <10 |
| D | 3H:1V | Bentofix NS | Stable | <35 | <15 |
| E | 3H:1V | Gundseal | Stable | <25 | <10 |
| F | 2H:1V | Gundseal | Slide occurred on March 24, 1996 (internal slide within GCL) | – | 750 |
| G | 2H:1V | Bentomat ST | Slide occurred on January 12, 1995 (interface slide between lower side of geomembrane and upper woven geotextile component of GCL) | – | 25 |
| H | 2H:1V | Claymax 500SP | Slide occurred on December 10, 1994 (interface slide between lower side of geomembrane and upper woven geotextile component of GCL) | – | 130 |
| I | 2H:1V | Bentofix NW | Large deformation in subsoil | <170 | <10 |
| J | 2H:1V | Bentomat ST | Large deformation in subsoil | <450 | <30 |
| K | 2H:1V | Claymax 500SP | Large deformation in subsoil | <750 | <75 |
| L | 2H:1V | Bentofix NW | Large deformation in subsoil | <280 | <125 |
| N | 2H:1V | Bentofix NS | Stable | <30 | <10 |
| P | 2H:1V | Gundseal | Stable | NA | NA |

**Note:** Total deformation is the total amount of downslope movement; differential deformation is the difference between downslope movement of the upper and lower surfaces of the GCL.

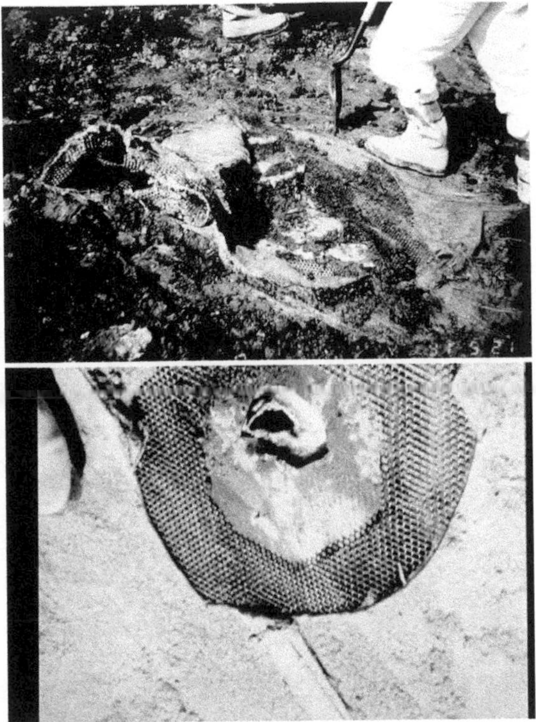

Figure 3. Mechanical damage to the GM/GCL primary liner at a Midwest United States Landfill (NRC, 2007).

Punctures, tears and holes due to construction and operations are well recognized mechanisms by which GCL integrity can be compromised. Daniel and Gilbert (1996) report on damage to a GM/GCL composite liner at a landfill located in the Midwest United States. The inflow to the leak detection system increased significantly during the third month of operation of a double-lined cell. Upon stripping the waste and soil on top of the liner, a total of 28 holes that penetrated both the GM and GCL of the primary liner were discovered. The authors suggest that the holes were caused by a backhoe used to place waste in the landfill. Figure 3 shows one of the holes in the liner system.

Peggs (2002) reports on leak location surveys at two surface impoundments that employed GCLs as barrier layer for liquid containment. In one case, the GCL was deployed on a subgrade containing stones as large as 250 mm and with only 10% of the soil finer than 0.2 mm. GCL was exposed at several locations identified by the leak detection survey. Peggs reports that defects in the exposed GCL ranged up to 40 mm in dimension and that around the larger holes there was no fine fraction evident in the subgrade. On the base of one of the ponds with an area of approximately 0.5 hectare 55 groups of holes were identified. Figure 4 shows one section of exhumed GCL with multiple holes in it.

In the second case the GCL was placed on a subgrade composed of compacted washed quarry stone with a maximum size of 20 mm. Again, the GCL was exposed at a number of locations identified by a leak location survey. Defects at the exposed locations included complete loss of bentonite through holes in the underside of the GCL, loss of bentonite from the cut edge of the GCL, a missing batten strip where the GCL was connected to a structural wall. Figure 5 illustrates the loss of bentonite through a hole in the GCL.

In 2001, engineers supervising the repair of an exposed portion of a geomembrane/GCL composite liner system on a relatively steep canyon side slope in southern California in the United States reported that the GCL was missing beneath one of the areas being patched. Investigation into the cause of the "missing" GCL revealed that the GCL panels had separated by as much as 30 cm along the overlapped seams after deployment. CQA records indicated that the GCL panels had been properly deployed with a seam overlap of between 150 and 300 mm. Subsequent investigation into this phenomenon revealed separation of lapped seams beneath exposed geomembranes

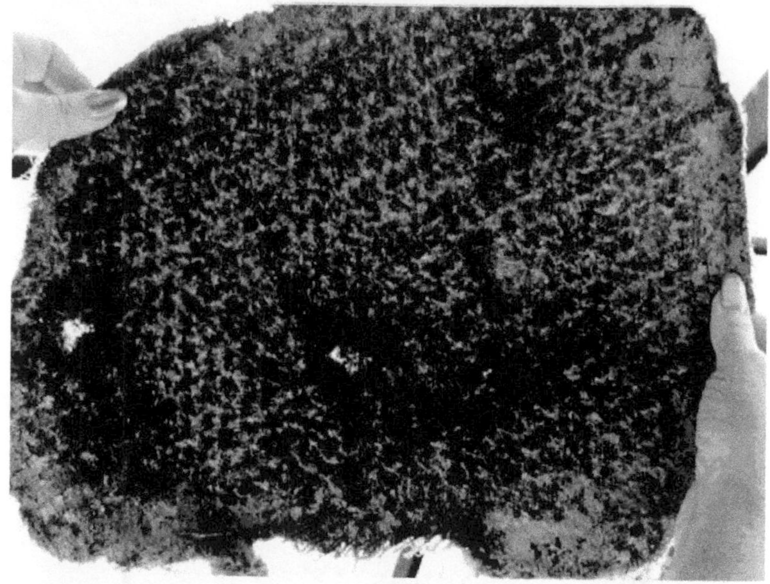

Figure 4. GCL with multiple holes due to puncture from stones in the subgrade (Peggs, 2002).

Figure 5. Loss of bentonite through hole in GCL (Peggs, 2002).

at 4 additional sites in the USA for GCLs placed upon relatively steep (1.5H:1V to 2H:1V) side slopes (Koerner and Koerner, 2005). In all cases the GCL was installed with a state-of-the-practice CQA program using overlaps between 150 and 300 mm at the seams and the separation occurred prior to placement of waste against the side slope. Separation between panels was typically between 0 and 300 mm from 2 months to 5 years after placement. Figure 6 shows GCL seam separation at California landfill with relatively steep (1.5H:1V to 2H:1V) side slopes. Separation of lapped seams beneath exposed geomembranes has also been reported on relatively flat slopes (Thiel and Richardson, 2005).

Figure 6. GCL seam separation at a California landfill (Koerner and Koerner, 2005).

Postulated GCL seam separation mechanisms include necking of the GCL panel under downward gravity loads and shrinkage of GCLs manufactured at moisture contents greater than the in situ equilibrium moisture content (Thiel and Richardson, 2005). The shrinkage mechanism is supported by observations of free water accumulating between the GCL and geomembrane at the toe of the steep side slope composite liners installed in climates with large diurnal temperature extremes (personal communications with J.P. Giroud, independent consultant and Jeff Dobrowolski, GeoSyntec Consultants) and by the results of laboratory tests in which GCL panels restrained at the top and bottom are subjected to a number of wetting and drying cycles (Thiel et al., 2006). To mitigate the potential for seam separation, Koerner and Koerner offer the following recommendations: 1) backfill the liner in a timely fashion and consider protecting the GCL with insulation when this is not possible; 2) use geotextiles to manufacture GCLs which minimize transverse contraction when they are longitudinally stressed (i.e., do not use double non-woven GCLs with no internal scrim); 3) increase the minimum overlap at GCL seams to 250–450 mm; and 4) control the manufacturing moisture content of the GCL to as low a value as possible (others have suggested it be limited to less than 20%). The use of a geomembrane with a white (or other pale-colored) surface to reduce diurnal temperature variation and reduce the tendency for moisture re-distribution beneath the geomembrane has also been suggested as a seam separation mitigation measure (Giroud, personal communication). It should be noted that these mitigation measures are unlikely to be effective for mitigating separation of GCL panels due to subgrade settlement, as reported by (Bonaparte et al., 2002).

### 3.2 *Containment Capability*

Field observations related to the containment capability of a GCL as a barrier layer include case histories of GCL infiltration performance, infiltration measurements on GCL test sections made using lysimeters, and observations and laboratory tests on GCLs exhumed from the field. Among the first case histories of GCL infiltration performance is the case history of the field performance of five renovated covered water reservoirs in which GCLs were used to seal the roofs presented by James et al. (1997). GCLs were used to seal the roofs of five Victorian-age covered fire water service reservoirs. The GCL was deployed over the roofs of the reservoirs and covered with 150 mm of gravel graded to drain laterally and 300 mm of native, generally calcareous top soil. Grass cover was established on top of this section. After two of the five reservoirs developed leaks, the GCL was uncovered by excavation for inspection at numerous locations and samples were exhumed

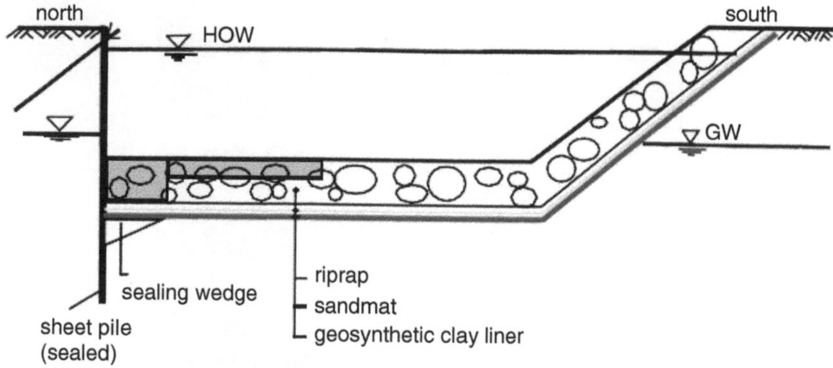


Figure 7. Cross section of the Oder-Havel navigation canal (Fleischer, 2002).

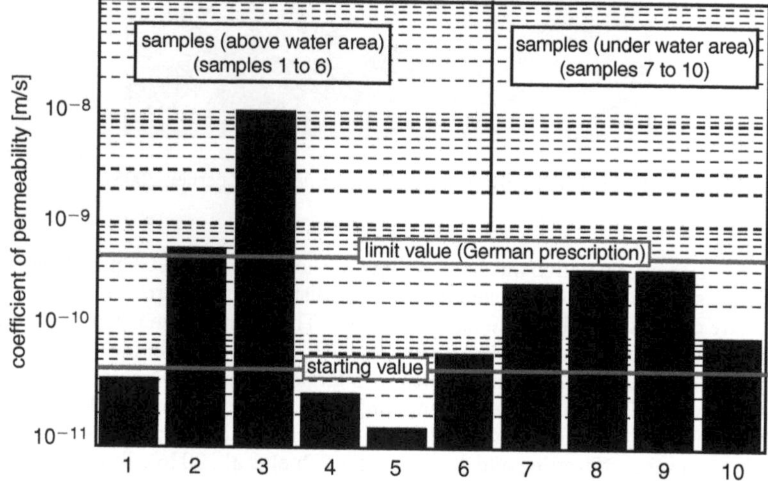


Figure 8. Saturated hydraulic conductivity on GCL samples from the Oder-Havel canal (Fleischer, 2002).

for laboratory testing. Visual observation of the exposed GCL showed frequent finely cracked zones that were judged to be responsible for the leaks. Testing of the exhumed GCL indicated that calcium had replaced sodium as the exchangeable ion in the bentonite of the GCL. The authors hypothesized that shrinkage of the bentonite due to ion exchange led to the observed cracking. The suspected source of the calcium was a small amount of calcite (2% by weight) contained in the as-manufactured GCL, although the overlying soil was also identified as a possible source.

Fleischer (2002) reports on the performance of GCL used as the lining for the Oder-Havel navigation canal in Germany. As illustrated in Figure 7, GCL with 4200 g/m$^2$ of sodium bentonite was deployed in 4 m of water and covered with a sand mat with a mass of 8000 g/m$^2$. A 700 mm layer of stone 150–450 mm in dimension was placed on top of the sand layer for scour protection.

Soil temperature and groundwater level measurements made adjacent to the canal over a three year period indicated minimal leakage of water from the canal. Figure 8 shows hydraulic conductivity measurements made on ten specimens recovered three years after construction based on an assumed thickness of 10 mm for the GCL. Testing of four samples recovered from blow the water level indicated an increase in saturated hydraulic conductivity of approximately one order of magnitude and replacement of approximately 75% of the exchangeable sodium ions with calcium. Testing of six samples recovered from above the water level indicated less ion exchange had taken place, yet two of the samples recovered above the water table had higher saturated hydraulic conductivity than any of the samples recovered from underwater. The high permeability of the above water samples was attributed to observed displacement of bentonite during rip-rap placement on the bank of the canal by dumping from an excavator.

Cover soil
Concrete ring
Nonwoven
GCL
Gravel 4/8 mm
Collection pot
and pipe
Gravel 2/8 mm
Slab

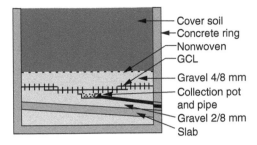

Figure 9. General configuration of the lysimeters at Lemförde (Blümel et al., 2002).

Figure 10. Construction of the lysimeter at the Aurach landfill (Henken-Mellies et al., 2002).

Blümel et al. (2002) report on lysimeter tests conducted in northern Germany. Infiltration rates were measured in three 2-m diameter lysimeters constructed in Lemförde, Germany. Figure 9 shows the general configuration of these lysimeters. GCLs composed of from 4500 g/m$^2$ to 5500 g/m$^2$ of sodium bentonite were covered by a layer of silty sand of about one meter in thickness. Measured percolation rates were on the order of 3 to 5 mm/year after three years of exposure. This percolation rate is less than 0.5 % of the precipitation. Observations and monitoring data suggest that the water content in the GCL never decreased enough during the summer to cause desiccation cracking. Testing on GCLs from other lysimeters at Lemförde exposed to similar conditions indicated showed that the exchangeable sodium ions in the GCL had been entirely replaced by calcium ions. However, the authors conclude that the measured infiltration shows that this ion exchange processes did not cause any significant increase in permeation.

Henken-Mellies et al. (2002) report on lysimeter tests conducted to model a landfill cover system in Aurach, approximately 60 km southwest of Nuremberg, Germany. The 520 m$^2$ lysimeter was constructed on a 5H:1V slope, as shown in Figure 10. A calcium bentonite GCL was employed to eliminate any effects due to ion exchange from the test. However, the GCL had a mass of 9500 g/m$^2$ to partially compensate for the higher saturated hydraulic conductivity of calcium bentonite compared to sodium bentonite. The GCL was overlain by a drainage geocomposite, 0.8 m of slightly loamy sand subsoil, and 0.2 m of loamy sand topsoil. Water balance measurements made over a 3-year monitoring period are presented in Figure 11. While annual precipitation was relatively constant over the three year period, both runoff an seepage through the GCL decreased after the first year, possibly due to establishment of vegetation in the top soil. Sensors installed in the GCL were not considered reliable enough to report moisture content data. However, the authors report that the sensor readings indicated the moisture content of the GCL did not decrease enough in the summer for desiccation to occur. The authors conclude that despite the calcium for sodium ion

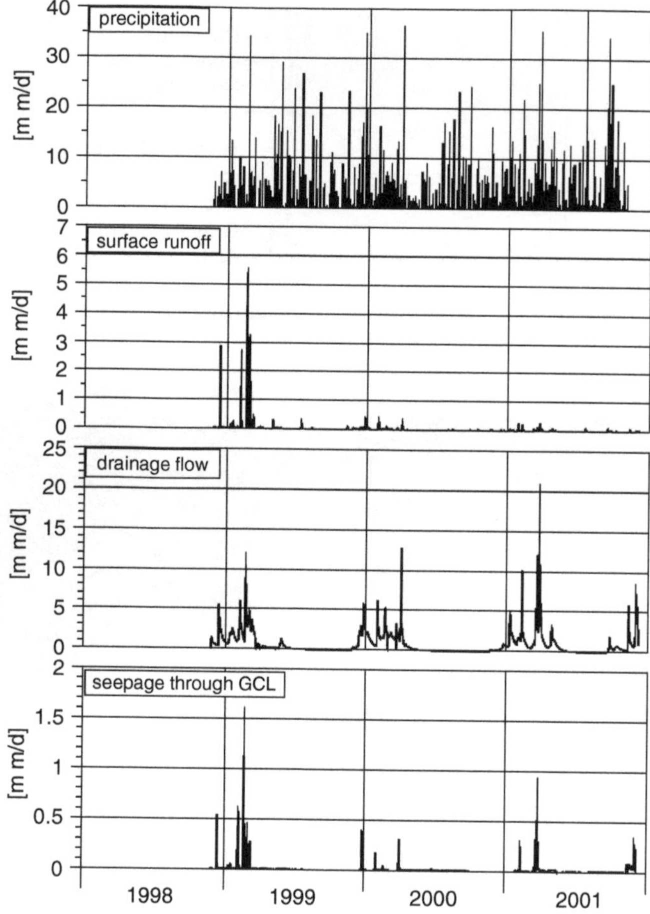


Figure 11. Water balance measurements from Nuremberg GCL test section (Henken-Mellies et al., 2002).

exchange, the GCL still provided excellent containment as the test section showed that only 0.5% of the total precipitation seeped through the GCL.

Melchior (2002) reports on the performance of GCLs exhumed from a landfill cover test site in Hamburg-Georgswerder. In 1994, two lysimeters with an area of 100 m$^2$ each, one with a needle-punched GCL and one with a stitch-bonded GCL, were integrated into the landfill cover. The GCLs were overlain by a 15 cm drainage layer of gravel and 30 cm of sandy loam topsoil. In 1995, 1996, 1998 and 1999 samples of the GCLs were recovered and evaluated visually using x-rays and thin section and electron microscopic imaging and by chemical, physical and mineralogical laboratory tests. Results of the laboratory tests showed that desiccation, cation exchange, plant root penetration and shrinkage increased the saturated hydraulic conductivity of GCLs significantly. Furthermore, re-wetting and swelling of the bentonite in the laboratory did not restore the saturated hydraulic conductivity of the desiccated GCLs. Figure 12 shows desiccation cracking of one of the GCLs as revealed by x-ray imaging. Laboratory testing indicated that the exchangeable sodium ions in the GCL had been completely replaced by calcium ions by 1999. Melchior (2002) concluded that the use of GCLs in capping systems could be problematical if they were not adequately protected against desiccation.

Sporer (2002) reported on testing of GCLs excavated from the cover systems at 8 landfills in Germany. The height of cover soil on top of the GCLs was between 0.5 and 1.0 m. The covers were 3 to 6 years old. Testing indicated that exchange of sodium ions for calcium ions was complete in all of the GCLs. X-ray imaging showed no evidence of tears or cracking in the GCLs. GCL permittivity measured in fixed wall cells varied from $5 \cdot 10^{-8}\ s^{-1}$ and $1 \cdot 10^{-7}\ s^{-1}$.

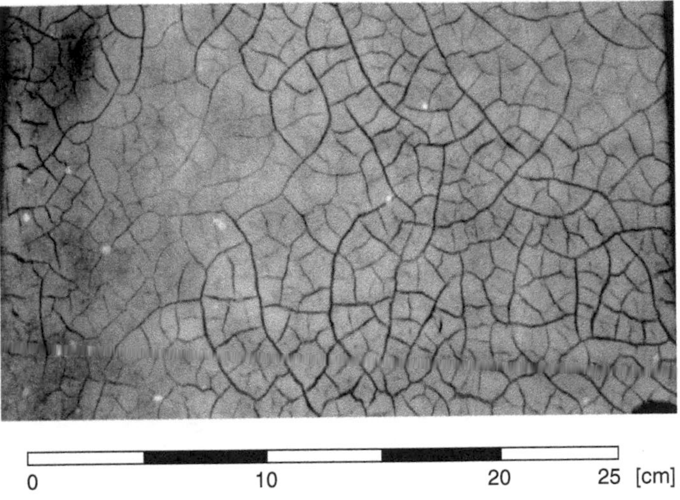


Figure 12. X-ray image of desiccation cracking in a GCL from the test section at Hamburg-Georgswerder (Melchior, 2002).

Heerten (2002) reports on excavation of GCLs at three landfills in Germany. The GCLs were subject to between 0.6 and 1 m of soil cover. Exchange of sodium ions for calcium was reported to be complete at all three sites. The results of triaxial and fixed wall tests gave permittivities between $10^{-9}\,s^{-1}$ and $10^{-8}\,s^{-1}$on these GCLs. Heerten (2004) describes excavation of a GCL after 5 years in service at a landfill at a site with mean yearly total precipitation 500 mm after an extremely hot summer. The GCL has been buried under 140 cm of top soil/ Again, exchange of sodium ions for calcium ions was reported to be complete and the permittivity was measured as $1.2 \cdot 10^{-8}\,s^{-1}$. No evidence of desiccation was observed despite the hot summer.

Meer and Benson (2007) report on laboratory testing of GCLs exhumed from four landfill covers in the United States. At three of the landfills, the GCL alone served as the infiltration barrier while at the fourth landfill the GCL was overlain by a geomembrane. Thickness of overburden soil on top of the GCL varied from 400 mm to 900 mm. The calcium carbonate content of these cover soils varied from 1 to 2.6%. Service life of the GCLs varied from 4 to 11 years. Laboratory testing for exchangeable cations indicated that calcium for sodium ion exchange was essentially complete in all of the exhumed GCLs, including the GCL protected by an overlying geomembrane. Swell index testing yielded values of between 7 and 11 mL/2g of bentonite, versus a typical value of 35 for new sodium bentonite GCLs. Hydraulic conductivity testing using a 10 mM $CaCl_2$ solution, presented in Figure 13, yielded hydraulic conductivities that varied over five orders of magnitude, from $1.6 \times 10^{-4}$ cm/s to $5.2 \times 10^{-9}$ cm/s at an effective stress of 20 kpa (versus $1.2 \times 10^{-9}$ cm/s and $1.7 \times 10^{-9}$ cm/s for new GCL).

GCL source S in Figure 13 is the GCL that was overlain by a geomembrane. Designations "-AB" and "-NP" for site O refer to adhesive bonded and needle-punched geomembranes, respectively. It should be noted that specimens were not back pressure saturated (to better simulate field conditions) and that testing on one of the exhumed GCLs using deionized water yielded a hydraulic conductivity from 3.6 to 10 times lower than the values measured using the $CaCl_2$ solution. While not offering a definitive explanation of the wide variation in hydraulic conductivity of the exhumed GCLs, the authors noted a strong dependence of hydraulic conductivity upon moisture content at the time of sampling as well as in controlled laboratory desiccation tests and suggest that calcium for sodium ion exchange prior to hydration and/or desiccation below a critical moisture content (hypothesized to be on the order of 100%, may led to significant (3-4 orders of magnitude) decreases in GCL hydraulic conductivity even after rehydration. Based upon the results for site S, where the GCL was protected by a geomembrane, the authors suggest that a decrease in hydraulic conductivity due to ion exchange may also be a concern for GCLs used in liner applications.

The above case histories suggest that almost all unprotected GCLs will undergo ion exchange in the field quickly (i.e. within three years) and suffer an associated decrease in saturated hydraulic

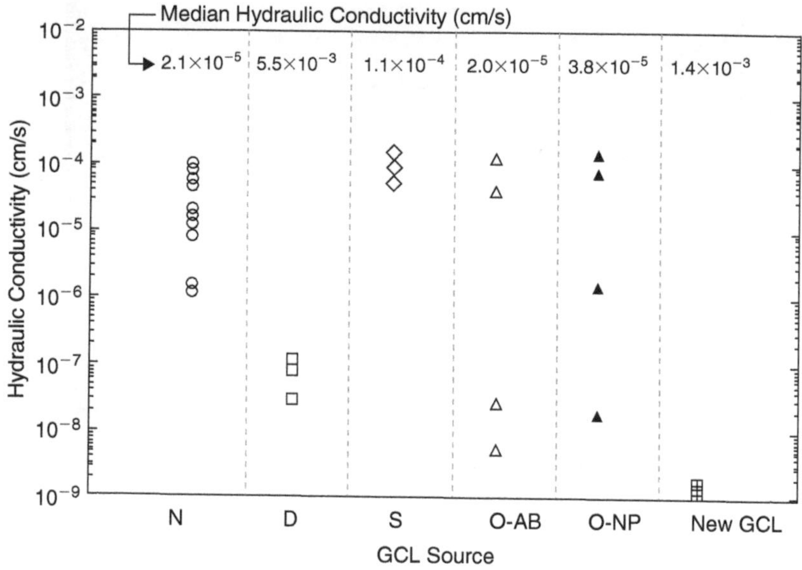


Figure 13. Hydraulic conductivity of exhumed GCLs at 20 kpa from four USA landfills (Meer and Benson, 2007).

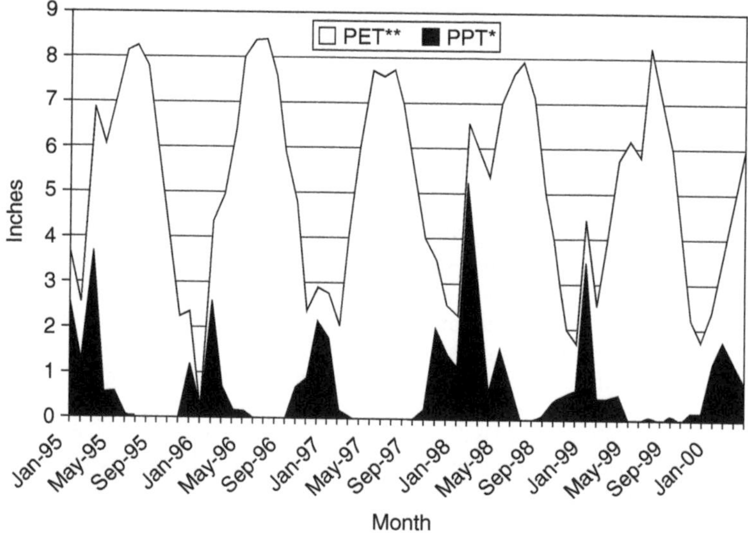


Figure 14. Climate conditions at the Bakersfield, California GCL test site (Mansour, 2001).

conductivity. However, Mansour (2001) reports that the saturated hydraulic conductivity of a needle-punch reinforced sodium GCL with a bentonite content of 4500 g/m$^2$ was unaffected after five years of burial under 0.66 m of vegetated cover soil despite exposure in a relatively extreme semi-arid climate at a site in Bakersfield, California, USA. The Bakersfield site had an annual precipitation (PPT) of less than 150 mm each year and annual potential evapotranspiration (PET) in excess of 1500 mm each year, as shown in Figure 14, and summertime temperatures routinely over 40°C, suggesting that drying out of the GCL during the summer was likely to have occurred. Hydraulic conductivity and swell index of the exhumed samples was essentially the same as measured on an archived sample of the original GCL, with a swell index of approximately 32 mL/2L and a hydraulic conductivity on the order of $2 \times 10^{-9}$ cm/s under an effective confining stress of 35 kPa. Laboratory testing indicated the soil above and below the GCL was relatively rich in sodium and deficient in calcium.

While most field studies of GCL performance looked at its behavior as a barrier to liquid migration, Spokas et al. (2006) evaluated the performance of a GCL as a gas migration barrier. These investigators compared measured methane emissions through the landfill cover and methane recovery rates from landfill gas collection system for a single GCL to those measured for a compacted clay liner (1 m thick) and a single geomembrane liner. The tests were conducted at three landfills in France in test cells that were filled with similar wastes but capped with the different cover configurations. All of the barrier materials were covered with 300 mm of top soil. The GCL was underlain by a sand layer and the geomembrane was underlain by a gravel layer. Emissions were significantly higher and the methane recovery rate was significantly lower for the GCL cover than for the other two covers.

Dwyer (2003) conducted a field study to compare the effectiveness of geomembrane- and GCL-based cover systems to soil-based evapotranspirative cover systems. While this study is frequently cited as evidence that soil-based cover systems are as effective, if not more effective than geomembrane and GCL-based systems, it must be recognized that Dwyer intentionally introduced 8 defects into his 0.13 hectare test section (over 60 defects per hectare) to simulate "typical" field conditions. The difference between this leak density of approximately 60 leaks/hectare and the leak densities at the end of installation for geosynthetic liner systems for good CQA reported by Forget et al. (2005) from electrical field leak detection surveys calls into question any conclusions on GCL effectiveness drawn by Dwyer from his test sections.

## 4 CONCLUSIONS

Field evidence from surveys of the performance of landfills in the USA indicated that GCLs perform at least as well, if not better, then compacted clay when used as the low permeability soil layer in composite liner systems. Large scale field tests indicate that laboratory interface shear testing provides a reliable basis for assessing the stability of containment systems employing GCLs. Field observations of mechanical damage to GCLs shows the importance of protecting the GCL against mechanical damage from subgrade soils, cover soils, and operating equipment as well as the prudence of providing a foundation soil with sufficient fines content to mitigate the potential for bentonite migration from holes in the GCL. Recent observations of seams separation for GCLs beneath exposed geomembranes (as well as laboratory testing to simulate these field conditions) suggest that it is desirable to place some overburden pressure on a GCL shortly after deployment to mitigate the potential for this phenomena to impact the continuity of a GCL barrier. Testing on exhumed GCLS indicates that replacement of the exchangeable sodium ion in a sodium bentonite GCL with calcium and an associated increase in the hydraulic conductivity of the GCL is likely to in a relatively short time (e.g. 3 years) after deployment of the GCL unless the soil in contact with the GCL is sodium rich. However, field observations of GCL performance in hydraulic applications (i.e. at the Oder-Havel Canal) and in landfill cover lysimeter test plots as well some of the laboratory test data on exhumed GCLs suggests that even after calcium ion exchange a GCL can remain an effective hydraulic barrier. Laboratory testing of exhumed GCLs suggests that prevention of GCL: desiccation below a critical moisture content of around 100% and possibly hydration of the GCL prior to ion exchange may be key factors in minimizing the increase in hydraulic conductivity associated with calcium ion exchange.

## REFERENCES


Bonaparte, R., Daniel, D., Koerner, R. 2002 *Assessment and Recommendations for Improving the Performance of Waste Containment Systems* Report No. CR-821448-01-0, Environmental Protection Agency, Washington, D.C., USA, 1039 pp.

Blümel, W., Müller-Kirchenbauer, A., Reuter, E., Ehrenberg, H., von Maubeuge, K. 2002. Performance of geosynthetic clay liners in lysimeters in Zanzinger, H., Koerner, R.M. and Gartung, E. (Eds.), *Clay Geosynthetic Barriers,* A.A. Balkema publishers, The Netherlands: 287–294.

Daniel, D.E., Gilbert, R.B. 1996. Practical methods for managing uncertainties for geosynthetic clay liners" in *Uncertainty in the Geologic Environment: from Theory to Practice, Proceedings of Uncertainty '96,* July 31 – August 3, Madison, Wisconsin, USA, ASCE publishers: 1331–1346.

Daniel, D.E., Koerner, R.M., Bonaparte, R., Landreth, R.E., Carson, D.A., Scranton, H.B. 1998. Slope stability of geosynthetic clay liner test plots. *Journal of Geotechnical and Geoenvironmental Engineering*, ASCE, 124(7): 628–637.

Dwyer, S.F. 2003. *Water balance measurements and computer simulation of landfill covers*. Dissertation submitted in partial fulfillment for the degree of Doctor of Philosophy, University of New Mexico, Albuquerque, New Mexico, USA.

Fleischer, P. 2002. Geosynthetic clay liners – first long time experience as an impermeable lining of a navigation canal. In *Proceedings of the* 7th *International Conference on Geosynthetics*, Delmas, P. Gourc, and Girard, H. (Eds.), Swets & Zeitlinger publishers: 823–826.

Forget, B., Rollin, A.L., Jacquelin, T. 2005. Lessons learned from 10 years of leak detection surveys on geomembranes. In *Proceedings of Sardinia '05: The Tenth International Waste Management and Landfilling Symposium*, Environmental Sanitary Engineering Centre, University of Padua, Italy, 9 pp. (on CD ROM).

Heerten,G. 2002. Geosynthetic clay liner performance in geotechnical applications. In Zanzinger, H., Koerner, R.M. and Gartung, E. (Eds.), *Clay Geosynthetic Barriers,* A.A. Balkema publishers, The Netherlands: 3–19.

Heerten, G. 2004. Bentonitmatten als mineralisches Dichtungselement im Umweltschutz. In *20th SKZ landfill conference "Die sichere Deponie"*, Würzburg, (in German), as reported in Kavazanjian, E. Jr., Dixon, N., Katsumi, T., Kortegast, A., Legg, P., and Zanzinger, H. (2006) "Geosynthetic barriers for environmental protection at landfills," in *Proceedings of the 8th International Conference on Geosynthetics,* Kuwano, J. and Koseki, J. (Eds.), Millpress Science Publishers: 121–152.

Henken-Mellies, W.U., Zanzinger, H., Gartung, E. 2002. Long-term field test of a clay geosynthetic barrier in a landfill cover system," in Zanzinger, H., Koerner, R.M. and Gartung, E. (Eds.), *Clay Geosynthetic Barriers,* A.A. Balkema publishers, The Netherlands: 303–309.

James, A.N., Fullerton, D., Drake, R. 1997. Field performance of GCL under ion exchange conditions. *Journal of the Geotechnical and Geoenvironmental Engineering*, ASCE, 123(10): 897–901.

Koerner, R.M., Koerner, G.R. 2005. In-situ separation of GCL panels beneath exposed geomembrane. *GRI White Paper No. 5*, Geosynthetics Institute, Folsom, PA, USA, April, 21 pp.

Mansour, R.I. 2001. GCL performance in semiarid climate conditions. In *Proceedings of Sardinia '01: the Eighth International Waste Management and Landfilling Symposium*, V. 3, October 1–5, Environmental Sanitary Engineering Centre, University of Padua, Italy: 219–226.

Meer, S.R., Benson, C.H. 2007. Hydraulic conductivity of geosynthetic clay liners exhumed from landfill final covers. *Journal of the Geotechnical and Geoenvironmental Engineering*, ASCE, 153(5): 550–563.

Melchior, S. 2002. Field studies and excavations of geosynthetic clay barriers in landfill covers. In Zanzinger, H., Koerner, R.M. and Gartung, E. (Eds.), *Clay Geosynthetic Barriers,* A.A. Balkema publishers, The Netherlands: 321–330.

NRC 2007. *Assessing the Performance of Engineered Waste Containment Barriers,* National Academies Press, Washington, DC, USA, 134 pp.

Peggs, I.D. 2002. Two leak location surveys on geosynthetic clay liners. In Zanzinger, H., Koerner, R.M. and Gartung, E. (Eds.), in Zanzinger, H., Koerner, R.M. and Gartung, E. (Eds.), *Clay Geosynthetic Barriers,* A.A. Balkema publishers, The Netherlands: 275–286.

Richardson, G.N., Thiel, R.S.,Marr, W.A. 1999. Lessons learned from the failure of a GCL/geomembrane barrier on a side slope landfill cover. pending publication in Giroud, J.P., *Lessons Learned from Failures Associated with Geosynthetics*: 11 pp.

Spokas, K., Bogner, J., Chanton, J., Morcet, M., Aran, C., Graff, C., Moreau-le-Golvan, Y., Bureau, N., Hebe, I. 2006. Methane mass balance at three landfill sites: What is the efficiency of capture by gas collection systems. *Waste Management*, 26(5): 516–525.

Sporer, H. 2002. Exhumed clay geosynthetic barriers. Unpublished presentation at the International Symposium on Clay Geosynthetic Barriers, Nuremberg, Germany, as reported in Kavazanjian, E. Jr., Dixon, N., Katsumi, T., Kortegast, A., Legg, P., and Zanzinger, H. (2006) "Geosynthetic barriers for environmental protection at landfills," in *Proceedings of the 8th International Conference on Geosynthetics,* Kuwano, J. and Koseki, J. (Eds.), Millpress Science Publishers: 121–152.

Thiel, R.,Richardson, G.N. 2005. Concern for GCL shrinkage when installed on slopes. In *Proceedings of Geo-Frontiers Congress05*, Rathje, E. (Ed.), Geotechnical Special Publications 130–142 and GRI-18; ASCE, 7 pp (on CD-ROM).

Thiel, R., Giroud, J.P., Erickson, R., Criley, K., Bryk, L. 2006. Laboratory measurements of GCL shrinkage under cyclic changes in temperature and hydration conditions. In *Proceedings of the* 8th *International Conference on Geosynthetics*, Kuwano, J. and Koseki, J. (Eds.), Millpress Science Publishers: 157–162.

# Author index